The Measurement of Health
and Health Status

The Measurement of Health and Health Status

Concepts, Methods and Applications from a Multidisciplinary Perspective

Paul F. M. Krabbe
University of Groningen,
University Medical Center Groningen,
Department of Epidemiology (Unit: Patient Centered Health
Technology Assessment)
Groningen, The Netherlands

AMSTERDAM • BOSTON • HEIDELBERG • LONDON
NEW YORK • OXFORD • PARIS • SAN DIEGO
SAN FRANCISCO • SINGAPORE • SYDNEY • TOKYO

Academic Press is an imprint of Elsevier

Academic Press is an imprint of Elsevier
125 London Wall, London EC2Y 5AS, United Kingdom
525 B Street, Suite 1800, San Diego, CA 92101-4495, United States
50 Hampshire Street, 5th Floor, Cambridge, MA 02139, United States
The Boulevard, Langford Lane, Kidlington, Oxford OX5 1GB, United Kingdom

Library of Congress Cataloging-in-Publication Data
A catalog record for this book is available from the Library of Congress

British Library Cataloguing-in-Publication Data
A catalogue record for this book is available from the British Library

ISBN: 978-0-12-801504-9

For information on all Academic Press publications
visit our website at https://www.elsevier.com/

Working together
to grow libraries in
developing countries

www.elsevier.com • www.bookaid.org

Publisher: Nikki Levy
Acquisition Editor: J. Scott Bentley
Editorial Project Manager: Susan Ikeda
Production Project Manager: Nicky Carter
Designer: Mark Rogers

Typeset by TNQ Books and Journals

Dedication

To my Swedish wife and all my children.

Contents

6. Health Status Measurement

7. Validity

Part III

Part IV

14. New Developments

15. Perspectives

Preface

You only see it once you get it [Je gaat het pas zien als je het doorhebt] Johan Cruijff (iconic Dutch soccer player and trainer, 1947−2016).

After working for almost 25 years on the evaluation of health interventions in the setting of university hospitals, I realized that the field of health outcomes measurement is partitioned off in segregated areas. This is particularly evident with respect to subjective phenomena such as perceived health status or (health-related) quality of life. Some of those who investigate or apply the instruments adhere to the framework of psychometrics; others call themselves clinimetricians and use different concepts and methods; another group consists of health economists and decision-science researchers who have their own framework. Most of them are unfamiliar with other frameworks and often not even aware of what these can offer. There is a widespread misunderstanding of the different approaches to conceptualize and measure health and health status. My goal here is to bring these frameworks together, as they are far more closely connected than generally recognized.

Inspiration for this book was drawn from my own teaching experience but also from my contacts with numerous other scientists and researchers in the field. In terms of its content, the book is a response to many presentations made by others at conferences—not a reflection of what I learned from their papers but rather of what seemed unclear or undefined. Another source of inspiration was the stimulating environment in which I have been actively engaged for more than 20 years: the international EuroQol Research Foundation. This group bundles researchers from all over the world with the same interest, namely to measure "health-related quality of life" and express this in a single metric figure. Because EuroQol members come from many different backgrounds, collaboration is often thought provoking. Despite its diversity, the group works under the umbrella of the health-economic doctrines that stipulate how health should be measured.

I hope this book will contribute to the field by unifying material that is currently dispersed over volumes that treat these measurement approaches separately. Because it crosses disciplinary boundaries, this book offers an illuminating perspective on the measurement of health. While I recognize that health outcomes research uses different types of measures—including clinical, health status, quality of life, and patient satisfaction—I have chosen to focus on health status. Thus, along with the measurement of health, this book covers the measurement of (perceived or experienced) health status, in particular health-related quality of life.

The book is intended to offer readers a set of directions in a sea of confusing concepts to help them select the best measurement framework. Which one is "best" depends on the situation, on what the analysts want, or which policy they are trying to assess. In that respect, this book complements but also distinguishes itself from the literature on this topic. First, it contains an integrative overview of material from various disciplines and fields. Second, it provides in-depth explanations of the pros and cons of specific measurement methods, highlighting their relationships and differences as well as their contributions. Third, it avoids detailed descriptions, though some of the most well-known and frequently applied instruments are explained and discussed to illustrate a point. Fourth, it provides professionals with an overview of the knowledge, language, and concepts that are used in this diverse and fragmented field.

This book cannot be expected to clear up all of the difficulties of measuring a subjective phenomenon such as health. Yet, because it is also intended to serve as an introductory handbook, some crucial principles are covered, though without presenting lengthy mathematical derivations or relying heavily on statistics. On the other hand, the interpretation of statistical measurement concepts such as reliability, validity, responsiveness, item-response theory, and factor analysis is treated more extensively. Practical applications and study questions, which are usually included in textbooks, were deliberately left out. Instead, the conceptual underpinning of current approaches to measurement is illustrated by empirical examples. For information on related issues, the reader is referred onward by citing other sources.

As such, this book presents a unifying perspective that is relevant to scientists and others working in several disciplines and in various fields. The target audience comprises academics and professionals: medical doctors, health researchers, health-care providers, clinical epidemiologists, health economists, and insurance providers. The book is also intended for people working in pharmaceutical companies, in the areas of health technology assessment, public health, nursing, pharmacy, dentistry, health services research, physiotherapists, and occupational therapists involved in developing and using health scales.

It is the elusiveness of health that ensures its enduring appeal to scientific inquiry. Interdisciplinary cooperation is the best way forward in many fields, and health measurement is no exception. Our endeavor to arrive at credible methods to quantify health calls for contributions from the fields of health economics, medicine, psychology, measurement theory, mathematics, and others, but also from patients, whose importance to research should not be underestimated. The issues and methods addressed here are too complex for the narrow focus of a single discipline.

Paul F. M. Krabbe
University of Groningen,
University Medical Center Groningen,
Department of Epidemiology (Unit: Patient Centered Health
Technology Assessment)
Groningen, The Netherlands
October 2016

I welcome and encourage readers to write their suggestions and comments to me at p.f.m.krabbe@umcg.nl.

Acknowledgments

Science is a cooperative enterprise, and as such my work on the measurement of health draws not only on my own ideas. Several other individuals have shaped my thinking and influenced my work. I should mention George Torrance first, as he was my primary inspiration when I started out in this field. Not only did he introduce new concepts but he could write clear scholarly articles, many as sole author. He was on the program of the first scientific conference I attended—in 1995 in Phoenix, Arizona—and I missed his plenary address because I was still driving near the Mexican border in my rental car (I had yet to learn that distances in the United States are incomparable to distances in the Netherlands). But later I was able to meet with George several times.

In 2004 I spent some time visiting an institute at Harvard University where I was hosted by Christopher Murray, who founded the Global Burden of Disease approach and developed the disability-adjusted life year concept. After a few days Joshua Salomon stopped by. I was struck by his sound and balanced reasoning and his multidisciplinary range of interests, impressed by his detailed knowledge of the origins of many scientific methods and ideas, and inspired by how he combined these in his own exploration of methods to measure health status.

But my greatest source of inspiration was Louis Thurstone (1887–1955). With a masters in engineering, Thurstone (originally his Swedish family name was Thunström) began his career as an electrical engineer. Thomas Edison (known for the phonograph, the motion picture camera, and the electric light bulb) had heard about Thurstone's invention of a flicker-free motion picture projector and offered him an internship. Thurstone was intrigued by Edison: did Edison's problem-solving capacity stem from his genius or did his genius stem from his ability to solve problems? For many years, Thurstone worked as a psychometrician at the University of Chicago. His groundbreaking work on the measurement of attitudes laid the foundation for measuring subjective and social phenomena.

English editing by Nancy van Weesep-Smyth.

Part I

Chapter 1

Introduction

Chapter Outline

INTRODUCTION

Ask any medical gathering whether 1 year of life is about the same as any other and you will surely hear a resounding chorus of "no." It is not just the quantity but also the quality of life that concerns people, medical professionals or otherwise. This applies to the life years saved in life-and-death situations as much as to interventions explicitly geared to changing morbidity and raising health status. So it can probably be agreed that the measurement of "quality of life" or "quality of health" is a necessary part of health evaluation (Brooks, 1995).

Although traditional health outcomes such as dead or alive or surviving are indisputable, these measures are often insufficient when studying chronic diseases or conditions such as pain, migraine, fatigue, mental status, or depression. That is mainly because survival does not play a major role in these types of problems. In addition, while we can measure a biological response, we may not be able to determine whether that response makes a noticeable difference to the patient. For example, when treating anemia (decrease in the amount of red blood cells), we can measure hemoglobin levels, but we should also know if the biologic response results in changes such as a perceived reduction in fatigue.

This realization also has led to changes in health policy. In evaluating health care, there is a noticeable shift in emphasis from traditional, simple-to-measure clinical indicators such as mortality and morbidity to more complex patient-based outcomes such as disability and "quality of life." For example, in the English National Health Service, health-status measurement is at the center of the reform effort, as seen in the development of PROMs (Devlin & Appleby, 2010) (Box 1.1, Fig. 1.1).

BOX 1.1

Concerning happiness, however, and what it is, they are in dispute, and ordinary people do not give the same answer as wise ones. For ordinary people think it as one of the plainly evident things, such as pleasure or wealth or honor—some taking it to be one thing, others another. And often the same person thinks it as different things, since when he or she gets a disease, it is health, whereas when he or she is poor, it is wealth.

From Ethica Nichomachea (translation Reeve, C.D.C., 2014. Hackett Publishing Company, Indianapolis, p. 14).

DEFINING HEALTH

Defining health is, at best, problematic. Its definition reflects the historical period and culture to which it applies. Over the past 150 years, rising expectations have changed the definition of health in the United States: from survival, it shifted to freedom from disease, to an ability to perform daily activities, and even came to embrace a sense of happiness and well-being.

FIGURE 1.1   Aristotle (384–322 BC).

Americans seem to expect their health to be not merely adequate but good, if not excellent. Expectations in other cultures are lower. In many African countries, for example, certain afflictions are so prevalent that they are not even considered diseases. Drawing the line between being sick or well may also depend upon age and sex. For instance, the elderly normally endure more sickness than the young or middle aged. Therefore, the definition of health for an elderly person may include pain and discomfort not experienced by younger persons. Pregnant women in developing countries do not get the medical attention that most women in developed countries get and may therefore undergo more pain and discomfort in pregnancy. Definitions of health are likely to reflect the ideology and culture of the most powerful groups in society. In modem societies, there is a tendency to consider more conditions as diseases, such as alcoholism, drug dependence, and delinquency. There is also a greater tolerance of the diseased person (Larson, 1991).

IMPORTANCE OF HEALTH

Abraham Maslow (1908–70) was an American psychologist who is best known for creating a *hierarchy of needs*. His theory is often represented as a pyramid with five hierarchical levels (Fig. 1.2). Based on motivational theory in psychology, it presumes that once people have met their basic needs, they seek to meet successively higher needs. Maslow called the bottom four levels of the pyramid "deficiency needs": a person does not feel anything if they are met but becomes anxious if they are not. Thus, physiological needs such as eating, drinking, and sleeping are deficiency needs, as are safety needs, social

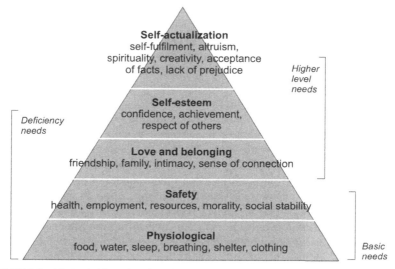

FIGURE 1.2 Maslow's hierarchy of needs.

needs such as friendship and sexual intimacy, and ego needs including self-esteem and recognition. Whether this is really the case for all deficiency needs is open to discussion, but it seems reasonable for health. In contrast, the fifth level of the pyramid is called a "growth need" because it enables a person to "self-actualize" or reach one's fullest potential as a human being. Although Maslow's hierarchy of needs has been criticized for being overly schematic and lacking in scientific grounding, it presents an intuitive and potentially useful theory of human motivation. Interestingly, it regards health as a basic condition of life. Good health is a requirement for attaining other important things in life or reaching the next Maslow level, although there are some exceptions. Sometimes, despite a bad health condition, people seem to be able to reach all the other Maslow levels. A good example is the physicist and cosmologist Stephen Hawking. Despite suffering from amyotrophic lateral sclerosis (characterized by stiff muscles, muscle twitching, and gradually worsening weakness), he was not only able to produce impressive theories about the universe but also got married, had children and—as the story goes—was dancing in his wheelchair to operatic music of Richard Wagner.

HEALTH MODELS

Epidemiology is largely concerned with the causes and mechanisms of poor health and disease, whereas physicians are more interested in interventions to improve health. A theoretical framework may provide guidance not only on how to improve health but also on who is best qualified to design and administer such interventions and when these might be most effective. As a poor health condition is reflected in a low health status, it is paramount to measure and monitor the health status of patients (Costa & King, 2013). A distinction should be made between the determinants (causes) and the manifestations (effects) of health (Chapter 6). Determinants are events or experiences that in some sense cause variation in the manifestations of health. This was recently noted by Costa and King (2013): "Which events and experiences are manifestations and which are determinants depends on the definition of health adopted. If responses to an instrument are aggregated across manifestations and determinants, resultant scores confound health with its potential causes. This problem also applies to the more general concept of health." Chapter 2 will elaborate on this issue. For now it is sufficient to mention the most prominent health model (Fig. 1.3), which was proposed by Wilson and Cleary (1995). Although they call it a conceptual model of patient outcomes, their model actually captures the factors affecting health as experienced by individuals. They posit that biological factors associated with disease cause symptoms, which influence functional status, which in turn influences general health perceptions, which finally impacts on overall "quality of life." Conceptual models such as that of Wilson and Cleary are attempts to elucidate the interrelationships between various measures and constructs.

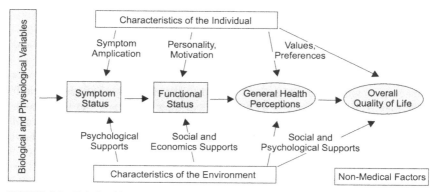

FIGURE 1.3   Relationships among measures of patient outcome in a health conceptual model. *Adaptation of Wilson, I.B., Cleary, P.D., 1995. Linking clinical variables with health-related quality of life: a conceptual model of patient outcomes. The Journal of the American Medical Association 273 (1), 59—65.*

PATIENT REPORTED

Over the past few years, there has been a fundamental shift in focus to emphasize the patients' involvement in the care they receive; an emphasis that is reflected in several policy and national initiatives. Patients also want to be involved in the decision-making process, and patient advocacy is now considered important in many aspects of health care. Against this background, the assessment of outcomes based on the patient's perspective using patient-reported outcome (PRO) measures increasingly accompanies traditional clinical ways of measuring health and the effects of treatment on the patient (Devlin & Appleby, 2010; Appleby et al., 2016).

Any assessment coming directly from patients, without interpretation by physicians or others, about how they function or feel in relation to their health condition is called a patient-reported outcome or patient-reported outcome measure (PROM). PROMs put together two terms (patient reported and outcome) and often refer to any assessment by patients about how they function or feel. PROMs are tools to gain insight from the perspective of the patient into how aspects of their health and the impact of the disease and its treatment are perceived to affect their lifestyle and subsequently their (quality of) life.

That acronym is proposed as an umbrella term to cover both single domains and multidomain measurement instruments. Common constructs measured by PROMs include health status, health-related quality of life, quality of life, well-being, treatment satisfaction, symptoms, functioning, adherence to treatment, and many other types of outcomes. PROMs can also differ in type of instrument (e.g., questionnaires, index, instrument, classification, Likert scales, visual analogue scales, item-response theory, multiattribute preference-based classifications). Although PROMs can be very simple, they may also be rather sophisticated (i.e., based on computer adaptive testing by having a connection to a central server to attain interactive and adaptive selection of next questions).

Access to the patients' perspective through the use of PROMs can have an impact on a wide range of aspects related to the delivery of effective health care. Identifying issues faced by patients and their families about living with an illness may generate knowledge that might impact treatment decisions and adherence. In general, use of PROMs may enhance the understanding of how health-practitioners can affect outcome.

PATIENT CENTEREDNESS

Another term that is currently used, and most likely better addresses the goal of PROMs, is patient-centered outcomes. These outcomes cover issues of specific concern to the patient. The idea of patient centeredness finds its origin in patient-centered medicine.

Patient-centered medicine is based on the claim that patients should be active participants in their care. This idea explains the emerging interest in PROMs. According to the National Institutes of Health (USA), patient-centered medicine is "health care that establishes a partnership among practitioners, patients, and their families (when appropriate) to ensure that decisions respect patients' wants, needs and preferences and solicit patients' input on the education and support they need to make decisions and participate in their own care." There are very few health-care providers, regulatory authorities, or clinical experts who do not publicly claim to provide patient-centered health care. Many agree with the definition of the Institute of Medicine (USA), which states that patient-centered means "providing care that is respectful of and responsive to individual patient preferences, needs, and values, and ensuring that patient values guide all clinical decisions" (Scholl et al., 2014). One important element of patient-centered health care is the cooperation of patients and health-care providers.

The Patient-Centered Outcomes Research Institute (PCORI) was established as a part of the US Patient Protection and Affordable Care Act of 2010 to fund patient-centered comparative clinical effectiveness research. The aim is to extend the concept of patient centeredness from health-care delivery to health-care research. In essence, the PCORI evaluates questions and outcomes that are meaningful and important to patients and caregivers. PCORI considers research to be patient-centered if the focus includes outcomes that matter to patients (Frank et al., 2014).

Patient-centered care is increasingly replacing the established physician-centered system with one that revolves around the patient. Effective care is generally defined by or in consultation with patients rather than by means of physicians' tools or standards. For instance, orthopedic surgeons employ the

Harris Hip Score to judge the success of a total hip replacement. The tool was designed for use solely by physicians and does not even ask patients to rate their satisfaction with the procedure. It answers questions that are likely to be important to doctors and thought to be important to patients. However, it is unknown whether any physician-derived tools, such as the Harris Hip Score, accurately reflect the patient experience, be it with hip replacement or other aspects of their medical care (Harris, 1969).

What distinguishes informative health outcomes from all other measures of health is the need to solicit and incorporate patients' values and preferences into the final assessment. Therefore, it makes sense to incorporate end outcomes such as health status and (health-related) quality of life in the evaluation of the overall effect of medical interventions. Such outcome measures become relevant when we place individual patients or patient groups at the heart of the health-care system.

DIGITAL REVOLUTION

Many methods that have been used to develop health measurement scales were first applied in the late 1970s. Since then, with the increasing availability of relatively inexpensive desktop computers followed by the spread of personal computers (PCs), researchers have gained access to a great number of modern methods to analyze data and construct measurement instruments. For example, the statistical technique of confirmatory factor analysis is now commonplace. But even the more straightforward exploratory factor analysis (Chapter 7), the estimation of Chronbach's α (Chapter 8), and not least the measurement models of item-response theory (Chapter 10) have all became widely available to almost every researcher only because mainframe computers were no longer necessary.

Take my own experience as a master student in 1988 at the University of Utrecht (The Netherlands). A simple factor analysis had to be written in hard-to-capture syntax, sent off from a terminal (monitor plus keyboard) to a mainframe (large computer operated by experts in a separate and distant building somewhere on campus), and if nothing went wrong (one dot missing or in the wrong place was enough to get no output the next day), then the output (results) would be generated on long zigzag scrolls in dot-matrix print.

BOX 1.2

September 1956: IBM launched the 305 RAMAC, the first SUPER computer with a hard disc drive (HDD). The HDD (Fig. 1.4) weighed over a ton and stored 5 megabyte of data. Data were stored on 50 rather large aluminum discs spinning at 1200 rpm. The discs were coated on both sides with a magnetic iron oxide.

Continued

Box 1.2 —cont'd

FIGURE 1.4   The volume and size of 5 MB memory storage in 1956.

In addition, the faculty was sent an invoice for every computation. Once the PC became a standard tool for everybody, user-friendly software was developed for almost every existing analytical method (Box 1.2).

REFERENCES

Appleby, J., Devlin, N., Parkin, D., 2016. Using Patient Reported Outcomes to Improve Health Care. John Wiley & Sons, Chichester.

Brooks, R.G., 1995. Health Status Measurement: A Perspective on Change. Macmillan Press, London.

Costa, D.S., King, M.T., 2013. Conceptual, classification or causal: models of health status and health-related quality of life. Expert Review of Pharmacoeconomics and Outcomes Research 13 (5), 631–640.

Devlin, N.J., Appleby, J., 2010. Getting the Most out of PROMs: Putting Health Outcomes at the Heart of NHS Decision-Making. The King's Fund. London.

Frank, L., Basch, E., Selby, J.V., 2014. The PCORI perspective on patient-centered outcomes research. The Journal of the American Medical Association 312 (15), 1513–1514.

Harris, W.H., 1969. Traumatic arthritis of the hip after dislocation and acetabular fractures: Treatment by mold arthroplasty. An end-result study using a new method of result evaluation. The Journal of Bone and Joint Surgery 51 (4), 737−755.

Larson, J.S., 1991. The Measurement of Health: Concepts and Indicators. Greenwood Press, Westport.

Scholl, I., Zill, J.M., Härter, M., Dirmaier, J., 2014. An integrative model of patient-centeredness - A systematic review and concept analysis. PLoS One 9, 9 e107828.

Wilson, I.B., Cleary, P.D., 1995. Linking clinical variables with health-related quality of life: A conceptual model of patient outcomes. The Journal of the American Medical Association 273 (1), 59−65.

Chapter 2

Health Outcomes

Chapter Outline

INTRODUCTION

Health is a construct encompassing a range of phenomena, and various factors define it differently (Levine, 1995; Huber et al., 2011). To physicians, guided by a biomedical model, it is predominantly a state that falls within acceptable biological norms. Accordingly, health outcomes have been overly objectified in terms of mortality, morbidity, or cure, though there is a clear tendency to move beyond the biomedical model. Morbidity measures, as expressed by laboratory tools, often deal mainly with the prognosis of a patient and less with the actual experienced health condition. The construct has been extended to cover psychological and even social factors, making subjective measures such as health status or "quality of life" (QoL) necessary—and rightly so, because the ultimate goal of all health interventions is to improve a patient's perceived health condition.

There is no clear and widely accepted classification of health outcome measures. As the field evolves, practitioners and researchers use different terms and classifications. This chapter describes and explains all those endpoint measures that potentially provide information about the current health condition of individual patients.

The now-famous WHO affirmation of 1948 (WHO, 1946) describing health as "a state of complete physical, mental, and social well-being and not merely the absence of disease or infirmity" is probably the most cited definition. Yet being so general, it is not providing an easy command. Critics call it Utopian, inflexible, and unrealistic, arguing that including the word "complete" makes it highly unlikely that anyone would be healthy for a reasonable period of time.

Others have noted that health is defined differently in different cultures, and therefore the WHO definition is too broad to apply at an international level. Though inspirational, the WHO definition is not very practical and seems more akin to "QoL" than to health (Saracci, 1997).

Nevertheless it covers many of the concepts discussed here. All generic health outcome measures—functional status, health status, health-related QoL (HRQoL), well-being, QoL—are closely interrelated. Different terms are often used for the same concept, having emerged from adjacent research communities (rehabilitation, medicine, psychology, epidemiology, clinimetrics, health economics, philosophy, medical decision making). QoL is often captured by well-being, satisfaction with life, even functional status and health status; the boundary between these concepts is elusive.

Health outcome measures are endpoints describing patients' current condition and must therefore be differentiated from risk factors (prognosis), accidents and events (causes), and process measures. Other names in use are patient-based and patient-centered outcomes, which denote outcomes that cover issues of specific concern to the patient. During the last 30–40 years, there has been a proliferation of questionnaire-type measures designed for self-completion by patients or their proxies.

BIOLOGICAL AND PHYSIOLOGICAL OUTCOMES

Given the unambiguous nature of death, its compulsory registration, and the availability of disease registries, event-based measures such as death provide objective, precise, and readily available estimates of health. Most event-based outcome measures (yes/no) stand out, as does survival (expressed in years), as special indicators of health: though crude, they are factual. The class of biological and physiological health outcomes is diverse. Diabetic patients, for example, are evaluated according to blood glucose; chronic obstructive lung disease patients are evaluated according to pulmonary function. Vital signs include height, weight, body temperature, blood pressure, pulse, respiration rate, and so on. Even though these measures do not concern outcomes (apart from death) and are mostly not reported by patients, they are advantageous to the clinician, who can use them to relate outcomes to specific diseases and as prognostic indicators. That advantage explains the wide array of morbidity outcomes.

SYMPTOMS AND DISORDERS

Another class of outcomes from clinical studies contains information that is less biomedical and more observational: the immense palette of symptoms and disorders. Although many symptoms are objectively observable, direct observation is not always possible for several others—notably pain, fatigue, anxiety, and depression. The issue of interest may be the patients' general appearance and specific attributes of disease (nutritional status, presence of jaundice, pallor, or clubbing). Alternatively, the focus could be on their neurological (consciousness, awareness, vision) or psychiatric status (orientation, mental state, evidence of abnormal perception or thought).

FUNCTIONAL STATUS

Functional status denotes an individual's ability to perform normal daily activities or tasks that are essential to meet basic needs and fulfill usual roles (Wilson & Cleary, 1995). The concept has two facets: capacity and performance. Functional capacity represents one's maximum ability to perform daily activities in the physical, psychological, and social domains of life, whereas functional performance refers to what people actually do in the course of their daily lives (Leidy, 1994). Both facets can be influenced by biological or physiological impairment, symptoms, and mood but also by health perceptions. For example, individuals who are well but think they are ill may have a low level of functional performance relative to capacity. Critically, measures of functional status focus on functional ability and overt behavior, as opposed to subjective experience. In other words, they assess what individuals can actually do rather than what they feel able to do.

Functional status is variously defined. Many clinicians include carrying out activities of daily living but also participating in life situations and society. The first range covers basic physical and cognitive activities such as walking or reaching, focusing attention, and communicating, as well as the routine activities of daily living, including eating, bathing, dressing, and toileting. The second embraces life situations such as school or play for children and, for adults, work outside the home or maintaining a household.

Functional status commonly informs rehabilitation medicine, physical and occupational therapy, and care in a nursing facility or at home. The concept guides therapy in areas such as hearing, speech, vision, cognition, and mobility. It is also used to design and coordinate services for children with special needs and to monitor the health of people with chronic conditions.

Functional status in different areas is measured by activities of daily living instruments. The Barthel Index and the Katz, for example, are generally applied to the elderly population, particularly to individuals in long-term care institutions (Katz et al., 1963; Mahoney & Barthel, 1965). Whereas these instruments take a functional approach (e.g., role and tasks), others take a broader view of functional status, referring to the activities of daily living but also to participation in life situations and society (Spector et al., 1987).

HEALTH STATUS

One of the first attempts to measure health by means other than the usual mortality and morbidity outcomes was the Karnofsky Performance Scale (KPS), developed in 1948 (Fig. 2.1). At that time, people started to question

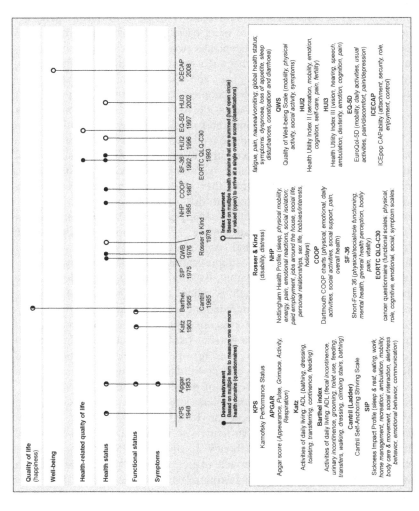

FIGURE 2.1   Illustration to show the various health outcomes (endpoints) represented by a selection of different but widely applied health-outcome instruments.

whether it was worth prolonging life if the treatments caused many side effects; this prompted efforts to measure health status or performance (Day & Jankey, 1996). The KPS, a 10-point scale to measure the ability of cancer patients to perform daily activities, was the precursor of today's health-status measures (Karnofsky et al., 1948). The neonatal Apgar score (evaluation of the condition of the newborn infant) was another early health-status measure (Apgar, 1953), though it can also be seen as a functional status or symptom measure. The Sickness Impact Profile is a behavior-based measure of sickness-related dysfunction, with attributes related to physical and social functioning and usual activities (Bergner et al., 1981). A simple measure of health status in primary care settings is based on the COOP/WONCA Charts (Nelson et al., 1987). The 36-item Short-Form survey (SF-36) is a compromise (created by the RAND corporation in a large-scale study) between lengthy instruments and single-item measures of health (Stewart et al., 1988; Ware & Sherbourne, 1992). Initially, it covered the health concepts that are usually included in widely used health surveys (physical, social, and role functioning; mental health; general health perception). Bodily pain and vitality were added in light of strong empirical evidence (Stewart & Ware, 1992). The SF-36 is indicated as an eight-scale profile of functional health and well-being, but also as a health-status instrument (Ware & Sherbourne, 1992) or as a QoL instrument. A wide range of health-status measures refer to particular disease groups, such as people with impaired hearing (Hinderink et al., 2000). Another essential disease-specific health-status instrument is the EORTC QLQ-C30 (Aaronson et al., 1993).

Health status is broader than functional status. Most measures include attributes such as physical function, sensation, self-care, cognition, pain, and discomfort (Rosser & Kind, 1987; Torrance et al., 1996; Feeny et al., 2002). These measures are directed "within the skin": focused on health and the capacity for living according to the individual patient. Many health-status measures have extensions that assess not only physical performance but also mental and social functioning. Some include metaconcepts such as "autonomy" and "dignity" (Fig. 2.2). It is important to recognize that health-status measures bear no relationship to the perceived impact of these attributes on individual patients and their values. That explains why the frequency and intensity of health outcomes (e.g., pain, mobility) are usually recorded in health-status instruments. As an important outcome measure in medical research, health status spurred the development of health-related QoL instruments.

HEALTH-RELATED QUALITY OF LIFE

Researchers associate HRQoL with an expanded concept of health status, one embracing social interaction as well as emotional and psychological well-being. In keeping with that broader view, many HRQoL instruments now reflect the WHO affirmation and are commonly aligned with "subjective" rather

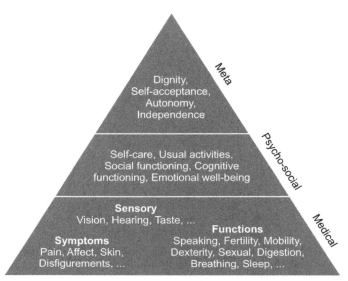

FIGURE 2.2   Examples of health attributes subdivided into medical, psycho-social, and meta levels.

than "objective" health status. Sometimes, researchers add to this that a measure of HRQoL needs to assess not only the frequency or level of complaints on certain outcome attributes but also the impact, expressed as the value of subjective perceptions and experiences of living in such a condition (Bonomi et al., 2000; Sullivan, 2003; Hamming & De Vries, 2007). As a case in point, patients may care little about major problems but a great deal about minor ones. With that in mind, HRQoL serves as a restricted definition of QoL (explained later), in the sense that it was designed to exclude externalities such as housing, neighborhood, and financial matters (Fig. 2.3).

The physical, mental, and social dimensions of health and HRQoL can be measured in terms of concepts that may vary in importance. To cover that range, several instruments have been developed (Kaplan et al., 1976). An early one was the Nottingham Health Profile (NHP), consisting of two parts. The first concerns health problems (sleep, physical mobility, energy, pain, emotional reactions, and social isolation). The second concerns how these affect daily life (employment, chores, social life, relationships, sex life, hobbies and interests, and holidays) (Hunt et al., 1985). Statements on each aspect are weighted for severity using the Thurstone method of paired comparisons, thereby transforming a health-status instrument into a value- or preference-based HRQoL measure. For the NHP this was done with a sample of 215 members of the general public (McKenna et al., 1981). The most widely applied preference-based HRQoL instrument is the EQ-5D, comprising five attributes (mobility, usual activities, daily activities, pain/discomfort, and anxiety/depression) (Dolan, 1997).

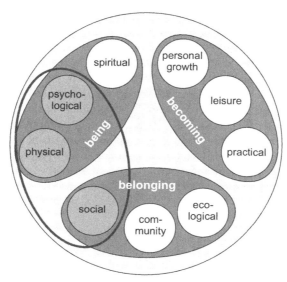

FIGURE 2.3 Health-related quality of life as a component of quality of life. *Reproduced with permission from Renwick, R., Brown, I., Nagler, M., 1996. Quality of Life in Health Promotion and Rehabilitation: Conceptual Approaches, Issues and Applications. Sage Publications, Thousand Oaks.*

Even though HRQoL is meant to express the value of subjective perceptions and experiences, many measures of HRQoL are actually generic health-status instruments. Some are intended to measure *perceived* or *subjective* health status and fall somewhere between health-status and HRQoL instruments. While these may address a patient's subjective perception by posing questions in a particular way, these instruments do not express the quality of the patient's health state. Having been developed under the classical test theory paradigm (Chapter 9), most are based on multiple domains and comprise multiple-rated Likert-item scales (Likert, 1932). Other instruments were developed under the clinimetrics paradigm (Chapter 13) to produce an index measure, but these are based on expert opinion (Feinstein, 1983, 1999). It takes special tasks (Chapters 11 and 12), together with dedicated measurement models, to produce genuine values for health states (Fayers, 2007; Krabbe, 2008; Salomon, 2014).

WELL-BEING

Ever since the WHO affirmation, the distinction between health and well-being has been under debate. Its critics have argued that health is a component of well-being, not identical to it, and that the WHO definition medicalized nonhealth elements of everyday life. The 1948 definition is said to have launched a lofty ideal, projecting health as an integral component of well-being and, furthermore, implying that good health is a necessary condition for attaining the highest possible levels in all other aspects of well-being (Salomon et al., 2003).

However, health is not the only outcome of interest. While the aim of conventional health care is to achieve health gains (or prevent deterioration), the focus of social care is to improve well-being. Elderly and disabled people are often supported by a combination of "conventional" health care and social care or long-term care (Coast et al., 2008). Here too, both physical and social well-being contribute to the QoL. Its perception seems to be captured by an instrument such as the quality of well-being scale, a preference-based measure combining three scales of functioning (physical activities, social activities, mobility) with a measure of symptoms and problems (Kaplan et al., 1979; Anderson et al., 1989). Given the descriptions of its attributes, however, it appears to be an instrument of HRQoL.

A similar argument has been put forward in the case of HRQoL, as viewed in terms of the contribution of an individual's health to his/her overall well-being. The conceptual problems emerge from the fact that well-being is not clearly separable into independent health and nonhealth components (Broome, 2002; De Charro, 2014).

QUALITY OF LIFE

Instead of HRQoL, the term "QoL" is often used. But the latter is a much broader construct, comprising aspects such as living conditions, spirituality, political system, and financial situation. Some researchers tend to conflate QoL with other concepts, using them interchangeably. Research on QoL has influenced health-status research in the context of medicine. In certain health fields (dementia, end-of-life), QoL in its broadest sense is an outcome measure equivalent to well-being.

QoL research began in the disciplines of sociology and psychology and soon became an issue within health care (Haas, 1999). For example, Neugarten et al. (1961) developed a self-assessment instrument to measure successful aging in individuals over 55. The research identified five components (zest vs apathy, resolution and fortitude, congruence between desired and achieved goals, self-concept, and mood tone) of the construct they named "life satisfaction." Other notable work was done by Campbell and colleagues (Campbell et al., 1976; Bergner, 1989) in a period of rising affluence but also of increasing violence and public disorder. In 1964, President Lyndon B. Johnson implicitly drew attention to QoL by placing it at the basis of a Great Society, stating that "these goals cannot be measured by the size of our bank balance. They can only be measured in the quality of the lives that our people lead" (Johnson, 1964). In that vision, figures on economic growth were insufficient to describe the quality of American life; policymakers started to express QoL in social attributes such as level of education and crime rates. As Campbell noted, objectively measured social attributes from the 1960s explained only 15% of the variance in an individual's QoL, whereas psychological attributes such as life perception, happiness, and satisfaction accounted for 50%.

CAPABILITIES

Another promising avenue is the capability approach, introduced as an alternative to standard utilitarian welfare economics (Sen, 1982). It posits that traditional outcomes in health should not be the sole object of welfare assessments; rather, one's capabilities [real opportunities (outcomes) to do and be what one has reason to value] should be included in the overall assessment of a person's well-being. The capability approach has been influential in development economics, and many health economists recognize its advantages. Compared to the traditional HRQoL approach, it has the potential to offer a richer theoretical evaluative space, particularly for assessing complex interventions in social care and public health (Simon et al., 2013). One of the first outcome instruments to appear under the heading of the capability approach is the ICECAP (Coast et al., 2008).

HAPPINESS

A widely used instrument is Cantril's self-anchoring striving scale (called the Cantril Ladder because it is often depicted as a ladder with 10 rungs, denoting steps from 0 to 10). This scale is used to assess satisfaction with life and reflects a general cognitive evaluation of a person's overall well-being, i.e., happiness (Cantril, 1965). Self-anchoring means that a person is asked to define the two extremes (often the worst and best possible life) of the spectrum on the basis of his/her own assumptions, perceptions, goals, and values. The self-defined continuum then serves as the measuring device. The question is, "How satisfied are you with your life as a whole these days?" Initially introduced as a public-opinion tool and later often used in satisfaction studies, this instrument has also been applied in health studies (Luttik et al., 2006; Peters et al., 2012).

Some researchers and health authorities consider "happiness" as a promising overarching outcome measure to express the status of a country and its citizens, and this outcome seems to involve the health-care sector too (Dolan, 2014; Kahneman et al., 2004). A famous pop group once sang that "happiness is a warm gun," obviously exercising poetic license (Wikipedia, 2016). Clearly, "health" and "happiness" are distinct life experiences whose relationship is neither fixed nor constant. Many people have poor health but are nevertheless happy, and perhaps even more are healthy but unhappy. Despite receiving the best health care, people may have disturbed and limited social relationships. Even more than HRQoL, concepts such as well-being, happiness, and satisfaction are befuddling in their breadth, abstractness, and ambiguity.

A good reason to distinguish health from happiness or satisfaction/dissatisfaction is that the former may vary quite independently of the latter. Let us turn to the interesting example given by McCall (1980). Consider a society blessed with every conceivable advantage: good schools, democratic government, delightful neighborhoods, public transportation, clean air, no crime, negligible unemployment, creative work opportunities, excellent health and longevity, a high level of affluence, and no poverty. And yet, sad to say, almost everyone in the society is, for various personal reasons, unhappy. A's mother has just died, B can't get along with his boss, C and D suffer the pangs of unrequited love, E is married to the wrong man, etc. Does this mean that the society's level of health is low?

DISCUSSION

Health outcomes are not clearly defined, and the conceptual basis for many of the instruments to measure them remains unclear. When describing a patient's health, investigators tend to substitute "QoL" for health status or functional status. Most refer to health status or HRQoL without defining the concept but then go on to operationalize it, leaving the reader to infer the definition. Certain research areas tend to prefer particular concepts. "Health status" is paramount in nursing; "HRQoL" is predominant in medical psychology and clinical epidemiology; "well-being" is privileged in sociology, psychology, and economics; and "QoL" is embedded and applied in sociology, psychology, policy science, and some specific areas in health. Disambiguation is essential to any meaningful application of concepts (Rescher, 1969). But because the concepts of health outcomes evolve with use, their definitions may vary according to time, place, discipline, and theoretical perspective.

The questions that comprise many health measures can be worded either in terms of performance ("I do not walk at all": Sickness Impact Profile) or in terms of capacity ("I'm unable to walk at all": Nottingham Health Profile). This distinction reflects to a large extent the contrast between objective and subjective measurement (Chapter 6). Performance may be recorded objectively whereas assessments of capacity tend to be subjective. Debate continues between those who favor performance wording and those who favor capacity wording. In general, couching the responses in terms of capacity gives an optimistic view of health, whereas recording performance is conservative. Proponents of performance wording argue that it gives a truer picture of what the person actually does, and not what they think they might be able to do on a good day if they try.

As noted, the WHO described health as a "state of complete physical, mental, and social well-being and not merely the absence of disease or infirmity." The ambiguity of this definition becomes less puzzling in light of its origin. Dr. Szeming Sze, a member of the Chinese delegation to the conference convened to draft the Charter of the United Nations in 1945, was appointed to a subcommittee. Sze had a background in health education and an interest in preventive medicine. Two other members were on the same subcommittee, one of whom was a psychiatrist. Together they came up with the wording that defines health as not merely the absence of illness and includes the mental domain (Anonymous, 1988).

Outcome instruments often contain domains and attributes that seem inappropriate. For example, many HRQoL instruments cover the "financial situation" or process variables, patient characteristics, or mechanisms. Instead,

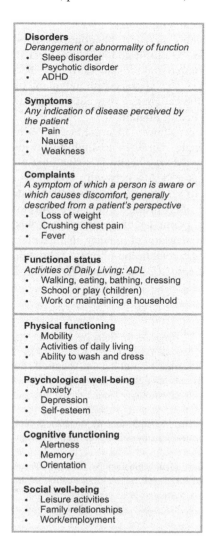

FIGURE 2.4 Overview of different types of health outcomes with examples of their facets.

to measure outcomes accurately, the instruments should take a "snapshot" of an individual's health condition and be as objective as possible. Some definitions of well-being include "self-acceptance," "pursuit of meaningful goals," "ability to cope," and "continued growth and development as a person." Obviously, such facets have more to do with character and personality than with health (Ryff, 1989). In addition, many measures do not reflect health outcomes but rather a constellation of interacting components; the vicious circle of depression is a good example. Sometimes it is hard to say whether "depression" is a result (outcome) of a disease or situation, or the cause of other elements of the health outcome.

Psychosocial problems need to be conceptualized to measure them. However, questions can be raised about the conceptualization of psychosocial aspects such as compassion, consideration for the patient's feelings, or maintenance of dignity. The appraisal or judgment of these humanitarian issues experienced by the patient is a very personal affair and may be affected by factors outside the health domain.

There is no consensus on how to measure health. Various metrics have been used, ranging from single measures to multidimensional instruments. The choice reflects not only the way health is defined but also the purpose of the measurement—to compare health between populations, inform policy, evaluate and plan services, or set priorities for research and development. Patients' reported outcomes are often deemed subjective and unreliable. Even Donabedian, the renowned public-health pioneer, had misgivings about using outcomes as the sole measure of quality. Nonetheless he concluded that "Outcomes, by and large, remain the ultimate validation of the effectiveness and quality of medical care" (Donabedian, 1966) (Fig. 2.4).

REFERENCES

Aaronson, N.K., Ahmedzai, S., Bergman, B., Bullinger, M., Cull, A., Duez, N.J., Filiberti, A., Flechtner, H., Fleishman, S.B., de Haes, J.C.J.M., Kaasa, S., Klee, M., Osoba, D., Razavi, D., Rofe, P.B., Schraub, S., Sneeuw, K., Sullivan, M., Takeda, F., 1993. The European organization for research and treatment of cancer QLQ-C30: a quality-of-life instrument for use in international clinical trials in oncology. Journal of the National Cancer Institute 85 (5), 365−376.

Anderson, J.P., Kaplan, R.M., Berry, C.C., Bush, J.W., Rumbaut, R.G., 1989. Interday reliability of function assessment for a health status measure: the quality of well-being scale. Medical Care 27 (11), 1076−1084.

Anonymous, 1988. WHO: from small beginnings. World Health Forum 9, 29−34.

Apgar, V.A., 1953. Proposal for a new method of evaluation of the newborn infant. Current Researches in Anesthesia & Analgesia 32 (4), 260−267.

Bergner, M., 1989. Quality of life, health status, and clinical research. Medical Care 27 (3), 148−156.

Bergner, M., Bobbitt, R.A., Carter, W.B., Gilson, B.S., 1981. The sickness impact profile: development and final revision of a health status measure. Medical Care 19 (8), 787−805.

Bonomi, A.E., Patrick, D.L., Bushnell, D.M., Martin, M., 2000. Validation of the United States' version of the World Health Organization Quality of Life (WHOQOL) instrument. Journal of Clinical Epidemiology 53 (1), 1–12.

Broome, J., 2002. Measuring the burden of disease by aggregating well-being. In: Murray, C.J.L., Salomon, J.A., Mathers, C.D., Lopez, A.D. (Eds.), Summary Measures of Population Health: Concepts, Ethics, Measurement and Applications. World Health Organization, Geneva, pp. 91–113.

Campbell, A., Converse, P.E., Rodgers, W.L., 1976. The Quality of American Life: Perceptions, Evaluations and Satisfactions. Russell Sage Foundation, New York.

Cantril, H., 1965. The Pattern of Human Concern. Rutgers University Press, New Brunswick, NJ.

Coast, J., Flynn, T.N., Natarajan, L., Sproston, K., Lewis, J., Louviere, J.J., Peters, T.J., 2008. Valuing the ICECAP capability index for older people. Social Science & Medicine 67 (5), 874–882.

Day, H., Jankey, S.G., 1996. Lessons from the literature: towards a holistic model of quality of life. In: Renwick, R., Brown, I., Nagler, M. (Eds.), Quality of Life in Health Promotion and Rehabilitation: Conceptual Approaches, Issues and Applications. Sage Publications, Thousand Oaks, CA, pp. 39–50.

de Charro, F., 2014. Happiness & Health and Health Status Measurement. Internal Report. EuroQol Research Foundation, Reigate.

Dolan, P., 1997. Modeling valuations for EuroQoL health states. Medical Care 35 (11), 1095–1108.

Dolan, P., 2014. Happiness by Design: Finding Pleasure and Purpose in Everyday Life. Penguin Group, New York.

Donabedian, A., 1966. Evaluating the quality of medical care. The Milbank Quarterly 83 (4), 691–729.

Fayers, P.M., 2007. Applying item response theory and computer adaptive testing: the challenges for health outcomes assessment. Quality of Life Research 16 (Suppl. 1), 187–194.

Feeny, D., Furlong, W., Torrance, G.W., Goldsmith, C.H., Zhu, Z., DePauw, S., Denton, M., Boyle, M., 2002. Multiattribute and single-attribute utility functions for the health utilities index Mark 3 system. Medical Care 40 (2), 113–128.

Feinstein, A.R., 1983. An additional basic science for clinical medicine: IV. The development of clinimetrics. Annals of Internal Medicine 99 (6), 834–848.

Feinstein, A.R., 1999. Multi-item "instruments" vs Virginia Apgar's principles of clinimetrics. Archives of Internal Medicine 159 (2), 125–128.

Haas, B.K., 1999. A multidisciplinary concept analysis of quality of life. Western Journal of Nursing Research 21 (6), 728–742.

Hamming, J.F., De Vries, J., 2007. Measuring quality of life. British Journal of Surgery 94 (8), 923–924.

Hinderink, J.B., Krabbe, P.F.M., van den Broek, P., 2000. Development and application of a health-related quality-of-life instrument for adults with cochlear implants: the Nijmegen Cochlear Implant Questionnaire. Otolaryngology-Head and Neck Surgery 123 (6), 756–765.

Huber, M., Knottnerus, J.A., Green, L., van der Horst, H., Jadad, A.R., Kromhout, D., Leonard, B., Lorig, K., Loureiro, M.I., van der Meer, J.W., Schnabel, P., Smith, R., van Weel, C., Smid, H., 2011. How should we define health? BMJ 343, d4163.

Hunt, S.M., McEwen, J., McKenna, S.P., 1985. Measuring health status: a new tool for clinicians and epidemiologists. Journal of the Royal College of General Practitioners 35 (273), 185–188.

Johnson, L.B., 1964. President of the United States: Remarks in Madison Square Garden. Downloaded on November 25, 2015. From: http://www.presidency.ucsb.edu/ws/?pid=26700.

Kahneman, D., Krueger, A.B., Schkade, D.A., Schwarz, N., Stone, A.A., 2004. A survey method for characterizing daily life experience: the day reconstruction method. Science 306 (5702), 1776—1780.

Kaplan, R.M., Bush, J.W., Berry, C.C., 1976. Health status: types of validity and the index of well-being. Health Services Research 11 (4), 478—507.

Kaplan, R.M., Bush, J.W., Berry, C.C., 1979. Health status index: category rating versus magnitude estimation for measuring levels of well-being. Medical Care 17 (5), 501—525.

Karnofsky, D.A., Abelmann, W.H., Craver, L.F., Burchenal, J.H., 1948. The use of nitrogen mustards in the palliative treatment of carcinoma. With particular reference to bronchogenic carcinoma. Cancer 1 (4), 634—656.

Katz, S., Ford, A.B., Moskowitz, R.W., Jackson, B.A., Jaffe, M.W., 1963. Studies of illness in the aged. The index of ADL: a standardized measure of biological and psychosocial function. JAMA 185 (12), 914—919.

Krabbe, P.F.M., 2008. Thurstone scaling as a measurement method to quantify subjective health outcomes. Medical Care 46 (4), 357—365.

Leidy, N.K., 1994. Functional status and the forward progress of merry-go-rounds: toward a coherent analytical framework. Nursing Research 43 (4), 196—202.

Levine, S., 1995. The meanings of health, illness, and quality of life. In: Guggenmoos-Holzmann, I., Bloomfield, K.P.H., Brenner, H.M., Flick, U. (Eds.), Quality of Life and Health: Concepts, Methods and Applications. Blackwell, Berlin, pp. 7—13.

Likert, R., 1932. A Technique for the Measurement of Attitudes. New York University, New York.

Luttik, M.L., Jaarsma, T., Veeger, N., van Veldhuisen, D.J., 2006. Marital status, quality of life, and clinical outcome in patients with heart failure. Heart & Lung 35 (1), 3—8.

Mahoney, F.I., Barthel, D.W., 1965. Functional evaluation: the Barthel index. Maryland State Medical Journal 14, 61—65.

McCall, S., 1980. What is quality of life? Philosophica 25 (1), 5—14.

McKenna, S.P., Hunt, S.M., McEwen, J., 1981. Weighting the seriousness of perceived health problems using Thurstone's method of paired comparisons. International Journal of Epidemiology 10 (1), 93—97.

Nelson, E., Wasson, J., Kirk, J., Keller, A., Clark, D., Dietrich, A., Steward, A., Zubkoff, M., 1987. Assessment of function in routine clinical practice: description of the COOP chart method and preliminary findings. Journal of Chronic Diseases 40 (Suppl. 1), 55S—63S.

Neugarten, B.L., Havighurst, R., Tobin, S., 1961. The measurement of life satisfaction. Journal of Gerontology 16 (2), 134—143.

Peters, L.L., Boter, H., Buskens, E., Slaets, J.P.J., 2012. Measurement properties of the Groningen frailty indicator in home-dwelling and institutionalized elderly people. Journal of the American Medical Directors Association 13 (6), 546—551.

Renwick, R., Brown, I., Nagler, M., 1996. Quality of Life in Health Promotion and Rehabilitation: Conceptual Approaches, Issues and Applications. Sage Publications, Thousand Oaks.

Rescher, N., 1969. On Quality of Life and the Pursuit of Happiness. RAND Corporation, Santa Monica.

Rosser, R., Kind, P., 1987. A scale of valuations of states of illness: is there a social consensus? International Journal of Epidemiology 7 (4), 347—358.

Ryff, C.D., 1989. Happiness is everything, or is it? Explorations on the meaning of psychological well-being. Journal of Personality and Social Psychology 57 (6), 1069—1081.

Salomon, J.A., 2014. Techniques for valuing health states. In: Culyer, A.J. (Ed.), Encyclopedia of Health Economics. Elsevier, Amsterdam, pp. 454—458.

Salomon, J.A., Mathers, C.D., Chatterji, S., Sadana, R., Üstün, T.B., Murray, C.J.L., 2003. Quantifying individual levels of health: definitions, concepts, and measurement issues. In: Murray, C.J.L., Evans, D.B. (Eds.), Health Systems Performance Assessment. World Health Organization, Geneva, pp. 301–318.

Saracci, R., 1997. The World Health Organization needs to reconsider its definition of health. BMJ 314 (7091), 1409–1410.

Sen, A., 1982. Choice, Welfare and Measurement. Harvard University Press, Cambridge, MA.

Simon, J., Anand, P., Gray, A., Rugkåsa, J., Yeeles, K., Burns, T., 2013. Operationalising the capability approach for outcome measurement in mental health care research. Social Science & Medicine 98, 187–196.

Spector, W.D., Katz, S., Murphy, J.B., Fulton, J.P., 1987. The hierarchical relationship between activities of daily living and instrumental activities of daily living. Journal of Chronic Diseases 40 (6), 481–489.

Stewart, A.L., Hays, R.D., Ware, J.E., 1988. The MOS short-form general health survey: reliability and validity in a patient population. Medical Care 26 (7), 724–735.

Stewart, A.L., Ware, J.E., 1992. Measuring Functioning and Well-Being: The Medical Outcomes Study Approach. Duke University Press, Durham.

Sullivan, M., 2003. The new subjective medicine: taking the patient's point of view on health care and health. Social Science & Medicine 56 (7), 1595–1604.

Torrance, G.W., Feeny, D.H., Furlong, W.J., Barr, R.D., Zhang, Y., Wang, Q., 1996. Multiattribute utility function for a comprehensive health status classification system: health utilities index Mark 2. Medical Care 34 (7), 702–722.

Ware, J.E., Sherbourne, C.D., 1992. The MOS 36-item short-form health survey (SF-36): I. Conceptual framework and item selection. Medical Care 30 (6), 473–483.

Wikipedia. "Happiness Is a Warm Gun" is a song by the Beatles on the double-disc album The Beatles, also known as the White Album, which was released on 22 November 1968. http://en.wikipedia.org/wiki/Happiness_Is_a_Warm_Gun.

Wilson, I.B., Cleary, P.D., 1995. Linking clinical variables with health-related quality of life: a conceptual model of patient outcomes. JAMA 273 (1), 59–65.

World Health Organization, 1946. Preamble to the Constitution of the World Health Organization as Adopted by the International Health Conference, 2. WHO, New York, p. 100 (June 19–July 22, 1946; signed on July 22, 1946 by the representatives of 61 States).

Chapter 3

Health Summary Measures

Chapter Outline

INTRODUCTION

In the areas of health policy and public health, the most frequently used measure of mortality is life expectancy. Mortality has generally been taken as the basic indicator of health loss. With life expectancy increasing and chronic diseases on the rise, however, mortality rates alone no longer fully describe the health of a population. Substantial resources are now invested not only in preventing premature death but also in promoting healthy living. But how effective have these efforts been? What is an appropriate metric to measure health? Summary measures that combine mortality and morbidity attempt to incorporate these two main components of health. The key challenges in developing summary measures are defining and measuring health. In this chapter, three of the most well-known health summary frameworks are presented. Each produces a measure that captures mortality and morbidity in a composite measure. Two of these are capable of expressing the effects of health interventions.

QUALITY-ADJUSTED LIFE YEAR

In many countries, regulatory authorities and governmental organizations require studies that demonstrate the contribution of specific health interventions (i.e., their benefit, added value). For example, the National Institute for Health and Care Excellence (NICE) provides guidance to the National Health Service in England on the clinical effectiveness and cost effectiveness of selected new and established technologies (drugs, medical devices, diagnostics, and surgical procedures). NICE strongly recommends that economic evaluations use a summary measure such as quality-adjusted life years (QALYs) as the unit of health benefit assessment in the reference case.

While the QALY is a simple technique for summarizing the impact of health-care interventions, it is not a rational tool on its own; it was developed to assist in decision making about health-care resource allocation. Though this remains as best-known application. From the very beginning, the QALY has been controversial, but when handled with care the QALY concept can be quite useful.

In a QALY, the quantity and quality of life are combined into a single measure (Fig. 3.1). Quantity (life years, survival, life expectancy) can be stated simply as the duration of certain health states or health conditions during the study period; alternatively, it can be extrapolated by imputing the life expectancy of patients. Central to the computation of QALYs is the "quality" component, which is expressed in a single metric health-status index score. (There is a widespread misspecification in this wording: an index is always a single number, an index is not a score but a number, a metric measure is always a single number. More on this terminology are discussed in Chapters 4, 5, and 6.)

The QALY approach is appealing because it provides a relatively simple means of reflecting the health effects of medical interventions. Accordingly, it enables comparison of interventions that have very different types of outcome. This approach is particularly useful if a patient's health status is unstable, in which case the research design has to include repeated measurements (at fixed time intervals or event related). Because QALYs are derived by multiplying two components—the quality of health (Q) and the length of life (L)—only

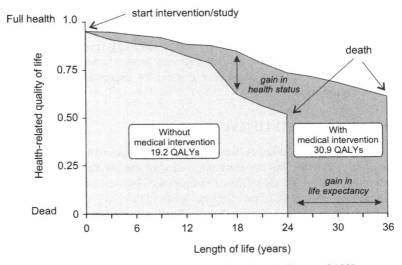

FIGURE 3.1   Graphical representation of the quality-adjusted life year (QALY) concept.

interventions with health-status consequences that are lasting and substantial may disclose significant effects. The required input to calculate QALYs is a set of values (preference scores, or utilities) for the health status of individual patients, expressed on a scale from 0.0 to 1.0, i.e., from dead to full health (Chapters 11 and 12).

The QALY approach is not feasible for all types of diseases or health interventions. Take treatments for infertility, such as in vitro fertilization (IVF), for example. When evaluating IVF, most studies explicitly rule out any consideration of the child's health status, using only "success" criteria (birth of a healthy child following treatment). A QALY-based evaluation of IVF, on the other hand, would presumably consider the mother's quality of life (not only her health status) together with that of the child and perhaps that of her partner or other close relatives. Deriving QALYs would not be appropriate in such cases because the treatment is not exclusively concerned with the improvement of an individual patient's health condition.

The QALY concept was developed in the early 1970s. It was first mentioned as "function years" in a compelling paper by Fanshel and Bush (1970) that evaluated a tuberculin skin testing program. This paper covered most of the aspects surrounding the measurement of health in general, though in terminology that has since changed. The use of Thurstone's model based on paired comparisons (Chapter 11) to estimate health-state values was first proposed in this publication. Around the same time, Torrance and colleagues (Torrance, 1970; Torrance et al., 1972) introduced the concepts of the index day and health day. Obviously unaware of these publications, Grogono and Woodgate (1971) published a similar approach to evoke indices that would measure health. They specified health according to 10 domains and arrived at "health years" after running a simple weighting algorithm. An earlier precedent may be in a publication by Herbert Klarman et al. (1968), who compared three options for treating patients with chronic renal disease in terms of life years gained. They also performed an analysis both with and without adjustments for "quality of life," but these were arbitrary adjustments in their epidemiological cohort model.

The term "QALY" was first used in 1976 by Zeckhauser and Shepard to indicate a health outcome measurement unit that combines duration and quality of life. It was for that paper that Shepard coined the acronym QALY. The term QALY appeared in the same year in a book by Weinstein and Stason (1976)—one of the contributors was Shepard—examining policies for control of hypertension. They used the QALY as a common scale that takes into account added years of life expectancy, decreased morbidity from myocardial infarctions (heart attack) and cerebrovascular accidents (stroke), and the discomforts associated with the side effects of medications. The following year, a major medical journal, *The New England Journal of Medicine* (NEJM), published their landmark article in which the QALY (gained) was described as the appropriate measure of effectiveness of treatments, thereby making this approach known to a broad audience in the fields of medicine and public health (Weinstein & Stason, 1977). They asserted that the outcome of these weights is ultimately subjective, but that subjectivity does not exclude "societal consensus." Furthermore, they argued that QALYs could be used as an outcome measure in cost-effectiveness analysis. The term "cost per quality-adjusted year of life saved" was explicitly taken as a cost-effectiveness measure.

The fact that an article on cost effectiveness was published in the NEJM reflects the expansion of interest in health evaluation beyond the medical field and undoubtedly reinforced this interest. That article is frequently cited as a milestone in the development of cost-effectiveness analysis in health and medicine. In England, *the British Medical Journal* carried a highly influential article by Alan Williams (1985) reporting an early instance of applying QALYs to the evaluation of coronary artery bypass grafting. Williams (founding father of the EuroQol-5D instrument, see Chapter 12) concluded, in his typically straightforward manner, that "The data on which these judgments are based are crude and in need of refinement. The methodology is powerful, far reaching, and open to comment."

DISABILITY-ADJUSTED LIFE YEAR

In 1993 the World Health Organization, in collaboration with the World Bank and the Harvard School of Public Health, published the Global Burden of Disease study (1993). Seeking to measure worldwide disability and death from a multitude of causes, this study marked the most comprehensive effort thus far to systematically measure the world's health problems. To carry out this important work, the Global Burden of Disease researchers adopted an internationally standardized form of the QALY, which they called the disability-adjusted life year (DALY).

DALYs combine years of life lost (YLL) to premature death and years lived with a disability (YLD) or disease, weighted for the severity of the health condition. With this measure, it is possible to compare health conditions that have different symptoms and outcomes using a standardized metric (Smith, 2015). The measure aggregates YLLs because of premature death and YLD of specified severity and duration (Fig. 3.2). YLD can also be described as years lived in less than ideal health. By definition, one DALY is one lost year of healthy life. DALYs allow us to estimate the total number of years lost due to specific causes and risk factors at the country, regional, and global levels (Murray & Lopez, 1997). DALYs cover conditions such as influenza, which may last for only a few days, or epilepsy, which could persist for a lifetime. A DALY is measured by taking the prevalence of the condition multiplied by its disability weight. To calculate the total DALYs for any disorder in any population, one sums that disorder's YLLs and YLDs (Murray et al., 2002).

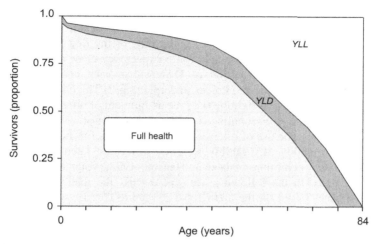

FIGURE 3.2 Graphical representation of the disability-adjusted life year (DALY) concept. *YLL*, years of life lost; *YLD*, years lived with a disability.

For example, to calculate the DALYs incurred through stroke in the Netherlands, one adds up all YLLs attributed to stroke. YLLs are calculated by subtracting the age at death from the longest possible life expectancy for a person at that age. For example, if the longest life expectancy for men is 85 years, but a man dies of stroke at 75 years, he would have lost 10 years of life due to stroke. Next, the total number of YLDs is calculated for those who have suffered a stroke in the Netherlands.

Disability weights reflect the severity of different conditions and are derived from surveys of the general public. In a formal exercise involving health workers from all regions of the world, the relative severity of 22 "indicator conditions," selected by the Global Burden of Disease researchers to represent distinct severities of disability, was weighted between 0.0 (full health) and 1.0 (equivalent to dead). To determine which disorders should be selected as indicator conditions, the researchers reached consensus on the weight of each one. These weights were then grouped into seven classes, where class 1 has a weight between 0.00 and 0.02 and class 7 a weight between 0.70 and 1.00.

One DALY can be thought of as one lost year of "healthy" life. The sum of these DALYs across the population, which amounts to the burden of disease, can be taken as a measurement of the gap between current health and an ideal situation where the entire population lives to an advanced age, free of disease and disability. Most importantly, they use the same "disability weight" for everyone who lives a year in a specified health state.

The Global Burden of Disease studies use the same "ideal" life expectancy for all population subgroups. For DALYs, the standard life expectancy has been set to match the highest observed national life expectancy. Japan has served as the standardized population for premature death, since it is the world's longest-surviving population. At the time of the first DALY studies, Japanese females at birth already had a life expectancy close to 82 years.

Since its inception, the notion of DALYs has been criticized for the methodological and normative choices underpinning it. The critique concerns four areas: the use of age weighting (i.e., the assignment of different weights to YLL at different ages); discounting (i.e., the assignment of a lower weight to YLL in the future); the use of different life expectancies for men and women; and the determination of disability weights, which are intended to capture the severity of a condition (Anand & Hanson, 1997; Voigt & King, 2014; Salomon, 2014). In the following, we will discuss the methodological and normative issues raised for the first Global Burden of Disease studies and then show that several of these points have been addressed in recent publications.

Age weighting is not a value-free, objective process. For example, a range of data indicate that most people value a year of life lived by a young adult more than a year of life lived by a very young child or an older adult. That is to say, there is a broad social consensus that the relative value of life rises rapidly from birth to peak in the early 20s and then steadily declines. Such age weighting was incorporated into the first bundle of DALY estimates.

Likewise, a year of life lived *now* is considered preferable to one lived in the future. In economics, waiting always entails cost in the form of lost opportunity. Initially, the Global Burden of Disease researchers discounted future life years by 3% per year when estimating DALYs. Thus, when we incorporate both discounting and age weighting, the death of a girl on her first birthday results in a loss of 34 years of life, while the death of a young woman on her 25th birthday results in a loss of 33 years. Needless to say, there was intense debate among ethicists and public-health workers on whether a year of healthy life is more preferable now rather than later, and more preferable at certain ages (Menken et al., 2000). In response to the first three criticisms, the developers made age weights uniform; dropped discounting, and used the same life expectancy for men and women (now 86 years) to estimate DALYs.

To assess disease severity, the Global Burden of Disease researchers employed the person trade-off method, which asks health workers to make judgments about quality versus quantity of life (Chapter 12). Many researchers were dissatisfied with the particular person trade-off method being used to determine disability weights. It consisted of asking respondents to choose between different hypothetical public-health interventions. Essentially, the method requires subjects to trade off number of persons lived healthily or with some defined disability. To address the fourth criticism, the team revised its method for determining disability weights.

These methodological and normative issues are not exclusive to DALYs. Discounting is also subject to ongoing discussion in cost-effectiveness analysis, where QALYs are taken as an effect component. Furthermore, several normative issues surround the techniques used by health economists to determine the "weights" for different health states. These techniques are notoriously susceptible to various biases (Chapter 12).

The Global Burden of Disease (2010) study was based on data from household surveys conducted in different countries—Bangladesh, Indonesia, Peru, the United Republic of Tanzania, and the United States of America—as well as from a Web-based survey. Thus, the disability weights reflected responses obtained in different settings and demographic groups, including individuals with little formal education. More important, health-state measures ("weights") in the Global Burden of Disease (2010) study were based on pairwise comparisons (Chapter 11) of different health states rather than on the person trade-off (Salomon et al., 2012). On the grounds of this study, the researchers were able to produce a list of leading causes of disease burden accompanied by a projection in time (Fig. 3.3). For informative and elegant graphs on global and population health, readers are referred to the website of the Institute for Health Metrics and Evaluation (www.healthdata.org).

Q-TWiST

In many clinical studies, especially in the field of cancer, the primary outcome of research is the occurrence of a particular event (e.g., death, relapse, toxicity). The outcome is often expressed as the duration of a particular health state that is associated with an event, using survival analysis. Conventionally, this statistical technique is applied to situations where the main outcome was

2004 Disease or injury	As % of total DALYs	Rank	Rank	As % of total DALYs	2030 Disease or injury
Lower respiratory infections	6.2	1	1	6.2	Unipolar depressive disorders
Diarrhoeal diseases	4.8	2	2	5.5	Ischaemic heart disease
Unipolar depressive disorders	4.3	3	3	4.9	Road traffic accidents
Ischaemic heart disease	4.1	4	4	4.3	Cerebrovascular disease
HIV/AIDS	3.8	5	5	3.8	COPD
Cerebrovascular disease	3.1	6	6	3.2	Lower respiratory infections
Prematurity and low birth weight	2.9	7	7	2.9	Hearing loss, adult onset
Birth asphyxia and birth trauma	2.7	8	8	2.7	Refractive errors
Road traffic accidents	2.7	9	9	2.5	HIV/AIDS
Neonatal infections and other	2.7	10	10	2.3	Diabetes mellitus
COPD	2.0	13	11	1.9	Neonatal infections and other
Refractive errors	1.8	14	12	1.9	Prematurity and low birth weight
Hearing loss, adult onset	1.8	15	15	1.9	Birth asphyxia and birth trauma
Diabetes mellitus	1.3	19	18	1.6	Diarrhoeal diseases

FIGURE 3.3 The 10 leading causes of disease burden in 2004 and 2030. *COPD*, chronic obstructive pulmonary disease; *DALY*, disability-adjusted life year. *Reprinted with permission from World Health Organization, 2008. The global burden of disease: 2004 update. p. 51. www.who.int/ healthinfo/global_burden_disease/2004_report_update/en.*

whether a patient was dead or still alive. But other mutually independent disease states can also be considered (e.g., disease free, severe side effects, relapse).

Several efforts were made in cancer research to integrate quality and quantity of life in a single analysis for the sake of treatment comparisons (Gelber & Goldhirch, 1986). Of particular interest is the quality-adjusted time without symptoms of disease and toxicity of treatment (Q-TWiST) method (Gelber et al., 1996). The development of the Q-TWiST was motivated by a medical controversy in 1986. The question was, which chemotherapy for breast cancer demonstrated improvement in disease-free survival, but no overall survival advantage, and had significant side effects. This inquiry prompted the TWiST approach (Price et al., 1987) and eventually the Q-TWiST. Initially, the Q-TWiST incorporated aspects of health status into operable breast cancer/adjuvant chemotherapy (Goldhirch et al., 1989). The approach has also been useful in other disease settings, such as acquired immune deficiency syndrome (Lenderking et al., 1994).

The underlying idea is to include the results of a survival analysis and health states/status in a single analysis. As in an ordinary survival analysis [known as survival analysis in the (bio)medical setting, but also called event-history by statisticians and others], the element of time is central to the Q-TWiST method. But instead of using single outcomes such as overall survival or disease-free survival, multiple outcomes corresponding to changes in health status are considered (Mills, 2001). A limited number of distinct clinical health states are distinguished, and each of them is assigned a value (i.e., quality of the health state). Estimation of the proportion of patients in each health state at a specific time period is based on survival curves. These clinical health states should be mutually exclusive and are selected to be relevant to the clinicians and patients. Each health state is assigned a value (from 0.0 to 1.0).

The first step in the analysis is to define health states that are relevant with respect to the disease under study. These should highlight specific differences between the treatments being compared. In the case of adjuvant chemotherapy for resectable breast cancer, the survival outcomes are defined as follows: the time with toxicity (TOX), represented by the period in which the patient is exposed to subjective side effects of therapy; disease-free survival (DFS), the time until disease recurrence or death, whichever occurs first; and overall survival (OS), the time to death from any cause. Survival is associated with progressive clinical health states: time spent with treatment toxicity (TOX); time without either symptoms of the disease or toxicity of treatment (TWiST = DFS−TOX); and time following the diagnosis of systemic spread of the disease or relapse (REL = OS−DFS). It is assumed that a patient will move from the first to the last health state, possibly skipping one.

The definitions of TOX and REL reflect the fact that these periods have a negative impact on the patient's overall quality of life. The defined survival outcomes (e.g., TOX, DFS, and OS) indicate transitions between the

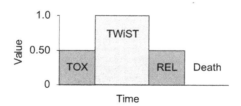

FIGURE 3.4 Components of the quality-adjusted time without symptoms of disease and toxicity of treatment (Q-TWiST): The division of overall survival into TOX (subjective toxic effects), TWiST, and REL (relapse), and the weighting of these time periods using values for TOX and REL.

progressive states of health (e.g., TOX, TWiST, and REL; Gelber et al., 1995). Fig. 3.4 displays the different time periods in this example according to the assumed values of 1.0 for TWiST and 0.5 for both TOX and REL. This diagram represents a scenario in which 1 month spent in TOX or REL is equivalent in value to one-half month spent with the better quality of life that characterizes TWiST.

The second step is based on survival analysis. Kaplan—Meier curves for the time to events that signal transitions between the clinical health states are used to partition the area under the overall survival curve separately for each treatment (Fig. 3.5). This allows partitioning of the overall survival into the amount of time spent in each state. The third step is to compare the treatment regimens using the weighted sum of the mean durations of each clinical health state as calculated in the second step.

The Q-TWiST is calculated as the weighted sum of the clinical health-state durations and the values attached to these health states. By definition, the value for the health state of dead is 0 and the weight for TWiST is generally assumed to be 1.

$$Q - TWiST = (V_{TOX} \times TOX) + (V_{TWiST} \times TWiST) + (V_{REL} \times REL).$$

The appeal of the Q-TWiST comes mainly from its strong clinical orientation. Given that its applications are based on randomized-controlled trials, implying a control group or a comparative treatment, the effect of external factors is accounted for. Moreover, the drop out (censoring) is captured by the statistical survival analysis part of the approach, though under certain conditions the estimates may be biased (Gelber et al., 1989). In addition, possible confounding factors may be controlled for by incorporating covariates into a more advanced survival analysis (Cox proportional hazard model). The fact that the method lends itself to graphical representation of the results enhances its appeal.

Extensions of the standard Q-TWiST have been suggested (Glasziou et al., 1998). For example, for each mutually exclusive state, rather than using values obtained from expert opinion, patient-derived values can be incorporated into a

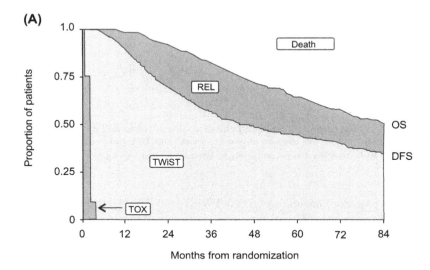

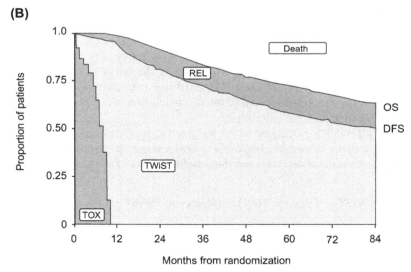

FIGURE 3.5 Partitioned survival for the long-duration treatment (A) and for the short-duration treatment (B) for IBCSG Trial V at 7 years of median follow-up. In each graph the area under the overall survival curve (OS) is partitioned by the survival curves for disease-free survival (DFS) and time with treatment toxicity (TOX). The areas between the survival curves give the average number of months spent in TOX, TWiST, and REL as indicated. *Reprinted by permission of the American Statistical Association. Based on Gelber, R.D., Cole, B.F., Gelber, S., Goldhirsch, A., 1995. Comparing treatments using quality-adjusted survival: the Q-TWiST method. The American Statistician 49, 161–169.*

Q-TWiST. Another extension determines appropriate time intervals for administering outcome assessments. These intervals should be selected so that any major change in the patient's health condition that is attributed to the treatment or the disease process can be captured in the analysis. Instead of taking measurements at fixed intervals, a protocol based on event-induced measurements could yield more reliable data.

APPLICATION

One of the clinical studies that used the Q-TWiST approach to investigate a new type of intervention was done in the setting of cancer (Wiering et al., 2011). The aim of this prospective study was to describe long-term health comprehensively by using health summary frameworks in patients undergoing surgical treatment of colorectal liver metastases. To illustrate the application of the Q-TWiST, the background and results of this clinical study are presented here. References are given to later chapters in this book where more detailed information can be found. In this example, the Q-TWiST was not intended to express differences between two different treatments or assess a new treatment against standard treatment. Rather, the aim was to graphically depict the dynamics of the transition of patients from one clinical state to another.

Colorectal cancer is the third most common cancer and the second most frequent cause of cancer-related death. Worldwide, approximately half a million people die of it every year. Liver metastases develop in 40% of all patients with colorectal cancer. If metastatic disease is limited to the liver, resection (removal) is the curative treatment of choice. Survival rates of 40−50% for the first 5 years after initial treatment have been reported.

Contraindications for surgical treatment of metastatic disease are increasingly being abandoned, making way for the development of local ablative techniques [operative procedure in which tissue is ablated (destroyed) by diathermy, cryotherapy, abrasion, or other means]. Given these upcoming treatment modalities, more and more patients are qualifying for aggressive surgical treatment. Before going on to our description, it should be noted that data on health in this patient category were scarce in the literature at the time this study was conducted (2009).

For the Q-TWiST analyses, the patients were distributed over four categories. These represent the main clinical states of patients with colorectal cancer after initial treatment of liver metastases: dead; the state after noncurative surgery; the state after curative surgery with recurrent disease; and the disease-free state after curative surgery. The patients could transition to a subsequent category in a certain order (Fig. 3.6). For each state, the proportion of patients was calculated for the entire follow-up period of 3 years. The EQ−5D, a generic preference-based health instrument (Chapter 12), was used to determine mean health values for each of the four clinical states. In the Q-TWiST, the element of time was expressed as the interval between two

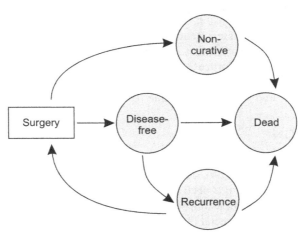

FIGURE 3.6   Transition diagram with possible clinical health states for patients with colorectal liver metastases.

measurements (note that, in this study, data were not registered upon the occurrence of an event but at fixed intervals). Finally, the mean duration of each clinical state was multiplied by the corresponding mean health value and the proportion of patients falling into that category. This yielded the number of QALYs accumulated for each clinical state during 3 years of follow-up.

A total of 136 patients were included in the study. The proportion of patients in each of the four clinical states and their distribution over time are depicted in Fig. 3.7, representing the Q-TWiST. Most patients were in the "disease-free" and "noncurative" states in the first few months. Not surprisingly, the percentage of patients with the states "recurrence" and "dead" rose in the course of the study. In total, 117 patients (86.0%) underwent successful surgical intervention (curative group), whereas 19 (14.0%) had a noncurative laparotomy because of inoperable disease at the time of surgery. The rate of disease-free survival among all 136 patients at 2 and 3 years was 32.4% (44 of 136) and 27.9% (38 of 136), respectively. The median time to disease recurrence was 8.5 (range 0−75) months. For patients with recurrent disease and no option for surgical reintervention, chemotherapy was started at a median of 12.0 (range 1−36) months after hepatic resection.

Three weeks after surgery, all patient groups showed a clear decline in health, as expressed in the EQ-5D values (Table 3.1). Thereafter, health values for three of the clinical groups showed distinct patterns over time (dead has a value of 0.0 by definition). In general, disease-free patients had the best health (0.78), whereas patients who had undergone noncurative surgery were doing worse than patients with recurrent disease. From the mean health values for each clinical state, it was clear that the disease-free group fared better than the group that had noncurative surgery or those who developed recurrent disease.

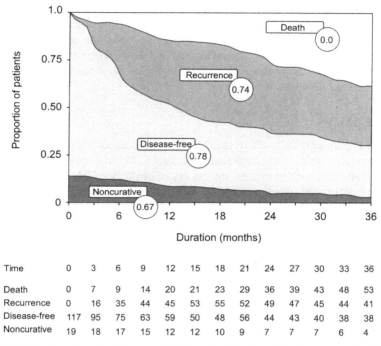

Time	0	3	6	9	12	15	18	21	24	27	30	33	36
Death	0	7	9	14	20	21	23	29	36	39	43	48	53
Recurrence	0	16	35	44	45	53	55	52	49	47	45	44	41
Disease-free	117	95	75	63	59	50	48	56	44	43	40	38	38
Noncurative	19	18	17	15	12	12	10	9	7	7	7	6	4

FIGURE 3.7   Proportion of patients ($n = 136$) in each of the four clinical states over time in the first 3 years after initial hepatic resection for colorectal metastases (patients may have moved from one state to another over the course of the study).

TABLE 3.1 Mean Health-Related Quality-of-Life Values for Each Clinical State, as Measured on the EQ–5D

State		Mean (s.d.) HRQoL	No of Observations	Maximum[a]	Minimum[a]
Death		0(0)	349	0	0
Disease-free		0.78(0.23)	891	0.92	−0.59
Noncurative		0.67(0.31)	162	0.92	−0.59
Recurrence		0.74(0.25)	450	0.92	−0.59
Chemotherapy	Without	0.82(0.17)	205	0.92	−0.43
	With	0.68(0.28)	245	0.92	−0.59

[a]Scale is from 0 (death) to 1 (perfect health).
HRQoL, Health-related quality of life.

Separate analyses were performed to express the number of QALYs gained for each clinical state. The disease-free group accumulated 0.78 QALYs annually over 3 years, the noncurative group 0.67, and the recurrence group 0.74. A total of 2.18 QALYs were collected in 3 years. Theoretically, a total of 3.0 QALYs could have been achieved if all patients had been in perfect health for the 3-year duration of follow-up.

DISCUSSION

One of the strengths of each of the three summary measures presented in this chapter is that they yield comprehensible results. Mainly, they are transparent because the measures to express the quality of health are scaled on a fixed unit (dead—full health). Such scaling is only possible if certain measurement frameworks are applied (Chapters 11 and 12). Because the values for the quality component (health state) of these summary measures are given as metrics, we can combine the states with their duration (i.e., survival, life years) to compute the summary statistic. In addition, a graphical representation can be drawn. Many common health-status measures or other clinical outcome measures either lack a clear unit of measurement or concern features that are only relevant to very specific disease areas.

An important question related to all three of these health summary measures is how to collect the quality components. Do we derive values from the literature, from expert opinion, from samples of the general population who place a value on hypothetical health states, or from actual patient judgments about their own health condition? It is clear that measuring the duration of health states may call for substantial logistic efforts, but in principle the task is rather straightforward. The problem, or rather the challenge—or perhaps the quest, to phrase it inspirationally—is the development of innovative methods to measure the quality of health.

In the standard QALY, health-state values are not derived from patients but are based on assessments collected from the general population (Chapter 12). For the DALY, "weights" for the disease conditions were originally supplied by selected experts. In the latest round of the Global Burden of Disease study, other methods were applied to elicit responses from large samples of the general population. For the Q-TWiST, expert opinions were initially used to attach values to specific health states. Eventually, however, the developers looked for other and better options. The Q-TWiST is usually applied in clinical studies by clinicians, who are less absorbed by measurement theory and do not hold strong convictions on how health should be measured. Sometimes this is a blessing, sometimes a curse.

Economic analyses of health care frequently apply QALYs, whereby costs are divided by the number of QALYs gained. Yet this measure is also useful in clinical settings when improved survival is associated with more toxic therapy, such as in the treatment of cancer. Methods for estimating QALYs can be

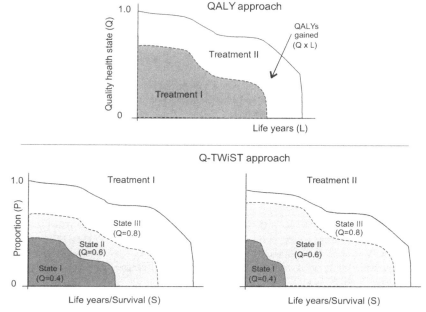

FIGURE 3.8 Conceptual and computational differences between the quality-adjusted life year (QALY) and the quality-adjusted time without symptoms of disease and toxicity of treatment (Q-TWiST).

complicated for various reasons. For example, a patient's time to death generally goes unnoticed, with some proportion surviving past the end of follow-up. Another reason is that the QALY is not applicable to every medical intervention. Indeed, we have already given an example of its drawbacks regarding IVF. But the QALY may also be unsuitable for diseases characterized by a very accidental course (e.g., migraine).

Basically, there are four main differences between QALYs and Q-TWiST (Fig. 3.8). First, QALYs presume reference to a wide range of optional health states, whereas the Q-TWiST refers to a limited number of distinct clinical health states. Second, for the QALY approach, valuations of health states are always based on preferences from a sample of the general population. For the Q-TWiST, in contrast, valuations of the distinct health states are based on expert opinion (physicians) or patient reports. Third, proportions of patients underpin the Q-TWiST, whereas the QALY puts more emphasis on quality values. Fourth, for QALYs, the element of time is expressed as the duration of the study or follow-up (in some instances extrapolated or combined with modeling). For the Q-TWiST, the element of time is expressed as survival (or time duration until event) and can be estimated through Kaplan–Meier or Cox regression analysis.

REFERENCES

Anand, S., Hanson, K., 1997. Disability-adjusted life years: a critical review. Journal of Health Economics 16 (6), 685–702.

Fanshel, S., Bush, J.W., 1970. A health-status index and its application to health-services outcomes. Operations Research 18 (6), 1021–1066.

Gelber, R.D., Cole, B.F., Gelber, S., Goldhirsch, A., 1996. Chapter 46: the Q-TWiST method. In: Spilker, B. (Ed.), Quality of Life and Pharmacoeconomics in Clinical Trials, second ed. Lippincott-Raven Publishers, Philadelphia.

Gelber, R.D., Cole, B.F., Gelber, S., Goldhirsch, A., 1995. Comparing treatments using quality-adjusted survival: the Q-TWiST method. The American Statistician 49 (2), 161–169.

Gelber, R.D., Gelman, R.S., Goldhirsch, A., 1989. A quality-of-life-oriented endpoint for comparing therapies. Biometrics 45 (3), 781–795.

Gelber, R.D., Goldhirch, A., 1986. A new endpoint for the assessment of adjuvant therapy in postmenopausal woman with operable breast cancer. Journal of Clinical Oncology 4 (12), 1772–1779.

Glasziou, P.P., Cole, B.F., Gelber, R.D., Hilden, J., Simes, R.J., 1998. Quality adjusted survival analysis with repeated quality of life measures. Statistics in Medicine 17 (11), 1215–1229.

Goldhirsch, A., Gelber, R.D., Simes, R.J., Glasziou, P., Coates, A., 1989. Costs and benefits of adjuvant therapy in breast cancer: a quality adjusted survival analysis. Journal of Clinical Oncology 7 (1), 36–44.

Grogono, A.W., Woodgate, D.J., 1971. Index for measuring health. Lancet 290 (7732), 1024–1026.

Klarman, H.E., Francis, J.O., Rosenthal, G.D., 1968. Cost effectiveness analysis applied to the treatment of chronic renal disease. Medical Care 6 (1), 48–54.

Lenderking, W.R., Gelber, R.D., Cotton, D.J., Cole, B.F., Goldhirsch, A., Volberding, P.A., Testa, M.A., 1994. Evaluation of the quality of life associated with zidovudine treatment in asymptomatic human immunodeficiency virus infection. New England Journal of Medicine 330 (17), 738–743.

Menken, M., Munsat, T.L., Toole, J.F., 2000. The global burden of disease study. Archives of Neurology 57 (3), 418–420.

Mills, M., 2001. Introducing Survival and Event History Analysis. SAGE Publications, London.

Murray, C.J.L., Lopez, A.D., 1997. Alternative projections of mortality and disability by cause 1990–2020: global burden of disease study. Lancet 349 (9064), 1498–1504.

Murray, C.J.L., Salomon, J.A., Mathers, C.D., 2002. A critical examination of summary measures of population health. In: Murray, C.J.L., Salomon, J.A., Mathers, C.D., Lopez, A.D. (Eds.), Summary Measures of Population Health: Concepts, Ethics, Measurement and Applications. World Health Organization, Geneva, pp. 41–51.

Price, K., Gelber, R., Isley, M., Goldhirsch, A., Coates, A., Castiglione, M., 1987. Time without symptoms and toxicity (TWIST): a quality-of-life-oriented endpoint to evaluate adjuvant therapy. Adjuvant Therapy of Cancer 5, 455–465.

Salomon, J.A., 2014. Disability-adjusted life years. In: Culyer, A.J. (Ed.), Encyclopedia of Health Economics. Elsevier, Amsterdam, pp. 200–203.

Salomon, J.A., Vos, T., Hogan, D.R., Gagnon, M., Naghavi, M., Mokdad, A., Begum, N., Shah, R., Karyana, M., Kosen, S., Farje, M.R., Moncada, G., Dutta, A., Sazawal, S., Dyer, A., Seiler, J., Aboyans, V., Baker, L., Baxter, A., Benjamin, E.J., Bhalla, K., Abdulhak, A.B., Blyth, F., Bourne, R., Braithwaite, T., Brooks, P., Brugha, T.S., Bryan-Hancock, C., Buchbinder, R., Burney, P., Calabria, B., Chen, H., Chugh, S.S., Cooley, R., Criqui, M.H., Cross, M.,

Dabhadkar, K.C., Dahodwala, N., Davis, A., Degenhardt, L., Díaz-Torné, C., Dorsey, E.R., Driscoll, T., Edmond, K., Elbaz, A., Ezzati, M., Feigin, V., Ferri, C.P., Flaxman, A.D., Flood, L., Fransen, M., Fuse, K., Gabbe, B.J., Gillum, R.F., Haagsma, J., Harrison, J.E., Havmoeller, R., Hay, R.J., Hel-Baqui, A., Hoek, H.W., Hoffman, H., Hogeland, E., Hoy, D., Jarvis, D., Karthikeyan, G., Knowlton, L.M., Lathlean, T., Leasher, J.L., Lim, S.S., Lipshultz, S.E., Lopez, A.D., Lozano, R., Lyons, R., Malekzadeh, R., Marcenes, W., March, L., Margolis, D.J., McGill, N., McGrath, J., Mensah, G.A., Meyer, A.C., Michaud, C., Moran, A., Mori, R., Murdoch, M.E., Naldi, L., Newton, C.R., Norman, R., Omer, S.B., Osborne, R., Pearce, N., Perez-Ruiz, F., Perico, N., Pesudovs, K., Phillips, D., Pourmalek, F., Prince, M., Rehm, J.T., Remuzzi, G., Richardson, K., Room, R., Saha, S., Sampson, U., Sanchez-Riera, L., Segui-Gomez, M., Shahraz, S., Shibuya, K., Singh, D., Sliwa, K., Smith, E., Soerjomataram, I., Steiner, T., Stolk, W.A., Stovner, L.J., Sudfeld, C., Taylor, H.R., Tleyjeh, I.M., Van Der Werf, M.J., Watson, W.L., Weatherall, D.J., Weintraub, R., Weisskopf, M.G., Whiteford, H., Wilkinson, J.D., Woolf, A.D., Zheng, Z.J., Murray, C.J.L., 2012. Common values in assessing health outcomes from disease and injury: disability weights measurement study for the Global Burden of Disease Study 2010. Lancet 380 (9859), 2129−2143.

Smith, J.N., 2015. Epic Measures: One Doctor. Seven Billion Patients. Harperwave, New York.

Torrance, G.W., 1970. A Generalized Cost-effectiveness Model for the Evaluation of Health Programs. McMaster University, Hamilton.

Torrance, G.W., Thomas, W.H., Sackett, D.L., 1972. A utility maximization model for evaluation of health care programs. Health Service Research 7 (2), 118−133.

Voigt, K., King, N.B., 2014. Disability weights in the global burden of disease 2010 study: two steps forward, one step back? Bulletin of the World Health Organization 92 (3), 226−228.

Weinstein, M.C., Stason, W.B., 1976. Hypertension: A Policy Perspective. Harvard University Press, Cambridge.

Weinstein, M.C., Stason, W.B., 1977. Foundations of cost-effectiveness analysis for health and medical practices. The New England Journal of Medicine 296 (13), 716−721.

Wiering, B., Oyen, W.J.G., Adang, E.M.M., van der Sijp, J.R.M., Roumen, R.M., de Jong, K.P., Ruers, T.J.M., Krabbe, P.F.M., 2011. Long-term global quality of life in patients treated for colorectal liver metastases. The British Journal of Surgery 98 (4), 565−571.

Williams, A., 1985. Economics of coronary artery bypass grafting. BMJ 291 (6491), 326−329.

World Bank, 1993. World Development Report 1993: Investing in Health. Oxford University Press, New York.

Zeckhauser, R., Shepard, D.S., 1976. Where now for saving lives? Law and Contemporary Problems 40 (4), 5−45.

Part II

Chapter 4

Measurement

Chapter Outline

INTRODUCTION

Early in the history of humankind, parts of the human body and the natural surroundings were used as measuring devices. Ancient records (Babylonian, Egyptian, and Biblical) indicate that length was first measured with the foot, forearm, hand, or finger. Time was measured by the periods of the sun, moon, and other heavenly bodies. With the invention of numbering systems and the development of mathematics, it was possible to create whole systems of measuring units suitable for trade and commerce, land division, taxation, and scientific research.

MEASUREMENT IN GENERAL

The Magna Carta (Latin for "Great Charter") set in train the long process that led to the rule of constitutional law as we know it today. Drafted in 1215 by the Archbishop of Canterbury to make peace between the king and a group of rebel barons, the latter were promised the upholding of church rights, protection from illegal imprisonment, access to swift justice, and limits on payments to the Crown, all to be implemented through a council. The king had to renounce certain rights, respect certain legal procedures, and accept that his will could be bound by law. But the charter also introduced a unit of measurement to represent a standardized quantity of a physical property:

> *Let there be one measure of wine throughout our whole realm; and one measure of ale; and one measure of corn, to wit, "the London quarter"; and one width of cloth (whether dyed, or russet, or "halberget"), to wit, two ells within the selvedges; of weights also let it be as of measures.*

The Magna Carta is still seen as an early initiative to set standards by imposing units of measurement, the first attempt to streamline the plethora of measurements that were used in Europe. Today, we see that the basic doctrine of "one measure" enshrined in the Magna Carta is almost universal practice. In commercial and official life, common measurements make our society more fair and safe. While it may be the English who started standardizing measures to make society more manageable, we have the French to thank for the metric system. Yet despite their taste for regulation, laissez-faire runs deep in France, as noted by the renowned military leader and statesman Charles de Gaulle: "How can you govern a country which has two hundred and forty-six varieties of cheese?" (1962).

The impact of dual measuring systems can be catastrophic, as in the following example. A news story in September 1999 reported how the National Aeronautics and Space Administration (NASA) lost a 125-million Mars orbiter in a crash onto the surface of Mars because a Lockheed Martin engineering team used the English units of measurement while the agency's team used the more conventional metric system for a key operation of the spacecraft. The unit's mismatch prevented navigation information from transferring between the Mars Climate Orbiter team at Lockheed Martin in Denver and the flight team at NASA's Jet Propulsion Laboratory in Pasadena. In short, nonstandardization of measuring units can have a dramatic impact on scientific endeavor.

Temperature

An early instrument to measure temperature, developed in the 1650s in Florence, depended on the expansion and contraction of alcohol within a glass tube. As temperature rises, alcohol expands rapidly though not at a regular speed, making accurate readings difficult. After the Florentine thermometer had been in use for half a century, a German glass blower and instrument maker working in Holland, Gabriel Fahrenheit, improved on its design by replacing alcohol with mercury. Mercury expands less than alcohol (about 7 times less for the same rise in temperature), but it does so in a more regular manner. Seeing the advantage of this regularity, Fahrenheit constructed the first mercury thermometer, which subsequently became the standard. The only practical method to turn this thermometer into a true (metric) measurement instrument was to choose two temperatures, which could be established independently, mark them on the tube, and divide the intervening length into equal degrees (Fig. 4.1). In 1701, Isaac Newton had proposed adopting the freezing point of water for the bottom of the scale and the temperature of the human body for the top end. However, Fahrenheit, who was accustomed to the cold winters that Holland was experiencing at that time, wanted to include temperatures below the freezing point of water. He therefore accepted the temperature of blood for the top of his scale but adopted the freezing point of salt water for the bottom. Then, in 1742, the Swede Anders Celsius proposed an early example of decimalization for his scale, which expressed the freezing and boiling temperatures of water as 0° and 100°, respectively. It took more than two centuries for English-speaking countries to embrace this less complicated system. Even today, the Fahrenheit unit of temperature is more prevalent in some countries, notably the United States.

Meter

After the French Revolution, the French Academy of Sciences determined (in 1799) that a meter would equal one ten-millionth of the distance from the north pole to the equator along the meridian running near Dunkirk in France and Barcelona in Spain (Robinson, 2007). Then, in 1875, a milestone was reached in international understanding and standardization; the International Treaty of the Meter established a basis for comparative measurement. While that agreement is still in force, it is now defined by the International Bureau of Weights and Measures as the distance that light travels in absolute vacuum in 1/299,792,458 of a second.

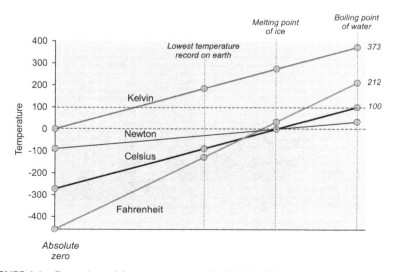

FIGURE 4.1 Comparison of four temperature scales (Celsius, Fahrenheit, Kelvin, Newton).

Kilogram

The kilogram was originally defined as the mass of 1 L of pure water at standard atmospheric pressure and at the temperature at which water has its maximum density (3.98°C). This definition was hard to apply accurately. In part, that is because the density of water depends slightly on the pressure, and pressure units include mass as a factor, thereby introducing a circular dependency into the definition. Since 1889 the magnitude of the kilogram has been defined as the mass of an object called the international prototype kilogram (Fig. 4.2). Today the kilogram is defined as a fixed number of Silicon-28 atoms.

MEASUREMENT DEFINITIONS

The literature on what measurement is (Table 4.1) and how measurements may be used comes from diverse backgrounds, with contributions from mathematicians and philosophers as well as researchers in the physical, social—behavioral, life, and other sciences (Hand, 2004).

The last definition on the list, taken from Wikipedia, is widespread on the web and captures the essence of measurement. Fundamentally, measurement is the act or the result of a quantitative comparison between a predefined standard and an unknown magnitude (Badiru, 2015). The first requirement is a standard to be used for comparison. A weight cannot simply be heavy; it can only be proportionately as heavy as something else, namely the standard. Unless a comparison is made relative to something generally recognized as the standard, the measurement can only have a limited meaning. As we shall see,

FIGURE 4.2 The International prototype of the kilogram, machined in 1878, is a cylinder of platinum—iridium alloy (Pt 90% and Ir 10% in mass) whose height (39 mm) is equal to its diameter. Six copies were designated as official copies and are kept in the same protective conditions as the international prototype. This is a replica for public display, shown as it is normally stored, namely under two bell jars.

TABLE 4.1 Different Definitions of (Subjective) Measurement

Source	Definition
Campbell (1938, p. 126)	The assignment of numerals to represent properties of material systems other than number, in virtue of the laws governing these properties.
Russell (1996, p. 176)	In its most general sense, any method by which a unique and reciprocal correspondence is established between all or some of the magnitudes of a kind and all or some of the numbers, integral, rational, or real as the case may be.
Stevens (1951, p. 22)	The assignment of numerals to objects or events according to rules.
Torgerson (1958, p. 14)	It involves the assignment of numbers to systems to represent that property.
Messick (1983)	May be viewed as loosely integrated conceptual frameworks within which are embedded rigorously formulated statistical models of estimation and inference about the properties of measurements and scores.
Nunnally & Bernstein (1994)	One part of successful quantification of health states involves choosing an empirical judgment procedure that is appropriate to the subjects' ability to respond; another part is to use a theory appropriate to the collected data.
Elshaw (2015)	The codifying of observations into data that can be analyzed, portrayed as information, and evaluated to support the decision maker.
Wikipedia	The assignment of a number to a characteristic of an object or event, which can be compared with other objects or events.

this certainly applies in the field of health, where a unit of measurement is often lacking or not clearly interpretable.

DATA THEORY

The problem of measurement was debated for 7 years by a committee of the British Association for the Advancement of Science. Appointed in 1932 to represent Section A (Mathematical and Physical Sciences) and Section J (Psychology), the committee was instructed to consider and report upon the possibility of "quantitative estimates of sensory events." In other words, is it

---

BOX 4.1

Stanley Smith (S.S.) Stevens built a renowned laboratory of experimental psychology in the basement of Memorial Hall at Harvard. His research focused on the relationship between the perceived magnitude of a stimulus and its objective physical magnitude. A pioneer in psychoacoustics, Stevens spent his entire career at Harvard shaping the department of psychology. His accomplishments also extended into the realm of quantitative data analysis, and in 1946 he published "On the theory of scales of measurement." He invented many gadgets, including a short downhill ski, which he passionately defended in many articles as superior to conventional skis.

---

possible to measure human sensation? Deliberation led only to disagreement, mainly about what is meant by the term measurement. In an interim report in 1938, one member complained that his colleagues "came out by the same door as they went in" and, hoping for another try at agreement, the committee begged to be allowed to continue for another year (Stevens, 1946).

In the early 1940s, the Harvard psychologist Stanley Stevens (Box 4.1) concluded that agreement can better be achieved if we recognize that measurement exists in a variety of forms and that scales of measurement fall into certain definite classes (Stevens, 1946). He then coined the terms nominal, ordinal, interval, and ratio to describe the hierarchy of measurement levels used in psychophysics. In addition, he classified the data transformations that were allowed for each of these measurement levels. This taxonomy, which was subsequently adopted for several important textbooks, has influenced the reasoning on measurement. In a later article, Stevens (1951) went beyond his simple typology to classify not just simple operations but also statistical procedures that were "permissible" according to the scales. Stevens' work provoked a great deal of controversy at the time (Nunnally and Bernstein, 1994). Since then, several authors have suggested alternative taxonomies of data. Nevertheless, the introduction of a hierarchy to describe the nature of information that lies within the numbers assigned to variables has had a major impact on thinking about such issues (Fig. 4.3).

Nominal

The lowest level of measurement is on the nominal scale, which classifies items (e.g., questions) into categories that are mutually exclusive and collectively exhaustive. That is, the categories do not overlap, and together they cover all possible characteristics being observed. For example, pain can be localized in the "head," "back," or "legs"; gender and the color of light are also measured on a nominal scale. The nominal scale groups variables such as the type of disease or country of birth. Since this is simply a method of grouping

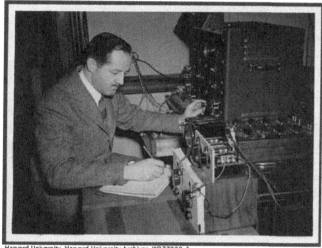

Harvard University, Harvard University Archives, W277900_1

FIGURE 4.3   S.S. Stevens (1906–73). *With permission from: HUP Stevens (4b), olvwork277900, Harvard University Archives.*

data, no mathematical operations are involved. For example, one might want to organize and describe types of disease in relation to the organs affected. We would collect data and then organize the information according to the diseases and their main effect on organs (i.e., lungs, heart, kidney, etc.). The groups we construct are just slots for organizing our data (i.e., classification) according to some predetermined criteria.

Ordinal

An ordinal scale is distinguished from a nominal scale by the order imposed on the categories. For example, the side effects of a treatment for a specific disease are classified as severe, moderate, or minor. We know that "first" is above "second," but we do not know how far above. Similarly, we know that "less" is preferable to "severe," but we do not know by how much. This type of scale simply refers to the order in which measurements are placed. It differs from a nominal scale in that the order of the categories actually matters in an ordinal scale.

A graded classification of chronic pain consists of an ordered set of categories. The ranking corresponds to qualitative differences in global severity. In many areas of medicine, graded classifications facilitate communication between researchers and clinicians, providing a simple way of describing the severity of the patients' condition and summarizing complex data. A graded classification may distinguish the stages of a disease that typically occur in sequence across time (e.g., the staging of tumor progression). Another example is a CT scan classification (Traumatic Coma Data Bank) of head injury that is used in clinical decision making (Vos et al., 2001).

In fact, many of the classifications used in hospitals and elsewhere have little to do with measurement. Moreover, many of these contain inconsistencies. While most register what clinicians consider relevant, this approach often diminishes their usefulness. For example, the Traumatic Coma Data Bank classification is meant to predict the development of intracranial hypertension or a fatal outcome. In that respect it does not seem to make sense that one of the categories is whether the patient has been operated on or not. Its inclusion implies that patients are differentiated not only by their CT characteristics (predictors) but also by treatment characteristics (outcome).

Interval

An interval scale differs from an ordinal scale by having equal intervals between the units of measurement. Temperature is a good example of an attribute that is measured on an interval scale. Even though there is a zero point on the temperature scale, it is an arbitrary and relative measure. Whether measured in Fahrenheit or Celsius, in both instances an interval scale is used, yet the unit of measurement is different. Note that the difference between 80° and 90° is the same as the difference between 20° and 30°. The interval provides information about the relative distance between measurements on such scales. Differences between two values on an interval scale have the same significance, no matter where the measured points are located. For instance, the difference on a health-status scale, as used in computing quality-adjusted life years (Chapters 3 and 12), between 0.4 and 0.6 must be understood as equal to the difference between 0.7 and 0.9. This requirement is necessary because the scores must reflect value in such a way that units along the scale have the same worth.

As noted earlier, the zero points on the common temperature scales (i.e., Fahrenheit, Celsius) were experimentally established and are not true points such as one might find in a counting system. Other examples of interval scales are IQ tests and blood pressure. An interval scale has equal distances between the obtained scores but does not include an absolute zero point. As the zero point is arbitrary, it does not mean anything; it is just another point on the scale. Another example is the Likert-type scale (Chapter 5), which is used to measure behavioral constructs but also health.

Ratio

A ratio scale has the same properties as an interval scale, except that the zero point is relevant here; it means something specific that the researcher can interpret. At zero, none of the phenomenon being measured exists. For example, an estimate of zero time units for the duration of a task is a ratio-scale measurement. Other examples of attributes measured on a ratio scale are cost, volume, length, height, and weight. Many aspects measured in engineering systems will be on a ratio scale. When computing quality-adjusted and disability-adjusted life years (Chapter 3), special measurement methods (Chapters 11 and 12) are applied to arrive at generic health measures that are on a ratio scale (e.g., dead = 0.0). Table 4.2 summarizes the four types of data classification described earlier.

Other Scale Types

Stevens' framework was later criticized by statisticians and others for its coarseness, and successive data theorists have introduced intermediate measurement levels. The original typology has been elaborated, especially by inserting "nonmetric" scales between the nominal and interval levels and by distinguishing between different types of ordinal data.

As part of his "theory of data," Coombs proposed a measurement level between the strictly ordinal and the interval, calling it the "ordered metric" level (Coombs, 1950). Coombs intended it to be used where an ordering exists between the scores and when information is available on the order of the magnitude of some of the constituent intervals. This type of scale is called "ordered metric" because such pairwise differences provide information on the order of the magnitude of the differences. Later, other measurement levels were introduced in the area of multidimensional scaling, notably by Shepard (1962a,b). It became clear that responses collected under specific conditions, though falling short on the interval criterion, nevertheless yielded more information than ordinal data (Fig. 4.4).

The distinctions between these various types of ordering elucidate some of the issues that arise in measurement approaches such as those worked out under choice models (Chapter 11) and in special methods such as nonmetric multidimensional scaling (Chapter 14).

MEASUREMENT THEORY

Counting is the basis for quantification. Measurement is deduced from well-defined sets of counts. The most elementary way to measure is to count the presence of a defined event. But more information is obtained when the conditions that identify countable events can be ordered into successive categories that increase in status along some underlying scale. It then becomes possible to count the number of steps in the ordered set of categories. So doing, we advance beyond knowing whether a phenomenon is present or not and are able to express the level of an event or condition.

TABLE 4.2 Four Types of Data Classification

Level of Measurement	Specific Characteristic	Basic Operation	Permissible Transformations	Permissible Statistics	Examples
Nominal	Classification	Equality or not ($=$, $\neq$)	Any one-to-one	Number of cases, mode	Gender, color
Ordinal	Order	Greater than or lesser than ($<$, $>$)	Monotonically increasing	Median, percentiles, order statistics	Hardness of minerals, class rank
Interval	Relative	Equality of difference ($+$, $-$)	General linear	Mean, variance, correlation	IQ, temperature (Celsius)
Ratio	True zero	Equality of ratios ($\times$, $\div$)	Multiplicative	Idem	Age, temperature (Kelvin)

Adapted from Stevens, S.S., 1951. Mathematics, measurement, and psychophysics. In: Stevens, S.S. (Ed.), Handbook of Experimental Psychology. Wiley, New York.

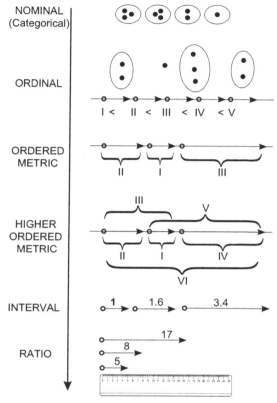

FIGURE 4.4   Diagrammatic classification of scale types and their data. *Adapted from Coxon, A.P.M., 1982. The User's Guide to Multidimensional Scaling. Heinemann Educational Books, London.*

Scores

Quantitative science begins with identifying conditions and events that are deemed worth counting. The resulting counts are sometimes called "raw scores" to distinguish them from "scaled" scores. The former is a score without any sort of adjustment or transformation; one example is the number of questions answered correctly. A scaled score is the result of some transformation applied to a raw score. Usually, "raw scores" are just called "scores" and are mere counts of observed events. While these may be essential for the construction of measures, they are not yet measures in themselves. Raw scores are often mistaken for measures and then misused as such.

Measures

To every scientist, cook, and carpenter, a "measure" means a number with which arithmetic can be done, a number that can be added and subtracted, and the differences among which can be multiplied and divided giving results that retain a numerical meaning. In any science, original observations are never measures in this sense. They cannot be, because a measure implies and requires the prior construction and maintenance of a necessarily abstract measuring system that has proven useful (Wright and Linacre, 1989).

However, not all physical measurement is direct, nor is it always concerned with a unique phenomenon. For example, density (mass divided by volume) cannot be measured directly because it is defined in terms of two separate properties, each with its own units of measurement. So the meaning of a measure should be expanded to include derived measures, which are based on established measures.

Subjective Measurement

When psychology was separated off from philosophy at the end of the 19th century, the new field sought to quantify subjective phenomena. The effort continues to this day, building on a succession of groundbreaking events (Barofsky, 2012). The milestones include Fechner's (1860) development of psychophysics, Campbell's (1920) and Bridgman's (1927) attempts to define measurement, the Ferguson Commission's (1940) inquiry into whether qualitative measurement is possible, Stevens' (1946) work on operational definitions and scales, and Luce and Tukey's (1964) development of conjoint measurement theory. Noteworthy reviews and comments on measurement include Fraser (1980), Michell (1986, 1997), Cliff (1992), and Borsboom (2005). Anyone not connected to this type of research will have trouble following the work of measurement theorists. It is formal, dense with definitions, annotations, and proofs of mathematical and logical propositions. We return to the issue of subjectivity in Chapter 6, where we explain how "objective" differs from "subjective" in the context of health measurement.

Representational Approach

Mathematical theories of measurement (often referred to collectively as "measurement theory") concern the conditions under which relations among numbers (and other mathematical entities) can be used to express relations among objects. One strong theory of (subjective) measurement is the representational position (or the "fundamentalist" position), named thus because it states that scale measures represent empirical relations among objects (Michell, 1986). In the mid-twentieth century, the two main lines of inquiry in measurement theory converged. Interest in the empirical conditions of quantification merged with interest in the classification of scales, thereby forming the representational theory of measurement. This remains the most influential mathematical theory of measurement to date. A rather complex mathematical work on fundamental measurement, it was published in three volumes by Krantz et al. (1971), Suppes et al. (1989), and Luce et al. (1990).

As a simple example of the representational theory, let us consider this approach to define the equivalence (" = ") of the quality of two health conditions. Equivalence requires transitivity, symmetry, and reflexivity. For example, "transitivity" means that the relation passes across health conditions. If patient A and patient B are considered eligible for a lung transplant and if patient C and patient D are also eligible, then patient A and patient D must both be classified as eligible. "Symmetry" means that the relationship extends in both directions. What has concerned the representationalists most is whether a particular measurement approach achieves interval-level status. Only then would comparisons between different measures be both meaningful and interpretable. And indeed, that same concern is also very important in most, if not all, of the health measurement methods discussed in this book.

MEASUREMENT PRINCIPLES

Although there are many principles that have to hold in formal systems of measurement, here we briefly present only three, highlighting those pertaining to health instruments in practical studies. These principles, or required conditions, are particularly relevant to instruments to measure health outcomes. Depending on the measurement framework that is used (Chapter 6), these principles may be applied for better or for worse.

Invariance

It is easy to imagine the chaos that would result if some physical measures lacked the invariance of ratio scales. Without invariance, a stick that is twice as long as another when measured in feet might be three times as long when measured in inches. The range of invariance of a scale determines the extent to which principles remain unaffected by expressing the scale in different units, e.g., feet rather than inches. This does not mean that the results of using the scale will not change. A mean temperature recorded in degrees Fahrenheit will be numerically different than a mean temperature in degrees Celsius even though calculating the mean is permissible for both.

It is important to consider the circumstances under which a particular type of scale maintains its measurement properties when the unit of measurement is changed. Following the scale taxonomy of Stevens, the higher the level in this hierarchy, the more powerful the mathematical operations, but the less free one is to change the measurement scale. Thus, the labels on a nominal scale may be changed in an almost unlimited manner as long as no two categories are given the same label. But at the other extreme, ratio scales lose their measurement properties when changed in any way.

Another consequence of the invariance principle shows up in the more general measurement frameworks such as item response theory (Chapter 10), choice models (Chapter 11), and valuation techniques (Chapter 12). According to this principle, the response on two (or more) attributes should be independent of the group of respondents who performed the task and judgments among the attributes should be independent of the set of attributes being presented.

The representation measurement theorists have their own way of expressing what invariance is about: "In Section 22.2.7 we argued that a necessary condition for a numerical relation T to represent an empirically meaningful relation is the invariance of T under permissible transformations of the numerical structure $\Re$. The reason is that if φ is a homomorphism and γ a permissible transformation, then $\gamma\varphi$ is also a homomorphism. For φ and $\gamma\varphi$ to induce the same relation on A, i.e., $S(\varphi, T) = S(\gamma\varphi, T)$, we must have γ $[\varphi(A)^k \cap T] \subseteq T$" (Luce et al., 1990).

Unit of Measurement

Measures of health-outcome instruments are generally unclear to most practitioners. This is particularly true for the more subjective instruments. A level of 210 mmHg is easily recognized as high blood pressure. But physicians typically have no idea whether or not a health score of 78 points on a certain health domain is meaningfully different from one of 69. For instance, what is the meaning of a two-point reduction over a 6-month period when anxiety is assessed on a 20-point scale? Is it a trivial or meaningful difference?

The difficulty of giving meaning to the findings has largely to do with the fact that most instruments measure on an interval level. This is a typical feature of almost all multiitem profile instruments (Chapter 9). Differences on an interval-level scale may be interpretable in the sense that an increase from 69 to 78 is more (better) than an increase from 69 to 74. But it is difficult and sometimes impossible to interpret these measures in an absolute sense, because such scales usually lack a meaningful upper and lower anchor. To obtain meaningful measures, the measurement scale must be on a ratio level (with fixed and meaningful lower and upper endpoints). If a measurement scale lacks such properties, a meaningful use is only possible when the scale is applied so often that, based on lengthy practical experience, one will be competent at interpreting the measures of the scale.

Dimensionality

Although there is no universally accepted definition for health or health status, the intrinsic multidimensionality of these concepts is well accepted. A single domain, e.g., physical functioning or fatigue, is not considered to capture a health status even though it may be patient reported. Nor does a single domain justify a claim of improvement in global health status. Within the field of health measurement there are two different and somewhat contradictory schools of thought. Both agree that health is a multidimensional phenomenon. Yet one side believes that this characteristic should be preserved at all costs and that health can only be represented as a profile of scores across distinct health domains. The other side, in contrast, believes that health can be projected on a unidimensional continuum.

A crucial assumption for deriving informative measures should be kept in mind, namely that the underlying phenomena can be represented on a unidimensional scale. But this may be questionable for quantifying a subjective phenomenon such as health, which has a rather broad scope. Nevertheless, from a measurement perspective, we may assume that generic health outcomes can be seen as unidimensional concepts. Of course, all data are multidimensional to some extent. However, conceptually and analytically, unidimensionality does not imply only one factor, domain, or dimension. Rather, it refers to the presence of a dominant factor and possibly minor ones that do not (substantially) affect the dominant one.

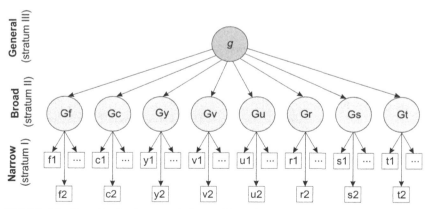

FIGURE 4.5   Second-order factor analysis—an illustration of John B. Carroll's three-stratum theory, an influential contemporary model of cognitive abilities. The broad abilities recognized by the model are fluid intelligence (Gf), crystallized intelligence (Gc), general memory and learning (Gy), broad visual perception (Gv), broad auditory perception (Gu), broad retrieval ability (Gr), broad cognitive speediness (Gs), and processing speed (Gt). Carroll regarded the broad abilities as different "flavors" of g. *Adapted from Bates, T., 2013. Own Work, CC BY-SA 3.0. From: https://en.wikipedia.org/wiki/Three-stratum_theory.*

A well-known example of such a hierarchical structure is the intelligence factor "*g*." The *g* factor (also known as general intelligence or general mental ability) is a construct that was developed during investigations of cognitive abilities and human intelligence (Neisser et al., 1996). It summarizes positive correlations among different cognitive tasks, reflecting the fact that an individual's performance on one type of task tends to be comparable to that person's performance on other types. The existence of the *g* factor was originally proposed by the English psychologist Charles Spearman in the early years of the 20th century. He suggested that all mental performance could be conceptualized in terms of a single general ability factor, which he labeled "*g*." Today's (common) factor models of intelligence typically represent cognitive abilities as a three-level hierarchy. It has a large number of narrow attributes at the bottom and a handful of broad, more general factors at the intermediate level. And at the apex it has a single factor, *g*, which represents the variance common to all cognitive tasks (Fig. 4.5).

REFERENCES

Badiru, A.B., 2015. Fundamentals of measurement systems. In: Badiru, A.B., Racz, L.A. (Eds.), Handbook of Measurements: Benchmarks for Systems Accuracy and Precision. CRC Press, Boca Raton, pp. 1–10.

Barofsky, I., 2012. Quality: Its Definition and Measurement as Applied to the Medically Ill. Springer, New York.

Bates, T., 2013. Own Work, CC BY-SA 3.0. From: https://en.wikipedia.org/wiki/Three-stratum_theory.

Borsboom, D., 2005. Measuring the Mind: Conceptual Issues in Contemporary Psychometrics. Cambridge University Press, Cambridge.

Bridgman, P.W., 1927. The Logic of Modern Physics. Macmillan, New York.

Campbell, N.R., 1920. Physics: The Elements. Cambridge University Press, Cambridge.

Campbell, N.R., 1938. Symposium: measurement and its importance for philosophy. In: The Proceedings of the Aristotelian Society. Supplementary, vol. 17. Harrison & Sons, London.

Cliff, N., 1992. Measurement theory and the revolution that never happened. Psychological Science 3, 186–190.

Coombs, C.H., 1950. Psychological scaling without a unit of measurement. Psychological Review 57 (3), 145–158.

Coxon, A.P.M., 1982. The User's Guide to Multidimensional Scaling. Heinemann Educational Books, London.

De Gaulle, C., 1962. Les mots du général. A. Fayard, Ernest Mignon, Paris.

Elshaw, J.J., 2015. Social science measurement. In: Badiru, A.B., Racz, L.A. (Eds.), Handbook of Measurements: Benchmarks for Systems Accuracy and Precision. CRC Press, Boca Raton, pp. 203–216.

Fechner, G.T., 1860. Elemente der Psychophysik. Breitkopf & Hartel, Leipzig.

Ferguson, A., Myers, C.S., Bartlett, R.J., Banister, H., Bartlett, F.C., Brown, W., Campbell, N.R., Craik, K.J.W., Drever, J., Guild, J., Houstoun, R.A., Irwin, J.O., Kaye, G.W.C., Philpott, S.J.F., Richardson, L.F., Shaxby, J.H., Smith, T., Thouless, R.H., Tucker, W.S., 1940. Quantitative estimates of sensory events: final report of the committee appointed to consider and report upon the possibility of quantitative estimates of sensory events. Report of the British Association Advancement of Science 2, 331–349.

Fraser, C.O., 1980. Measurement in psychology. British Journal of Psychology 71, 23–34.

Hand, D.J., 2004. Measurement Theory and Practice: The World through Quantification. Arnold, London.

Krantz, D.H., Luce, D.R., Suppes, P., Tversky, A., 1971. Foundations of Measurement. In: Additive and Polynomial Representations, vol. I. Dover Publications, Mineola, NY.

Luce, D.R., Krantz, D.H., Suppes, P., Tversky, A., 1990. Foundations of Measurement. In: Representation, Axiomatization and Invariance, vol. III. Dover Publications, Mineola, NY.

Luce, R.D., Tukey, J.W., 1964. Simultaneous conjoint measurement: a new type of fundamental measurement. Journal of Mathematical Psychology 1 (1), 1–27.

Michell, J., 1986. Measurement scales and statistics: a clash of paradigms. Psychological Bulletin 100, 398–407.

Michell, J., 1997. Quantitative science and the definition of measurement in psychology. British Journal of Psychology 88 (3), 355–383.

Messick, S., 1983. Assessment of children. In: Mussen, P.H. (Ed.), Handbook of Child Psychology, History, Theory and Methods, vol. 1. Wiley, New York, pp. 477–526.

Neisser, U., Boodoo, G., Bouchard, T.J., Boykin, A.W., Brody, N., Ceci, S.J., Halpern, D.F., Loehlin, J.C., Sternberg, R.J., Urbina, S., 1996. Intelligence: knowns and unknowns. American Psychologist 51 (2), 77–101.

Nunnally, J.C., Bernstein, I.H., 1994. Psychometric Theory, third ed. McGraw-Hill, New York.

Robinson, A., 2007. The Story of Measurement. Thames & Hudson, London.

Russell, B., 1996. (Reprint From 1938; Originally Published 1903). Principles of Mathematics. Norton, New York.

Shepard, R.N., 1962a. The analysis of proximities: multidimensional scaling with an unknown distance function: I. Psychometrika 27 (2), 125–140.

Shepard, R.N., 1962b. The analysis of proximities: multidimensional scaling with an unknown distance function: II. Psychometrika 27 (3), 219–246.

Stevens, S.S., 1951. Mathematics, measurement, and psychophysics. In: Stevens, S.S. (Ed.), Handbook of Experimental Psychology. Wiley, New York.

Stevens, S.S., 1946. On the theory of scales of measurement. Science 103 (2684), 677–680.

Suppes, P., Krantz, D.H., Luce, D.R., Tversky, A., 1989. Foundations of Measurement Volume II: Geometrical, Threshold, and Probabilistic Representations. Dover Publications, Mineola, NY.

Torgerson, W.S., 1958. Theory and Methods of Scaling. John Wiley & Sons, New York.

Vos, P.E., van Voskuilen, A.C., Beems, T., Krabbe, P.F., Vogels, O.J., 2001. Evaluation of the traumatic coma data bank computed tomography classification for severe head injury. Journal of Neurotrauma 18 (7), 649–655.

Wikipedia. From: https://en.wikipedia.org/wiki/Measurement.

Wright, B.D., Linacre, J.M., 1989. Differences between scores and measures. RASCH Measurement Transactions 3 (3), 63. http://www.rasch.org/rmt/rmt33a.htm.

Chapter 5

Constructs and Scales

Chapter Outline

INTRODUCTION

Inspired by early work in Germany by Weber (1795−1878) and Fechner (1801−87), many empirical experiments have been conducted and several theories have been developed to quantify physical and especially sensory phenomena such as weights, tones, and flashes of light (Stevens, 1966). However, intelligence, psychopathology, and quality of health, like many other abstract phenomena, are neither physical nor psychophysical, because no physical element underlies these concepts.

Psychologists and educators have been grappling with the issue of measuring abstract phenomena for many years, dating back to European attempts at the turn of the 20th century to assess individual differences in intelligence. Particularly since the 1930s, much has been accomplished; we now have a more sound methodology for the development and application of instruments to measure subjective health outcomes. This chapter explains many topics that play a fundamental role in several of the measurement frameworks that will be introduced in subsequent chapters.

CONCEPTS

Historically, clinicians have been most confident when measuring their patients' observable physical aspects of health. However, in the current environment of outcomes research, many aspects of interest are not directly observable. Therefore, phenomena such as quality of life, patient satisfaction, depression, and even diseases such as diabetes mellitus or Alzheimer's disease can be thought of as concepts.

Concepts provide a common language and shared meaning, allowing us to communicate clearly and precisely. Broadly speaking, concepts are the building blocks of theories, helping to explain how and why certain phenomena behave the way they do. Concepts are mental abstractions that we use to indicate the ideas, people, organizations, events, and things that we are interested in. Concepts bring theory down to earth, thereby helping to explain the different components. Concepts can refer to ideas (self-esteem, social capital), people (ethnicity, fatigue), and events (cultural revolution, urban regeneration), among other things (Laerd Dissertation, 2016).

We often consider concepts as mental abstractions because they are seldom directly observable. We cannot observe depression directly, even though we may associate it with certain signs: a person who often cries, engages in self-harm, has mood swings, and so forth. Since concepts are broad and abstract, conceptual clarity has become one of the cornerstones of good research. Concepts vary significantly in their complexity, by which we mean the relative difficulty that people have in understanding and measuring them. Some concepts can be very easy to understand and therefore to measure (e.g., gender, ethnicity), but others are more difficult (e.g., sexism, self-esteem). Another complex concept is health, and our primary objective here is to show that measuring this concept can help us understand it.

CONSTRUCTS

Measuring a multifaceted and partially subjective concept such as health involves capturing many features that are not directly observable. Concepts often lack clarity and precision. Because they are ambiguous, they need to be translated from their abstract form (i.e., ideas) to a concrete one (i.e., made measurable in the form of variables). A construct is a concept that has been operationalized so it can be empirically studied and measured. (Some researchers consider the terms concept and construct as interchangeable.) For example, health status can be defined as the set of levels of symptoms and complaints that affect overall health and the capacity for living according to an individual patient.

Constructs can be represented by a wide range of attributes (Chapter 6). For example, happiness could be associated with love, financial security, cigarettes, puppies, a song, and so on (Elshaw, 2015). The usefulness of one conceptualization over another depends largely on a construct's validity (Chapter 7). An interesting example of getting control over all kinds of constructs can be found in the Grid-Enabled Measures Database of the National Cancer Institute (www.gem-measures.org/public).

LATENT VARIABLES

Some aspects of health will have clear, precise, and universally agreed definitions. Others may be less apparent, and many of these are not directly and reliably measurable. Because such aspects are inferred, they are commonly described as postulated constructs, latent traits, or latent factors. These hypothetical constructs, which are believed or postulated to exist, are measured by latent variables. In statistics, a variable is latent (from Latin: lateo "lie hidden") when it is not manifest but rather inferred from other variables that are observable. Latent-variable models are used in many disciplines, ranging from psychology to economics, medicine, physics, bioinformatics, natural-language processing, and management. Some of the most widely applied latent-variable models are factor analysis and structural equation modeling (Chapter 7).

Reflective Versus Formative

Two measurement models are generally used to represent latent variables: the reflective and the formative model. It is important to distinguish between them when creating statistical models to analyze data, especially in the context of structural equation modeling or other explicit data modeling techniques. There is considerable disagreement in this area of research. To add to the confusion, the two models often yield similar results. One researcher advocating reflective measurement may end up with the same conclusions as another researcher adhering to formative measurement. Nevertheless, the distinction is important as it corresponds to two different approaches to health measurement.

Reflective Constructs

In a reflective construct, the correlation among the variables is caused by, and therefore reflects, variation in the underlying latent variable. Any changes in the latent variable should cause comparable changes in all the manifest (observable) variables (Fig. 5.1). Any single manifest variable or combination of these variables should be equally valid in representing the latent variable (Bollen and Lennox, 1991). Cognitive ability is a good example of a reflective model. It refers to the capability associated with knowledge in problem solving and is typically assessed using variables that evaluate quantitative ability, verbal reasoning, spatial reasoning, and so on. Scores on each of these

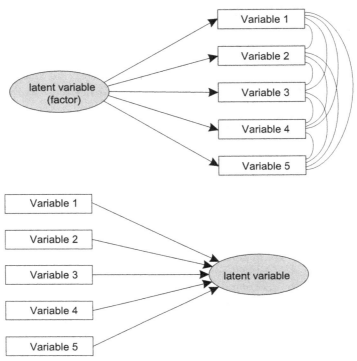

FIGURE 5.1   Reflective model (top) and the formative model (bottom) with the construct to be measured (*oval*) and the manifest variables (*rectangles*).

variables reflect the underlying construct of cognitive ability. Consequently, cognitive ability is assumed to influence all of these, and any change in cognitive ability should be reflected in all these variables (Elshaw, 2015).

Formative Constructs

A formative measurement model works with the arrows of causality pointing in the opposite direction (Fig. 5.1). Any changes in the observed manifest variables are hypothesized to cause changes in the underlying latent variable. Therefore, the manifest variables are formative in creating the latent variable of interest, and the manifest variables jointly determine the conceptual and empirical meaning of the latent variable. It is assumed that all the scores on the manifest variables contribute to the information of the latent variable, but none of the individual manifest variables themselves are actually caused by the latent construct. A change in one manifest variable does not necessarily correlate to a change in another manifest variable, and any of the manifest variables may or may not be correlated. The implication is that since the manifest variables define the characteristics of interest (i.e., the construct), the conceptual domain of the construct (i.e., latent variable) is sensitive to the number and types of manifest variables the researcher selects. Adding or removing a manifest variable can change the conceptual domain of the construct significantly. This issue is highlighted if we discuss measurement approaches that are based on preferences (Chapters 11 and 12) and can also be observed in the index measures used in the setting of clinimetrics (Chapter 13).

ATTITUDES, PREFERENCES, JUDGMENTS, AND CHOICES

There are many different opinions about what a preference, a choice, or a judgment is, depending on which field of research people are working in. I have even found definitions of "judgments" that completely contradict the one used in this book. Several notions are relevant to measuring the quality of health and play a crucial role in specific disciplines. In psychology, desirability is often classified under attitude, defined as a disposition to respond favorably or unfavorably to an object, person, institution, or event. Alternatively, a "preference" may refer to an evaluative judgment in the sense of liking or disliking an object (Dirksen et al., 2013). These two notions, attitude and preference, seem to be closely connected.

Attitudes

Psychologists distinguish between perceptions, beliefs, attitudes, motives, and preferences (McFadden, 1999). Central to the appraisal of subjective phenomena such as perceived health status is that these are regarded as attitudes (Eagly and Chaiken, 1993; Kahneman et al., 1999). Attitudes denote a psychological tendency that is expressed by evaluating a particular entity with some degree of favor or disfavor. Objects of attitudes include anything that people can like or dislike, wish to protect or to harm, want to acquire or to reject. Attitudes are often defined as latent variables in statistical models (Lichtenstein and Slovic, 2006).

Preferences

A dictionary defines preference as liking one thing better than another; a tendency to choose. The term generally refers to the (relative) "desirability" of something or someone but is conceptualized and measured differently across disciplines. In economics, preferences relate to hedonic components (i.e., things we want: consumption, leisure, health). The desirability of a good or service in economics is understood in relation to its utility, referring to a measure of satisfaction gained from the consumption of a good or service (such as health care) (Drummond et al., 2005). As scarcity lies at the core of economic reasoning and is closely related to choices between goods and services, making a choice includes the notion of sacrificing one alternative to obtain another (opportunity costs or trade-off; see Chapter 12).

In economic evaluations of health care, states of illness or disability are commonly projected on a scale from zero to unity. A value or utility of 0.0 is assigned to "being dead," whereas a value or utility of 1.0 is assigned to "full health." Gold et al. (1996) refer to preferences as numerical judgments of the desirability of a set of outcomes, and the term "preference" (or preference score) is used interchangeably with "utility" or "value" (Chapter 12).

Judgments

Allwood and Selart (2001) distinguish between intellectual tasks, such as those for which there is a demonstrably correct answer, and judgmental tasks, which call for evaluative preferences (behavioral, ethical, aesthetic). Other definitions have been used in the literature: "Judgment is the evaluation of evidence to make a decision"; "It is the process of forming an opinion or evaluation by discerning and comparing." In another definition, "judgment refers to the process by which people make decisions and form conclusions based on available information and material combined with mental activity (thought) and experience."

Nunnally and Bernstein (1994, p. 51) make the following distinction between what they call "judgments," where there is a correct response, and "sentiments," which involve preferences. Thus, there are right and wrong answers to "How much is two plus two?" and "Which of the two weights is heavier?" In contrast, sentiments cover personal reactions, preferences, interests, attitudes, values, and likes and dislikes. A person is neither right nor wrong for preferring one medical treatment to another. Consider some examples of sentiments: rating how well your present health status is on a seven-category Likert item; answering the question, "Which of these two health-state descriptions is better?"; and rank-ordering 10 different results after plastic surgery in terms of aesthetics. Still, according to Nunnally and Bernstein, judgments also tend to be cognitive, involving "knowing," whereas sentiments tend to be affective, involving "feeling."

Although I agree with Nunnally and Bernstein on most topics, in regard to the meaning of judgments I take a different stance. In this book I apply the term judgment to situations in which a person has to make a choice, conclusion, or decision on the basis of an individual evaluation. In my view, the outcome of that evaluation is neither right nor wrong. It seems that what Nunnally and Bernstein describe as "sentiments" would be considered judgments by many others.

Choices

Choice involves mentally making a decision: assessing the merits of multiple options and selecting one or more of them. One can make a choice between imagined options or between real options followed by the corresponding action. While choice plays a role in many scientific disciplines, in some it is a key concept. In philosophy and political science, for example, the concepts of "autonomy" and "freedom of choice" are important. Choices are also closely associated with making decisions.

In economics, it is important to understand people's choice behavior to establish and confirm economic theories. The underlying assumption is that there are unlimited human wants that are to be met by limited resources. Essentially, this is what economists call scarcity. Scarcity is considered the fundamental economic problem, and all economic activities revolve around trying to solve it. Choice comes about as a result of scarcity. People (as consumers) have to make choices among the things they want, as they cannot have them all.

Explanations and predictions of people's choices, in everyday life as well as in the social sciences, are often founded on the assumption of human rationality. The definition of rationality has been much debated (Tversky and Kahneman, 1981). But then even outside the field of economics, the connection between choice and rationality is not always straightforward either. In the field of psychology, for example, "choice" is not necessarily rational. Glasser (2007) postulated that unhappiness results from bad relationships with relatives and friends, which in turn suffer as a result of making misguided choices. He believed that by making better choices, we can maintain better relationships, which in turn will help us to lead more fulfilled lives.

MEASUREMENT AND SCALING MODELS

A measurement model is a framework consisting of a particular way of assessing an object of interest that will produce a typical type of response data, which will subsequently be processed in a specified statistical model. These two steps (response task, statistical model) together create a measurement model. This means that researchers are not free to apply whichever measurement task they deem appropriate, nor whichever statistical model they consider most relevant. In a measurement model certain elements are specified because that model is based on a theory that makes certain assumptions. So, for any measurement model, we accept that certain assumptions hold, that the mode of response or data collection is specified, and that a predefined analytical technique is obligatory.

To solve the problem of measuring in the absence of a standard or a unit of measurement, methodologies have been developed to measure phenomena that are unobservable (latent). Such methodologies, known as scaling models, can translate subjective perceptions into quantitative measures. Scaling models are nothing more than measurement models and are purpose-built to measure attitudes. These models are based on theoretical concepts. They stipulate that the data must be collected through judgment tasks that generate data with certain characteristics, and that data collection must be followed by specific analytical steps to arrive at the final quantification (a measure). Most scaling models also provide a statistic to express the goodness of fit. Nearly all scaling models were developed by psychologists (psychometricians) and mathematicians during the period 1927–64. The more advanced scaling models are rarely encountered in the health sciences. Yet they have had considerable success in many other research areas, including educational measurement, personality assessment, and marketing.

SCALING OBJECTS AND SUBJECTS

Scaling or measurement models may be used to scale subjects, objects, or both. Most have been developed in the field of psychology, which explains why these models refer to "subjects" and "stimuli." Other sciences may use different terms for the same entities. In the setting of this book, it is convenient to replace the terms subjects and stimuli by patients and health.

Because it is more difficult to scale objects (e.g., health) than subjects, most of the complex scaling models have been developed for objects. This difference in orientation has influenced the language used to describe the research. "Scaling" and "scaling methods" usually denote the scaling of items (objects). For methods used to scale people (subjects), the terms "measurement" and "test construction" are more common. Those who are interested in the details of scaling may consult the classical works of Guilford (1954) and Torgerson (1958) and the very readable article of Froberg and Kane (1989).

Scaling Subjects

Likert scaling (see this chapter) is employed only to scale subjects (patients). In Likert scaling, patients are presented with a list of statements about a single topic (e.g., complaints, functioning). They are instructed to respond to each statement (Likert item) in terms of their degree of agreement or disagreement. Thus, this scaling model involves a single type of object (complaints, functioning) and a single type of response (Chapter 9). The problem is how to combine responses for each patient in such a way that valid and reliable differences among patients can be represented. Likert scaling is considered a subject-centered or person-centered approach, since only subjects (persons) receive scale scores (Torgerson, 1958).

Scaling Objects

Instead of subjects, however, researchers sometimes need to scale objects (e.g., health). Consider, for example, a group of people who are asked to judge the severity (impact) of a set of health conditions. In this instance, the objects (health states) are quantified with respect to one or more particular health characteristics (attributes). Then it is the objects (health states) rather than the subjects that are scaled (Chapters 11 and 12). The same would be true if the group were asked to rate the loudness of tones, the sweetness of candy, or empathy for their general practitioner. In each of these examples, the task set for the subject (person) is to evaluate the object with respect to some designated attribute(s) (Coxon and Davies, 1982).

Scaling Subjects and Objects

What if the researcher wants to scale both subjects and objects? This calls for a third approach to scaling, the response approach (Chapter 10). As Torgerson (1958, p. 48) observes, "the task set for the subject is to respond to an object on the basis of the position of the object in relation to the subject's own position with respect to the attribute." For example, a person might be asked to pick a particular candidate from a list that is closest to his or her ideal candidate or pick a particular statement from a set of statements with which he/she is in closest agreement. Or to stay closer to the content of this book: a patient is asked to select from a set of health-state descriptions the state that is most similar to his/her own health condition. This approach includes the multi-attribute preference response model presented in Chapter 14.

SCALING METHODS

Thurstone and Likert can be credited for their contributions to the measurement of attitudes, opinions, and other noncognitive constructs. Both investigators worked out scaling methods for the purpose of measuring noncognitive traits. However, their methods differ in important ways, particularly in the assumptions about how people respond to items and in the criteria for item selection during the scale construction process (McIver and Carmines, 1981). Likert's method is much more widely used in practice because it is less laborious and easier to understand.

In the following section, we will briefly introduce two classes of scaling methods: direct and indirect. The main disadvantage of direct methods is that these methods are more susceptible to various types of bias. By contrast, indirect scaling methods require only simple qualitative judgments and are more robust. However, indirect methods presume far more knowledge on the part of the developers and the analysts. All of the advanced scaling models to measure subjective phenomena are based on indirect scaling methods.

Direct Scaling Methods

A classical example of a simple direct measurement procedure is measuring the physical attribute "length." One picks a particular object to serve as a unit of length and determines the length of any other (larger) object by the number of duplicates of that unit required to match it. This direct measurement procedure, based on the notion of concatenating duplicate objects, is widely used in physics. Of course, the fact that something can be measured directly does not mean that it has to be that simple. Even length and distance are now measured by much more sophisticated indirect processes. For instance, an estate agent takes a room's dimensions using a laser range finder. Measuring length by the concatenation process is almost entirely a representational measurement strategy, requiring only that a unit must be chosen on external grounds.

Another direct measurement procedure, the visual analogue scale (VAS), is simple and pragmatic. While VASs come in several variants, one common format consists of a horizontal 10-cm line. Its end points are clearly marked and a word at each end indicates the extremes of the attribute to be measured. With a direct scaling method such as the VAS, the attributes of an object are measured directly, in the sense that no intermediate measurements or calculations are needed. In many situations, however, direct measurement is not possible or not sufficiently robust. Direct scaling methods have in common that they convert patients' judgments directly into numerical scores, assuming that the resulting measures are meaningful.

Absolute and Comparative Responses

Another noteworthy distinction is that between absolute and comparative responses. In general, an absolute response concerns a particular judgment of one object (e.g., health condition), whereas a comparative response relates two or more objects. These two different types coincide almost perfectly with direct and indirect scaling methods.

One of the important findings is that people are almost invariably better (more consistent and/or accurate) at making comparative responses than absolute responses. This points up the frame-reference problem in absolute responses, a problem that is avoided in comparative responses. Asking someone "Is this cola sweet?" raises the question of how sweet is sweet, provoking a deliberation that is avoided when one is asked to judge which of several colas is the sweetest. Not surprisingly, people rarely make absolute judgments in daily life, since most choices are between some competitive alternatives and are thus inherently comparative. Scientific research has shown that in many situations discrimination is a basic operation of judgment and of generating knowledge. Therefore, the core activity of measurement in many scaling models (Chapter 11) is to compare two or more objects so that the data provide compelling information.

An extremely powerful way to gather informative data is through paired comparisons (Chapter 11). A research application of this method could involve preferences among N health states. The subject is presented with two states in succession and asked to state a preference (make a choice in favor of one of two health states). This query is repeated for all possible pairs of health states. Unfortunately, the procedure requires $N(N-1)/2$ pairs to be judged. The number of comparisons rises rapidly with N. For example, if there are 20 health states in the study, 190 (20 × (19/2)) assessments are needed. However, it is much quicker to have subjects rate each health state individually. Conversely, paired comparisons generally give much more reliable results, when applicable.

Absolute responses are useful because they are much easier to obtain than comparative responses. Noncomparative scaling is frequently referred to as monadic rating or measurement. In monadic rating, the respondent is asked to rate a health condition on a scale without any reference or comparison to another health condition. Respondents evaluate only one object at a time and the resulting data are assumed to be either at the interval or ratio scale.

Indirect Scaling Methods

Many indirect measurement methods have been developed. Their differences lie in the nature of the information collected about the relationship between objects. Indirect scaling methods require only simple qualitative judgments from respondents and provide built-in tests of the validity of the underlying theoretical construct. Indirect scaling procedures arrive at the numerical scale (the representation of the attributes) by modeling the respondents' judgments. Much of the research on indirect scaling methods has been reported in the behavioral science literature.

For each of the various indirect scaling methods, the tasks that respondents have to perform do not in themselves produce measures. These can only be derived from additional data transformations. The successful quantification of health depends on choosing an empirical judgment procedure that is appropriate to the subjects' ability to respond; it also requires the use of a measurement theory that is appropriate to the collected data (Nunnally and Bernstein, 1994). The tasks for these indirect scaling methods may differ. For example, the task for the Thurstone paired comparison model is to compare two objects (e.g., scans, faces, descriptions of health) and to choose in favor (better) of one. In item response theory (Chapter 10), the patient usually has to indicate whether he or she is capable of performing a certain task.

DIRECT-RESPONSE SCALES

Several sophisticated scaling approaches exist. So far, these models have been used infrequently in the area of health measurement (Kind, 1982; Hadorn et al., 1992; van Agt et al., 1994). They have been worked out in a different framework than the most prominent scaling model, which is classical test theory (Chapter 9). Therefore, the three frameworks of indirect measurement will be discussed separately in, respectively, Chapters 10, 11, and 12. In the following section, the most frequently used and more simple direct-response scales will be introduced.

Rating

A rating scale is a set of categories designed to elicit information about a quantitative or qualitative attribute. Common examples are the 1–10 rating scales in which a person selects the number that is considered to reflect the perceived quality of a product or the level of experienced pain. Categories can be organized in many ways: for example, by statements, by numbers, by icons, or combinations of these (Fig. 5.2). Some variants of this technique go by another name: category rating. The term "VAS" is often also used when a rating scale is employed, although this term is not confined to this method (Wewers and Lowe, 1990; McCormack et al., 1998).

Visual Analogue Scale

The VAS, a renowned direct measurement technique, originated in the social sciences and has been popular among psychologists. This type of scale has a long history and was initially called "graphic rating" (Hayes and Patterson, 1921). Aitken (1969) and Zealley and Aitken (1969) were among the first to apply a VAS in medicine, using it to construct single-item mood scales. Ever since, the VAS has been a common research and clinical tool in psychological medicine, especially for measuring pain. To my knowledge, Patrick et al. (1973) were the first to use a VAS to derive values for descriptions of hypothetical health states. Priestman and Baum (1976) were probably the first to use a VAS ("linear analogue self-assessment") to assess quality of life among cancer patients. It is generally accepted that VASs are highly feasible and show moderate-to-good test–retest reliability (Siegel et al., 1997; Salomon, 2003). Therefore, they are often used to assess patients' health in longitudinal studies.

Essentially, a VAS (also sometimes called a "semantic differential") is simply a straight line of a specified length with verbal descriptors at each end (anchors) consisting of short and easily understood phrases that describe the

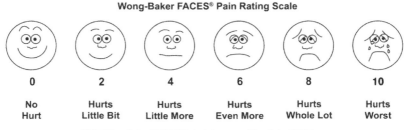

FIGURE 5.2   Rating scale based on a combination of three distinct elements: numbers, phrases, and icons (Wong-Baker FACES Foundation, 2015).

extremes of the attribute being measured. However, markers are often added to the line, usually with numbers attached (formally it is then a rating scale). Respondents indicate their position between these two extremes with a mark on the line. The length, in suitable units, from one end of the scale to the mark constitutes the measure of the attribute. There are variants of the VAS, but a common format consists of a horizontal 10-cm line, though the millimeter is also a common unit (Fig. 5.3).

It is clear that several pragmatic decisions are made when constructing such an instrument. Chief among these is the choice of descriptive terms for the end points. Regarding the measurement of health, these end points are often "fixed" in the sense that one end of the scale should represent the least preferred or "worst" health to the respondent and the other end point should indicate the most preferred or "best." Another issue concerns an appropriately standardized instruction that introduces the scale and specifies what the respondent is expected to do (Hand, 2004).

In principle, a VAS can provide fine discrimination since the investigator may choose to measure the positions of the response very precisely. In practice, however, there is some doubt as to whether respondents can really discriminate between fine differences of position along the line. The VAS is also said to be methodologically flawed. It has been criticized for being prone to end aversion (people are averse to use the ends of the scales) and having context bias. In addition, there is no clear underlying theoretical measurement framework. Moreover, the anchors are potentially ambiguous and therefore may not be comparable across populations (Bleichrodt and Johannesson, 1997; Krabbe et al., 2006).

Likert Scaling

Let us begin our consideration of Likert scales by briefly reviewing the contributions of Louis Thurstone, a psychologist at the University of Chicago. Not only was Thurstone one of the first psychologists to propose systematic procedures for measuring attitudes but also his research led to the development of three related but distinct methods of scaling: paired comparisons, successive intervals, and equal-appearing intervals (Thurstone, 1927, 1928, 1929a, 1929b; Thurstone and Chave, 1929). Thurstone scaling techniques were quite popular during the 1920s and 1930s, but they are not employed widely today because of a number of practical limitations.

No scaling approach has more intuitive appeal than the scale introduced by Rensis Likert (1932). Likert was especially interested in simplifying the procedures for constructing attitude scales and in using intensity-scaled responses for each item. He developed his approach explicitly to avoid the rigorous and time-consuming procedures for scaling that had been introduced by Thurstone. Typically, in Likert scaling, each item is scored on a five-point categorical scale, ranging from one extreme to the other (e.g., strongly

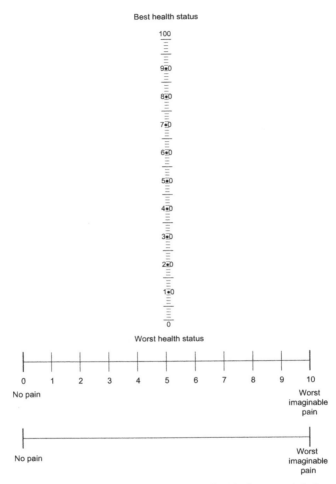

FIGURE 5.3   Different examples of visual analogue scales (the bottom one is the only official visual analogue scale).

disagree, disagree, undecided, agree, strongly agree), with these categories scored from one to five.

The reason why Likert-style scales are easier to work with than Thurstone scales is that they make the response process cumulative. By implication, the scores can be summed, yielding a summary score. A Likert scale is never an individual item; it is always a set of several items, with specific format features. The responses to these items are added or averaged to produce an overall score or measurement. Generally speaking, any scale obtained by adding up the response scores on a related set of items is referred to as a Likert or "summative" scale. The Likert method is most successful when the response scale for each item covers a range of scale levels. Likert items are extensively used in many instruments that are developed under classical test theory (Chapter 9).

BOX 5.1

Rensis Likert, a sociologist at the University of Michigan, is best known for survey research methods and for the Likert scale; a psychometric scale commonly involved in research using questionnaires. His original report entitled "A Technique for the Measurement of Attitudes" was published in the Archives of Psychology in 1932. After retirement in 1970, he was an active researcher in management styles; he also developed the linking pin model. Likert was known for his support of interdisciplinary collaborations and emphasis on using social science research to effect positive change (Fig. 5.4).

FIGURE 5.4   Rensis Likert (1903–81).

In brief, Likert scaling may be described in the following manner. A set of items, if possible composed of approximately an equal number of favorable and unfavorable statements concerning the attitude object, is given to a group of subjects. They are asked to respond to each statement in terms of their own degree of agreement or disagreement (Fig. 5.5). The specific responses to the items are combined so that individuals with the most favorable attitudes will have the highest scores while individuals with the least favorable attitudes will have the lowest scores. The weighting scheme is reversed for unfavorable

Strongly Disagree	Disagree	Undecided	Agree	Strongly Agree
1	2	3	4	5

Not at All Interested	Not Very Interested	Neutral	Somewhat Interested	Very Interested
1	2	3	4	5

Not at All	Not Really	Undecided	Somewhat	Very Much
1	2	3	4	5

Not at All Like Me	Not Much Like Me	Neutral	Somewhat Like Me	Very Much Like Me
1	2	3	4	5

Not at All Happy	Not Very Happy	Neutral	Somewhat Happy	Very Happy
1	2	3	4	5

Never	Sometimes	Regularly	Usually	Always
1	2	3	4	5

FIGURE 5.5 Examples of response options in Likert scales in which the categories are scored with successive integers and where a person's measure is simply the sum of the scores for the set of items.

statements so that higher scores always indicate more favorable attitudes toward the object or phenomenon. Thus, a Likert scale has to be considered as a specific type of rating scale. Though eminently practical, and with more than two categories of response making it potentially more precise, Likert's procedure is not as rigorous as Thurstone's original methods.

The Likert procedure is "subject centered"; that is, its purpose is to scale respondents, not items or objects. That purpose differs from, for instance, item response theory (Chapter 10), in which both respondents and items are scaled. In Likert scaling, all systematic variation in the responses to the items is attributed to differences among the respondents. Underlying these response tasks is a series of assumptions about the nature of the items and the final scale. First, the items are considered replications of one another. We assume that the items as a set, which only measure the attribute under investigation. In other words, all items should be related to a single common factor (Chapter 7). The sum of these items is expected to contain all the important information contained in the individual items. Second, it is assumed that each item is monotonically related to the underlying attitude continuum (see next section). In other words, the more favorable (unfavorable) a respondent's attitude, the higher (lower) his or her expected score for the item would be. Third, a defining feature of a Likert scale is that it consists of a number of response categories that have distinct cut-off points and assume linearity and equal intervals between these points. This means, in the example of Fig. 5.5, that the steps or differences in response from category "Never" to category "Sometimes" are the same as from "Usually" to "Always." Furthermore, the step from "Never" to "Regularly" is assumed to be twice as big as the step from "Never" to "Sometimes." So, the distance between each of the response categories is assumed to be linear, and the level of agreement with each statement is also assumed to be equal and linear.

Having designed an initial set of items to measure a particular aspect of health, the next phase of scaling requires evaluation of the item as a set. Are these items related to one another and do they in conjunction represent the construct they are supposed to measure? Which items may be combined to form the best single measure? Which items apparently fail at their given task and should be dropped from the final set of items that will comprise the scale? It makes little sense to combine unrelated items into an overall sum, since undifferentiated items contribute little useful information to the big picture. Indeed, they may actually decrease the reliability and/or validity of the scale. Consequently, such items should not be retained in the final measuring instrument. This process of item selection is described in detail in Chapters 7 and 8.

RESPONSE FUNCTIONS

Measurement models can be distinguished by their trace lines. A trace line is simply a curve describing the relationship between the probability of a specific response to an item (e.g., the probability of agreeing with a question or a statement) and the attribute that the item is supposed to measure. Each model discussed in this book is associated with a particular type of trace line, as these lines represent the basic theoretical difference among the measurement models. The cumulative model and the unfolding model are based on the two main response processes in the measurement of subjective aspects of health. The distinction between these two types of measurement models is best represented by the shape of their trace lines (or item characteristic curve): monotone increasing in the cumulative model and single-peaked in the unfolding model (Fig. 5.6).

For each of these two processes, there are two basic data collection designs: direct response and pair comparison. In the direct-response design, persons respond directly to items; in the pair-comparison design, they compare items in pairs (Andrich, 1988).

Cumulative

In the cumulative response model, the probability of a positive response is a monotonic function of the relevant parameters. Cumulative models are well

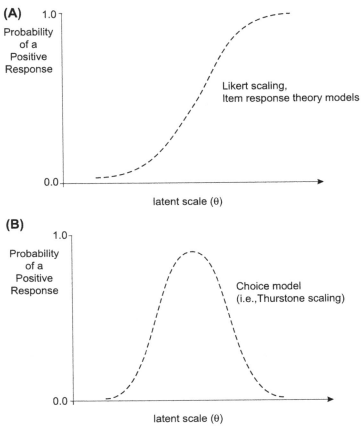

FIGURE 5.6 Response functions for different scaling models: (A) cumulative and (B) ideal point.

suited for modeling dominance relations. The top trace line (Fig. 5.6A) corresponds to items suitable for Likert scales and item response theory models (Chapters 9 and 10). Given the location of an item on the latent dimension, an ascending trace line denotes the increasing probability of responding positively on an item.

Ideal Point

Coombs (1964) referred to a person's location as the ideal point, reflecting the assumption that a person will tend to agree with statements that are close to the person's own attitude and disagree with those statements (i.e., categories of the item) that are far from the person's location on the scale in *either* direction.

The response function for this response process is single-peaked and based on proximity relations (Fig. 5.6B). Such proximity relations and their single-peaked response functions are part of the choice models (Chapter 11).

Unfolding

This special variant of scaling, in which the response is characterized as an ideal-point process, is based on the work of Coombs (1950, 1964). Coombs coined the term "unfolding" for the simultaneous process of locating persons and items on a scale using the agree/disagree responses, and this term is still found in the literature. Unfortunately, Coombs' procedure was also extremely cumbersome for a larger set of statements, items, health states, or other entities. Therefore, his method did not challenge other scaling approaches as the favored practical procedure. However, that was in the early days of computers with their limited capacity. Nowadays, there are probabilistic (statistical) unfolding models with more flexibility. Health attributes as well as patients can now be scaled on a unidimensional or multidimensional scale. It may be argued that unfolding theory has not received the attention it deserves. Unfolding has proved to be a very constructive data-analytic principle and a source of inspiration to many scientists (Bossuyt, 1990).

Reflection

Starting in the 1990s, several researchers have put forth the argument that many binary or graded disagree—agree (Likert scale) responses to attitude statements generally result from an ideal-point process. In such a response process, an individual endorses an attitude statement to the extent that the sentiment expressed by the statement adequately matches the individual's own opinion. This argument implies that such responses are best analyzed with some form of model that implements a single-peaked response function (Andrich, 1995; Luo, 2001).

The cumulative response model and the ideal-point response model are mutually incompatible. If one of them correctly describes how the probability of endorsing an item is related to an underlying attribute, then the other cannot. Both models make another assumption about the underlying data structure. This difference is subtle but may have strong repercussions if different types of responses are analyzed under the wrong model (Andrich, 1995; Luo, 2001). Usually cumulative models are used to measure latent abilities and unfolding models to measure latent attitudes and preferences. Many in-depth scientific discussions are dealing with this kind of theoretical reflections. We shall, if necessary, return to the typical characteristics of these different measurement models in Chapters 9—12.

DISCUSSION

There are various ways to scale items or subjects, depending on the purpose(s) the resulting scale has to serve. Each of these ways is associated with a specific scaling or measurement model, which may be defined as an "internally consistent plan for developing a measure" (Nunnally and Bernstein, 1994, p. 33). Measurement methods range from the very simple to the exceedingly complex, from the almost purely mathematical to the almost purely pragmatic, from direct to indirect, from univariate to multivariate (Hand, 2004).

There is much discussion, if not controversy, about the competence of Likert scales. In particular, the assumptions underpinning this measurement method are questioned. In Likert scales, numbers are used as labels for response categories and their assignment is arbitrary, except that the numbers reflect the increasing quantity of the characteristic being measured. Responses are ordered in terms of magnitude and a sequential code is assigned to each. "Mild," "moderate," and "severe" disability might be coded 1, 2, and 3. Because people use adjectives in different ways, we cannot assume that "mild" implies the same thing to everyone, nor that "often" implies the same frequency when referring to common health problems as when referring to rare ones. Thus, there is no intrinsic meaning to the actual value of the numbers in an ordinal scale and the distance between each increment. Of course, different interpretations as part of a linguistic/cognitive appraisal process may have less effect in a homogeneous sample of people. However, if health instruments are used, for example, in international comparisons, and they are based on simple direct scaling, there is a serious danger of bias in the measures (Barofsky, 2012). In most cases, adjective (linguistic) categories are not characterized by nicely uniform distances between them, as demonstrated by the more refined but complex indirect scaling methods (which also have other drawbacks). Sometimes the differences in interval may be substantial. Nevertheless, Likert scales are widely used. Purists may criticize the simple direct measurement methods based on Likert scales for making these strong assumptions, but others argue that the errors produced are minor (McDowell, 2006).

Another criticism of Likert scales is that some items may be more important than others. Nunnally and Bernstein (1994, p. 332) argue that, in most cases, the summation of raw scores in which each item is scored identically and each item contributes equally to the total scale score is perfectly adequate. They find it difficult to defend arbitrary weighting systems and see little reason to try: unweighted and weighted sum scores tend to correlate quite highly. Differentially weighting items might increase the reliability slightly, but reliability might also be increased by adding two or three new items. If the reliability is undesirably low, their advice is to increase the number and quality of the items (Chapter 8).

Finally, it is worth mentioning that for formative measurement, the set of variables defines the construct, which is not true for reflective measurement.

Bollen and Lennox (1991, p. 308) note that "omitting a variable is omitting a part of the construct" under formative measurement. This conclusion is highly relevant, being closely related to the specific measurement approaches discussed in Chapters 11 and 12. There is no validity check in a formative measurement model, such as to some extent exists for reflective measurement. Although the ontologies and the structures of the two approaches for measurement differ, either one can be appropriate, depending on the researcher's purposes (Bagozzi, 2011).

REFERENCES

Aitken, R.C.B., 1969. Measurement of feelings using visual analogue scales. Proceedings of the Royal Society of Medicine 62 (10), 989–993.

Allwood, C.M., Selart, M., 2001. Decision Making: Social and Creative Dimensions. Kluwer Academic Publishers, Dordrecht.

Andrich, D., 1988. The application of an unfolding model of the PIRT type for the measurement of attitude. Applied Psychological Measurement 12 (1), 33–51.

Andrich, D., 1995. Hyperbolic cosine latent trait models for unfolding direct-responses and pairwise preferences. Applied Psychological Measurement 19 (3), 269–290.

Bagozzi, R.P., 2011. Measurement and meaning in information systems and organizational research: methodological and philosophical foundations. MIS Quarterly 35 (2), 261–292.

Barofsky, I., 2012. Quality: Its Definition and Measurement as Applied to the Medically Ill. Springer, New York.

Bleichrodt, H., Johannesson, M., 1997. An experimental test of a theoretical foundation for rating-scale valuations. Medical Decision Making 17 (2), 208–216.

Bollen, K.A., Lennox, R., 1991. Conventional wisdom on measurement: a structural equation perspective. Psychological Bulletin 110 (2), 305–314.

Bossuyt, P., 1990. A Comparison of Probabilistic Unfolding Theories for Paired Comparisons Data. Springer-Verlag, Berlin.

Coombs, C.H., 1950. Psychological scaling without a unit of measurement. Psychological Review 57 (3), 145–158.

Coombs, C.H., 1964. A Theory of Data. John Wiley & Sons, New York.

Coxon, A.P.M., Davies, P.M., 1982. The Users Guide to Multidimensional Scaling: With Special Reference to the MDS(X) Library of Computer Programs. Heinemann Educational Books, London.

Dirksen, C.D., Utens, C.M.A., Joore, M.A., van Barneveld, T.A., Boer, B., Dreesens, D.H.H., van Laarhoven, H., Smit, C., Stiggelbout, A.M., van der Weijden, T., 2013. Integrating evidence on patient preferences in healthcare policy decisions: protocol of the patient-VIP study. Implementation Science 8 (1), 1–7.

Drummond, M.F., Sculpher, M.J., Torrance, G.W., O'Brien, B.J., Stoddart, G.L., 2005. Methods for the Economic Evaluation of Health Care Programmes, third ed. Oxford University Press, New York.

Eagly, A.H., Chaiken, S., 1993. The Psychology of Attitudes. Harcourt, Brace & Jovanovich, Forth Worth T.X.

Elshaw, J.J., 2015. Social science measurement. In: Badiru, A.B., Racz, L.A. (Eds.), Handbook of Measurements: Benchmarks for Systems Accuracy and Precision. CRC Press, Boca Raton, pp. 203–216.

Froberg, D.G., Kane, R.L., 1989. Methodology for measuring health-state preferences — II: scaling methods. Journal of Clinical Epidemiology 42 (5), 459—471.

Glasser, W., 2007. Choice Theory: A New Psychology of Personal Freedom. Harper Perennial, New York.

Gold, M.R., Patrick, D.L., Torrance, G.W., Fryback, D.G., Hadorn, D.C., Kamlet, M.S., Daniels, N., Weinstein, M.C., 1996. Identifying and valuing outcomes. In: Gold, M.R., Siegel, J.E., Russel, L.B., Weinstein, M.C. (Eds.), Cost-effectiveness in Health and Medicine. Oxford University Press, New York, p. 83.

Guilford, J.P., 1954. Psychometric Methods, second ed. McGraw-Hill, New York.

Hadorn, D.C., Hays, R.D., Uebersax, J., Hauber, T., 1992. Improving task comprehension in the measurement of health state preferences: a trial of informational cartoon figures and a paired-comparison task. Journal of Clinical Epidemiology 45 (3), 233—243.

Hand, D.J., 2004. Measurement Theory and Practice: The World Through Quantification. Arnold, London.

Hayes, D.G., Patterson, M.H.S., 1921. Experimental development of the graphic rating method. Psychological Bulletin 18 (1), 98—99.

Kahneman, D., Ritov, I., Schkade, D., 1999. Economic preferences or attitude expression? An analysis of dollar responses to public issues. Journal of Risk and Uncertainty 19 (1), 203—235.

Kind, P., 1982. A comparison of two models for scaling health indicators. International Journal of Epidemiology 11 (3), 271—275.

Krabbe, P.F.M., Stalmeier, P.F.M., Lamers, L.M., Busschbach, J.J., 2006. Testing the interval-level measurement property of multi-item visual analogue scales. Quality of Life Research 15 (10), 1651—1661.

Leard Dissertation, 2016. Constructs in Quantitative Research. Dissertations and Theses: An Online Textbook, From: https://dissertation.laerd.com.

Lichtenstein, S., Slovic, P., 2006. The Construction of Preference. Cambridge University Press, New York.

Likert, R., 1932. A technique for the measurement of attitudes. Archives of Psychology 22 (140), 5—55.

Luo, G.Z., 2001. A class of probabilistic unfolding models for polytomous responses. Journal of Mathematical Psychology 45 (2), 224—248.

McCormack, H.M., de Horne, D.J., Sheather, S., 1998. Clinical applications of visual analogue scales: a critical review. Psychological Medicine 18 (4), 1007—1019.

McDowell, I., 2006. Measuring Health: A Guide to Rating Scales and Questionnaires, third ed. Oxford University Press, Oxford.

McFadden, D., 1999. Rationality for economists? Journal of Risk and Uncertainty 19 (1), 73—105.

McIver, J.P., Carmines, E.G., 1981. Unidimensional Scaling. Sage Publications, Beverly Hills.

Nunnally, J.C., Bernstein, I.H., 1994. Psychometric Theory, third ed. McGraw-Hill, New York.

Patrick, D.L., Bush, J.W., Chen, M.M., 1973. Toward an operational definition of health. Journal of Health and Social Behavior 14 (1), 6—23.

Priestman, T.J., Baum, M., 1976. Evaluation of quality of life in patients receiving treatment for advanced cancer. Lancet 1 (7965), 899—900.

Salomon, J.A., 2003. Reconsidering the use of rankings in the valuation of health states: a model for estimating cardinal values from ordinal data. Population Health Metrics 1 (1), 1—12.

Siegel, J.E., Torrance, G.W., Russell, L.B., Luce, B.R., Weinstein, M.C., Gold, M.R., 1997. Guidelines for pharmacoeconomic studies. Recommendations from the panel on cost effectiveness in health and medicine. PharmacoEconomics 11 (2), 159—168.

Stevens, S.S., 1966. A metric for the social consensus. Science 151 (3710), 530—541.

Thurstone, L.L., Chave, E.J., 1929. The Measurement of Attitude: A Psychophysical Method and Some Experiments with a Scale for Measuring Attitude toward the Church. University of Chicago Press, Chicago.

Thurstone, L.L., 1927. A law of comparative judgment. Psychological Review 34 (4), 273–286.

Thurstone, L.L., 1928. Attitudes can be measured. American Journal of Sociology 33 (4), 529–554.

Thurstone, L.L., 1929a. Theory of attitude measurement. Psychological Review 36 (3), 222–241.

Thurstone, L.L., 1929b. Fechner's law and the method of equal appearing intervals. Journal of Experimental Psychology 12 (3), 214–224.

Torgerson, W.S., 1958. Theory and Methods of Scaling. John Wiley & Sons, New York.

Tversky, A., Kahneman, D., 1981. The framing of decisions and the psychology of choice. Science 211 (4481), 453–458.

van Agt, H.M.E., Essink-Bot, M.L., Krabbe, P.F.M., Bonsel, G.J., 1994. Test-retest reliability of health state valuations collected with the EuroQol questionnaire. Social Science and Medicine 39 (11), 1537–1544.

Wewers, M.E., Lowe, N.K., 1990. A critical review of visual analogue scales in the measurement of clinical phenomena. Research in Nursing and Health 13 (4), 227–236.

Wong-Baker FACES Foundation, 2015. Wong-Baker FACES® Pain Rating Scale. From. http://www.WongBakerFACES.org.

Zealley, A.K., Aitken, R.C.B., 1969. Measurement of mood. Proceedings of the Royal Society of Medicine 62 (10), 993–996.

Chapter 6

Health Status Measurement

Chapter Outline

INTRODUCTION

For decades, practitioners in the health sciences—physicians, psychologists, sociologists, and epidemiologists—have tried to formulate an overarching definition of health or health status, one that could be widely accepted. Despite their relentless effort, a meaningful description of the specific components of health remains elusive (Chapter 2). The World Health Organization's (WHO) version is clearly the most popular alternative to viewing health as the absence of disease or infirmity. Its definition of health as "physical, mental, and social well-being, and not merely the absence of disease or infirmity" concerns both positive and negative elements but accentuates the positive elements of health. In the 1950s, that definition was criticized for not being measurable. Then other criticisms were raised: that it was too narrow and did not include other aspects, particularly personal/spiritual health, well-being, etc. As discussed in Chapter 2, this latter criticism emanates from a perspective that may well go beyond health. Still, the WHO definition has been widely adopted by researchers as a starting point for the development of health-outcome instruments. But widespread acceptance of the WHO definition does not mean that consensus has been reached in operational terms. On the contrary, the question of how to measure the three principal domains of the WHO definition has been difficult to answer (Larson, 1991).

CONTENT

In its constitution, the WHO described health as a "state of complete physical, mental, and social well-being." Although physical and mental health are distinct concepts, they are also interrelated; the state of the one often affects the state of the other. The notion of social well-being extends the concept of health beyond the individual to include the quantity and quality of social contacts and social resources. In a model of health containing social variables, a change in social support would thus implicate a change in personal health outcome.

Ware et al. (1981) object to the inclusion of social well-being, however. They find that evidence "supports restriction of the definition of personal health status to its physical and mental components, rather than including social circumstances as well." Another argument has been advanced in defense of a restricted definition. It poses that even though the experience of a particular condition (e.g., paraplegia) can differ vastly depending on where a person lives and the level of care and social support available, the underlying health state remains the same (Voigt and King, 2014).

On the other hand, because people are essentially social beings, it has been argued that a health-outcome measure should take the impact of physical disfigurement into account. Diseases and disabilities affect social function, so social function may be considered a central component of health outcomes (Kaplan and Anderson, 1990). Seen from another angle, social support facilitates the ability to cope with severe health conditions. Therefore, social well-being is supportive of health, not an aspect of health itself. It is clear that there are divergent perspectives on how to measure the three principal domains of health.

QUALITY OF HEALTH

From this point on we will use the word "health" as an umbrella term for "health-related quality of life" and "perceived health status." Of course, there are subtle differences between these two concepts. But both refer to the endpoint and outcome of the human condition as understood in the medical or health setting. Another reason to put aside the cacophony of terminology and abbreviations is to make this book more readable. Moreover, this book deals mainly with the measurement of health conditions as perceived by individuals, and less with concepts and constructs specifying how perceived health is worked out. From my perspective, instead of simply using the term "health," a better alternative would be a term that might make all the other ones obsolete, and that could be "quality of health."

MEASUREMENT IN HEALTH

Measurement is essential to research in the health sciences, as in other fields. In the laboratory disciplines, however, measurement is not inherently difficult.

There, subjective judgment plays a minor role; any issue of reproducibility or validity is amenable to a technological solution. Measurement is then a reasonably straightforward process. Most of the research problems approached by clinical investigators—cardiologists, epidemiologists, and the like—do not depend on subjective assessment. Trials of therapeutic regimens focus on the prolongation of life or other objective measures. Unlike the clinical setting, however, diagnostic investigators are confronted with subjective human judgment, as evidenced by the errors that occur in radiological diagnosis, for instance. The same is true for medical care, where no objective standards have been formulated to express the quality of its delivery.

In the past few decades, the situation in clinical research has become more complex. The effects of many new drugs and surgical procedures have a marginal impact on the quantity of life. Instead, they are targeted toward improvement in the quality of health. Meanwhile, there is heightened awareness of the impact of health and health care on the quality of human life. Therapeutic efforts in many areas of medicine such as psychiatry, rheumatology, and rehabilitation are directed equally if not primarily to the improvement of the quality, not quantity of life. If efforts in these disciplines are to be placed on a sound scientific basis, methods must be devised to measure what was previously thought to be unmeasurable.

The fundamental problem is that, unlike length or weight, health cannot be measured directly. Instead, its measurement is indirect and stepwise. Health is not an objective phenomenon that can be readily observed in the outside world. Health and health status are abstract concepts, constructs involving many interrelated factors and influences. The measurement of health may even be considered a sociomedical activity. Debate will surely continue about how best to measure it, given the complexity and abstract nature of health itself.

ATTRIBUTES

One of the most challenging tasks in the development of a measurement instrument is to determine which attributes (variously also called items, domains, dimensions, or indicators) should be incorporated to capture health. Attribute is used here as a generic term that reflects a specific aspect or feature of health. The word can be used synonymously with health item or even health domain/dimension. In this book an attribute is more general than an item. An attribute can be any reference (couched in a phrase, picture, movie, explanation, term, question, etc.) to a specific health aspect, whereas an item is a verbal description (most often with response categories). Health attributes may be overarching and abstract concepts (e.g., independence, dignity) or indicators that are very concrete (ability of vision) and detailed (ability to open a can). Furthermore, certain aspects of health can be described on a generic or a specific level. The description might take the form of a hierarchical structure of attributes that eventually could be depicted graphically as clusters of

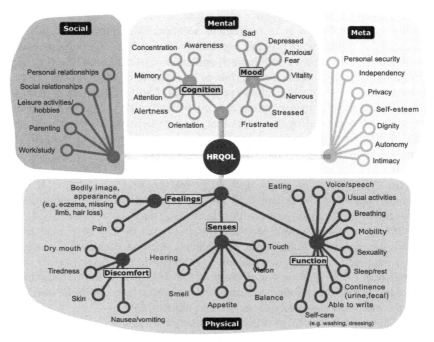

FIGURE 6.1   Health attributes depicted in a graphical template (HealthFan©) reflecting different domains (social, mental, physical, meta) and subdomains (discomfort, function, senses, feelings, cognition, mood).

different domains (Fig. 6.1). Once we identify which attributes we are interested in studying, we must decide how to describe them.

Selection of Attributes

In existing instruments, health attributes are selected predominantly on the grounds of consensus and expert opinion. However, there is some concern that those instruments do not sufficiently reflect the perspective of the individual patient. This concern has been raised in particular with regard to certain patient groups (Leplège and Hunt, 1997; Carr and Higginson, 2001; Ridgeway et al., 2013). When researchers decide which attributes to measure, they risk omitting ones of great relevance to the patients. Conversely, they risk including aspects that have little relevance. The field is becoming more aware of a better strategy: to select attributes based on patient input (Cella et al., 1993; Carr and Higginson, 2001; Hamming and De Vries, 2007; Ridgeway et al., 2013; Krueger and Stone, 2014).

To generate the content of a health instrument, either individual interviews or group discussion with patients (typically focus groups) will be required in most circumstances. Which approach is more appropriate depends on the sensitivity of the topic to be raised and whether there is a need for participants to react and feed off each other when discussing it. Focus groups need strong moderation to facilitate the conversation and ensure that no single person dominates the session. Individual interviews allow more in-depth exploration and are generally easier to analyze, but it may take more time to collect the data. Generally, a mixture of the two approaches is beneficial (Cappelleri et al., 2014). More information about qualitative research as a means to generate and select attributes, procedures for conducting such research, and considerations about the wording of the items can be found in Cappelleri et al. (2014) and in Streiner et al. (2015).

A recent study (Reneman et al., 2016) concerns the development of an outcome instrument to measure the overall impact of complaints on people with chronic pain. The researchers set out to systematically determine which aspects (attributes) are considered to be most important for people with chronic pain. From a large set of candidate attributes, people with chronic pain were asked to select the ones (a limited number) they deemed most important. The 20 that were most frequently chosen were compared to attributes in widely applied instruments that measure pain. Surprisingly, several of those 20 were absent in instruments that based the selection of attributes largely on expert judgment. It turned out that fatigue, by far the most important attribute, was not included in many current (chronic) pain instruments.

OBJECTIVE VERSUS SUBJECTIVE

The distinction between objective and subjective measures reflects the differentiation between mechanical methods based on laboratory tests and reasoning methods in which a person (e.g., clinician, patient, family member) makes a judgment. There is a difference between a subject response (by the person who is evaluating) and its subjective content. If the content is subjective (e.g., feelings, beliefs), then we are dealing with a personal perspective; assessments, judgments, and other such responses are by definition subjective. Observations done by a person and followed by a response, however, may be objective.

Ratings that involve judgments are generally called "subjective" measurements. By contrast, objective measurements involve no human judgment in the collection and processing of information (although judgment may be required for its interpretation). This distinction is often not clear-cut, however. Mortality statistics are commonly considered "objective," although judgment may be involved when assigning a code to the cause of death. Similarly, observing behaviors only constitutes an objective measure if the observations are recorded without subjective interpretation. Thus, difficulty in climbing stairs may be considered an objective indicator of disability if it is observed, whereas it is subjective if it is reported by the person. Note that the distinction between "subjective" and "objective" measurement does not refer to who makes the rating: objectivity is not bestowed on a measurement merely because it is made by an expert. Nor should we assume that subjective measures are necessarily "soft.". In longitudinal studies, subjective self-ratings of health are consistently found to predict subsequent mortality as well as, or better than, physical measures do. The terms objective and subjective are difficult to define, but the key criterion is the involvement of personal judgment.

Objective measurement is the estimation or determination of extent, dimension, or capacity in relation to some standard such as Quetelet index and forced expiratory volume score or in relation to some unit of measurement, e.g., blood pressure (mm Hg) and temperature (Celsius or Kelvin). Many of the basic measures in health are objective and thus correspond to these principal measurement properties. For example, the interval between a specific point in time and a person's death (where the interval is called "survival") can be precisely recorded. The number of complaints and type of side effects can easily be counted. (Note that a measurement is usually distinguished from a count. A measurement is expressed as a real number and is never exact, whereas a count is expressed as a natural number and may be exact.) Laboratory tests are more problematic. On some occasions, the test itself may lack precision (i.e., reliability). More often, the problem may be the variability of metabolic processes within and between patients (e.g., blood pressure, cholesterol level). Another branch of health-outcome measurement uses subjective measures. Some examples are pain, functional disability, severity of side effects, seriousness of symptoms and complaints, and speech clarity.

Measurement error occurs in both subjective and objective assessments; neither is necessarily more accurate nor precise under all circumstances. Some of the biomedical endpoints that we consider objective can include a demonstrably high degree of measurement error (for example, blood pressure), misclassification by experts, or have poor predictive and prognostic validity.

It is hard to say whether patients should assess only objective or only subjective attributes. Some authors advocate the combined use of objective and subjective health attributes, as the limitations of each type are different (Barofsky, 2012). Using both has the advantage of breaking down a false dichotomy between objective and subjective approaches. Similarly, Schalock (1996) argues for using a core set containing both objective and subjective attributes.

The difference between subjective measurement and objective measurement has been clearly demonstrated by Stensman (1985). This study asked individuals to rate their own health on a scale from 0.0 to 10.0. One group consisted of 36 persons with a severe mobility disability, while the other group, also of 36 persons, was nondisabled ("healthy"). Stensman depicted the results simply but clearly (Fig. 6.2). If we sum up all the scores of the 36 individuals in each group, we see that the mean score is comparable for both

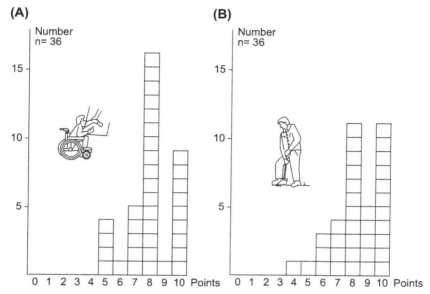

FIGURE 6.2 Self-reported quality of life on a 0–10 point scale for 36 subjects with severe mobility disability (A) compared with 36 nondisabled controls (B) (Stensman, 1985).

groups (disabled = 8.0, healthy = 8.3). What does this tell us? Asking people to rate their own health produces scores that truly reflect the perception of how individuals experience their health condition in relationship to their own internal standards. This approach is what is usually called subjective measurement. Other (indirect) measurement methods (see Chapters 11 and 12) also try to measure the health of (individual) patients. However, these methods disregard the adaptation mechanisms (see one of the following sections) that play a major role in how people respond in direct (subjective) measurement. The patients still have to make judgments, but some of these entail comparing one or a set of health attributes that serve as a reference standard.

CAUSE VERSUS EFFECT

Some authors have tried to come to grips with the relationships between health-outcome measures and their attributes. Leidy (1994), noticed that illness-related symptomatology is not considered an element of health but rather a precursor or effect of it. Pain, for instance, may hinder mobility and lead to loss of muscle function, movement, and/or flexibility, thereby restricting an individual's functional capacity. But pain may also increase the subjective suffering associated with a given activity, resulting in a choice to function at a lower level. However, we can also reason the other way around. When patients develop serious symptoms, their overall health is changed by those symptoms. In fact, the rationale for including symptoms in health instruments is that symptoms are believed to affect health. The person with the highest functional ability has the greatest probability of answering all questions positively, whereas a patient in poor health will not necessarily be suffering from all symptoms.

Symptoms and similar signs can be worked out as variables that measure an attribute. Thus elaborated, these can be entered into statistical models as causal indicators (Fayers and Hand, 1997). Symptoms and side effects serve as causal indicators of overall health. For example, few patients will experience all possible symptoms and side effects, but one serious symptom, such as pain, suffices to reduce overall health. Health attributes and the variables used to measure them may be partly indicative and partly causal. The terms "causal indicator" and "effect indicator" are widely used in structural equation modeling (Bollen, 1989), a specific area of research that combines regression analysis and various types of factor analysis. What we call causal indicators and effect indicators here may in another context be referred to as reflective and formative constructs or measurement models (Chapter 5). This distinction between causal indicators and effect indicators has become well known, even beyond the field of structural equation modeling. However, its implications are rarely recognized in health instrument development. They should be, though, since these two types of variables behave in fundamentally different ways in measurement scales. Moreover, their differences have a considerable impact upon the design of scales and statistical models.

Here, the latent variable (i.e., the concept denoted as θ in the literature on item—response theory and as η in the literature on structural equation modeling) is shown as a circle that represents the unobservable construct, depicted as the Greek letter η (eta). The rectangles stand for the observable items (e.g., variables). In the reflective model these are written as Y, because they are the consequences of η, whereas in a formative model the rectangles are the determinants of η and are written as X. This convention corresponds to Y as the conventional notation for dependent variables and X for independent variables (Fig. 6.3). Each Y is accompanied by an error term ε (the Greek letter epsilon). In the formative model there is only one error term, ζ (the Greek letter zeta), often called the disturbance term. A theory of how the scores generated by the items represent the construct to be measured is thus based on the relationships between the Xs and η, or between the Ys and η (de Vet et al., 2011).

A key difference between the formative and reflective models is the treatment of measurement error. An assumption underlying the reflective model is that all error terms (δ) associate with the observed scores (X) and, therefore, represent measurement error in the latent variable. The formative model does not assume such a correlational structure. For the formative model, the disturbance term (ζ) associates with neither the individual variable nor the

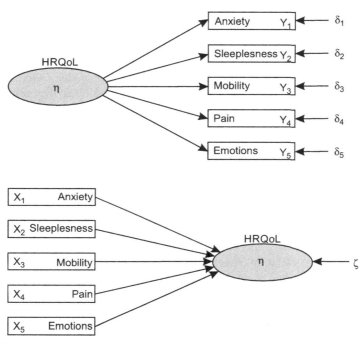

FIGURE 6.3 Structural relationship for reflective (top) measurement model (effect indicators) and formative (bottom) measurement model (causal indicators). *HRQoL*, health-related quality of life.

set of attributes as a whole. This term therefore does not represent measurement error. In the case of reflective models, researchers can identify and eliminate measurement error for each attribute using common factor analysis. In the case of formative models, however, that strategy is not feasible. Two well-known measurement models, namely classical test theory (Chapter 9) and item–response theory (Chapter 10), are reflective models. In contrast, choice models (Chapter 11), the valuation techniques from health economics (Chapter 12), and the index measures developed in the setting of clinimetrics (Chapter 13) are grounded in formative measurement models.

The distinction between formative and reflective models is not always clear-cut. Take the Apgar score, which was developed by Apgar (1953) to rate the clinical condition of a newborn baby immediately after birth. It consists of five variables to score five attributes: color (appearance); heart rate (pulse); reflex response to nose catheter (grimace); muscle tone (activity); and respiration—hence the eponymous acronym. According to Feinstein (1987), the Apgar score is a typical example of a measurement instrument in which the attributes refer to five different clinical signs that are not necessarily related to each other. In that sense, the instrument seems to correspond to the constraints of a formative model. However, it is questionable whether it is actually based on a formative model. If we consider the Apgar score as an indication that a baby was premature, then it may be based on a reflective model. Given that all organ systems in premature babies will be less well developed, the infant may show signs of problems in all these systems. Thus, depending on the underlying hypothesized conceptual model, we may see the Apgar score as either a formative or a reflective model. This illustrates the importance of specifying the underlying conceptual model (de Vet et al., 2011).

Since reflective measurement models have positive intercorrelations between the variables, researchers can use statistical routines such as factor analysis and Cronbach's α to empirically assess their reliability. However, reliability statistics assume internal consistency—that is, high intercorrelations among the variables in question. Such statistical routines are inappropriate for formative measurement models, where no theoretical assumption is made about interitem correlation. One of the key operational issues in the use of formative models is that no simple, easy, and universally accepted criteria exist for assessing their reliability.

These are rather complicated issues. There is still considerable discussion among scientists on how to deal with them, both conceptually and analytically. Moreover, the representation of the structure of the (assumed) relationship between variables (i.e., the construct) is important. That relationship can be expressed verbally, it can be represented graphically (Fig. 6.3), and it can be shown in mathematical notation. But in the end, we are scoring certain health characteristics that have to be processed in a specific statistical model to arrive at the health measures.

ADAPTATION, COPING, AND RESPONSE SHIFT

Coping and adapting are facts of life. People tend to adjust to new situational states. They develop coping mechanisms such as skill enhancement or activity adjustments. Those who become disabled can usually mobilize specific responses to deal with their condition. Their responses include adaptation, coping, adjustment, and other mechanisms.

When asked which features of life make the most important contribution to its quality, a "healthy" individual is likely to say "my social and family life, my work, my health, and my financial circumstances." But a year later, the same individual, after being diagnosed and treated for laryngeal cancer, may answer the same question differently: "my ability to speak and eat, my family and social life, and my spiritual happiness." The terms of reference (and thereby the standard) this individual uses have completely changed. It is important to recognize two factors involved in this change. The domains or attributes contributing to that individual's evaluation of his/her own quality of life have changed, but so has their relative importance (Allison et al., 1997).

Adaptation

The meaning of adaptation that is most appropriate in the context of health is to adjust oneself to new or changed circumstances. Patients may find it difficult to admit how poor their objective, functional health really is. Though denial is largely a cognitive matter, it may still express a person's need to accommodate a new reality and can therefore easily be taken for adaptation. A related cognitive deficiency that comprises part of the adjustment to chronic, long-term illness occurs when patients cease to realize what full health is like or what it would enable them to do. Over time, chronically ill or disabled persons may develop greater skill in using whatever physical or mental capacities they retain. People simply improve their ability to reach prior goals in their current activities, beyond what they could ever have imagined possible. Realizing that a disease or disability is likely to be chronic, people may adjust their activities—not only the activities they select to pursue their goals, but the direction of the goals themselves.

Patients with chronic illness often adapt physically and emotionally to their health states. Adaptation is partly physical. A person with blindness may learn how to read Braille and how to ambulate with a cane. Adaptation is also psychological. When people lose the ability to perform certain activities, they start preferring alternative ones. People may also lower their expectations. A patient with severe emphysema may forget that long walks in the park used to be an important part of his/her life (Ubel et al., 2003). Adaptation may involve altering one's plans to give greater importance to activities in which performance is less diminished by disability. For example, one might spend less time listening to music and more time reading as a result of severe hearing loss.

Coping

The concept of coping is similar to that of adaptation. However, coping suggests the presence of stressors (e.g., a biological agent, environmental conditions, external stimulus, or an event that causes stress) and the individual's attempts to deal with them. Lazarus and Folkman (1984) defined coping as "constantly changing cognitive and behavioral efforts to manage specific external and/or internal demands that are appraised as taxing or exceeding the resources of the person." So coping can be seen as a process, a strategy, and a response to all the elements (e.g., environment, individual disposition) that play a role in the effort to adapt. Coping is better understood as an underlying individual process than as an outcome. The product of a coping process may be seen as an "overt" adaptation.

Social support is an external factor that may directly affect health, but being an external factor, it is a circumstance that does not define personal health. Therefore, neither coping nor social support can be considered part of health outcomes. Their external position can also be discerned in the health model (Chapter 2) of Wilson and Cleary (1995).

Coping involves lowering one's expectations for performance so as to reduce the self-perceived gap between those expectations and one's actual performance. For instance, coping might pertain to one's ability to climb stairs when living with chronic obstructive pulmonary disease. Disabled persons who have gone through a process of coping typically report lower levels of distress and limitation of opportunity but a higher perceived health with their disability than the nondisabled do when evaluating the same condition.

Response Shift

Medical models of health traditionally assume that as more symptoms arise, health declines and disease progresses. Contrary to that assumption, many persons with chronically invalidating diseases keep reporting a high level of health. Caregivers initially rate the patients' health lower than the patients themselves do, and caregivers also rate health lower than patients during disease progression (Boyer et al., 2004; de Wit et al., 2000).

Such mechanisms, which can affect health measures, are collectively known as "response shift." In general, this concept denotes a change in the measurement of health over time, which may result from a change in individual internal standards of health. Response shift covers the psychological changes in perception of health following a change in health and thus can affect the results of health-outcome measurements.

People's interpretations and subsequent responses as registered on scales may change as result of their changing circumstances. That phenomenon is known as scale recalibration (Sprangers and Schwartz, 1999; Wilson, 1999; Ubel et al., 2010). It is one of three possible mechanisms that allow for "response shift." In addition, a response shift can reflect a change in the relative importance to a respondent of the domains of health (reprioritization) or it can reflect a redefinition of one's meaning of health itself (reconceptualization) (Schwartz and Sprangers, 1999).

Several methods have been proposed to estimate response shift: for instance, the then-test, ideal scale approach, anchor recalibration, structural equation modeling, and rating of vignettes. The then-test is a retrospective judgment of preintervention health and is used in combination with a pretest—posttest design. The then-test is one of the least complicated methods to measure response shift. However, the retrospective aspect is a serious drawback: it presumes correct recall of the preintervention health state (Visser et al., 2005). In anchor recalibration, shifts whereby individual patients place the scale anchors over time are assessed (Sprangers and Schwartz, 1999).

In structural equation modeling, response shift is deduced from mathematically defined changes by conducting longitudinal confirmatory factor analysis. It is claimed that structural equation modeling can detect reconceptualization, reprioritization, and recalibration (Visser et al., 2013; Schwartz et al., 2013).

If methods are used that measure an individual's perceived and subjective appraisal of his or her own health, and if that patient has adapted to a deteriorated health condition, then the report given by that patient will provide responses that will result in measures that are counterintuitive to an observer. This discrepancy derives from the subjective nature of the report and the type of measurement instruments used. In fact, these types of measurement instruments do not measure the "objective" health status of a patient; rather, they "measure" the patient's subjective appraisal. Had we used other types—namely, instruments that were specifically developed to arrive at more "objective" measures—we would have obtained different results. In that case, our measures would not be affected by response shift.

QUESTIONNAIRES AND INSTRUMENTS

Instruments for assessing health usually comprise multiple items (questions). A few instruments rely on a single global question such as, "Overall, what has your quality of life been like over the last week?" Some health instruments combine items; for example, the item scores might be averaged to produce an overall score for health. Most developers of instruments recognize that health has many domains. Therefore, they generally group items into separate "scales" corresponding to the different domains.

A distinction can be made between questionnaires with items that measure separate variables (e.g., age, length, or distinct health outcomes) and questionnaires with items that are aggregated into either a scale or an index. The former is usually seen as a component of surveys, whereas the latter is considered as instruments in themselves. Nevertheless, these aggregate instruments are commonly called "questionnaires" in the literature. As a matter of fact, most of these are questionnaires in the sense of a specific type of measurement tool.

In survey research, a questionnaire consists of a set of questions to be posed to the participants. Sir Francis Galton, an English polymath, introduced its use in surveys. The questions are usually phrased to elicit ideas and behaviors, preferences, traits, attitudes, and facts. A survey is a method to gather data. It is intended to collect, analyze, and interpret the views of a group of people drawn from a target population. Surveys have been extensively applied in various fields of research, particularly sociology, marketing, politics, and psychology. Today, questionnaires are often administered in face-to-face, telephone, paper-and-pencil, and computerized modes.

In an early definition, an instrument was said to include the questionnaire, the method of administration, instructions for administration, the method of scoring and analysis, and the interpretation (Guyatt et al., 1991). A later one stated that if the measurement of health outcomes is based on a set of items covering the phenomena of interest, and if the phrasing of the items and the response collecting are performed on the basis of an underlying measurement theory or model, we can speak of an instrument (Testa and Simonson, 1996; Feinstein, 1999). A recent definition of the term "instrument" as used in the field of patient-reported outcomes was given by Cappelleri et al. (2014): "A means to capture data (i.e., a questionnaire) plus all the information and documentation that supports its use. Generally, that includes clearly defined methods and instructions for administration or responding, a standard format for data collection, and well-documented methods for scoring, analysis, and interpretation of results in the target patient population." Whether all research is conducted in such a way that it can be classified as "well documented" remains to be seen. But what makes this definition novel is that it treats a questionnaire as just the appearance of one of the components of an instrument. Therefore, on many occasions a questionnaire will be part of a larger conceptual framework: an instrument.

In the following discussion of health measurement, the terms "instruments" and "scores" are used generically. "Instruments" can refer to questionnaires, tests, rating scales, observation procedures, and other devices or techniques used to measure or assess specific characteristics or attributes of people, objects, or events. "Scores" are the numbers resulting from the application of measurement instruments. We refer the reader to the Glossary at the end of this book for an explanation of specific terms; that is where most of the terms introduced in this chapter are defined.

TYPES OF INSTRUMENTS

Health instruments differ in several ways: for instance, in their emphasis on objective or subjective dimensions; their coverage of certain domains; and the format of their questions.

Sometimes health is regarded as a single concept, which can be assessed by asking patients about their health status. Health can then be quantified as a single overall quality-of-life score. This can be done by scaling on one type of scale. However, most investigators agree that health is a multifaceted concept. This implies that its aspects are best evaluated as distinct domains, such as physical, psychological, and social. A major task in scale design is to determine the number of items that should comprise a particular scale. If more than one item seems required, it is important to assess how consistently these items work together.

Measurement Approaches

Three contrasting approaches have been proposed to measure health. The most simple and straightforward one is to use a single scaling method with only one item. There is considerable disagreement on whether it is meaningful to ask a patient questions such as "Overall, what would you say your quality of life has been like during the last week?" One of the main objections to the use of single items is that concepts such as health, role functioning, and emotional functioning are complex and ill defined. People have different ideas about their meaning. Measurement based on a global single-item instrument assumes that respondents will consider all aspects of a phenomenon, ignore aspects that are not relevant to their situations, and differentially weigh the other aspects according to their values and ideals to arrive at a single rating. Responses to a single item that was chosen to represent one aspect of a complex concept are likely to be unreliable and will probably also lack validity. Yet, most OECD (Organisation for Economic Cooperation and Development) countries conduct regular health surveys that allow respondents to report on different aspects of their health. In these surveys, no multiitem or sophisticated item—response theory measurement models are applied. Instead, the most commonly asked question relates to self-perceived health, of the type, "How is your health in general?" (OECD, 2015).

Most of the existing instruments can be classified as descriptive profile instruments. This approach is used to assess one or more health domains, and each of these domains is covered by several items. Supporters of the multiitem profile approach argue that health is inherently multidimensional, so scores on each domain should be presented separately. Psychometric theory also favors multiitem instruments because they tend to be more reliable and less prone to random measurement errors than single-item instruments.

In a third approach, a set of health attributes is quantified as a single metric, often referred to as an index measure. In general, index measures are based on clinical practice and expertise instead of statistical procedures. The items are rated and then summed up to obtain a total score. Almost all of these instruments are unweighted index measures (e.g., Apgar score). Another category of instruments was developed along the same lines but with a preference-based methodology (Feinstein, 1983; Boyle and Torrance, 1984; Gold et al., 1996). Under this measurement framework, the distinct attributes are differently weighted on the basis of empirical studies that assign relative importance to the attributes.

Scope

In general, there are three types of instruments for measuring health: generic; disease-, disorder-, or condition specific; and symptom- or domain specific. Generic instruments are designed to cover a wide variety of diseases and medical treatments, allowing comparison of conditions and interventions. However, generic instruments may fail to capture specific aspects of patients' experience, which means they lack sensitivity. Generic instruments, such as the sickness impact profile, permit comparison across disease categories.

TABLE 6.1 Types of Health Measurement Instruments

	Single Item	Multiitem Profiles	Index	Preference Based
Generic	VAS, Cantril's ladder, rating scale (various)	SF-36	Apgar	EQ-5D
Disease- or condition specific	VAS, Cantril's ladder, rating scale (various)	EORTC QLQ C30 (cancer)	Barthel	DQI (dementia)
Symptom or domain specific	VAS (pain)	McGill Pain questionnaire	Zung (anxiety)	Pain

Therefore, generic tools are convenient for evaluating broader topics, such as whether the care a nurse practitioner delivers is comparable to that of a family physician.

Disease-specific instruments narrow in on the consequences of certain diseases. They contain domains or attributes that are relevant to a patient group. Apart from being designed for a particular disease, they can also be specific to a particular type of person (women's health) or to an age group (child health).

Symptom-specific instruments focus exclusively on symptoms produced by a particular disease or condition. One such instrument is the McGill Pain Questionnaire (Melzack, 1975). Symptom-specific instruments are generally intended for clinical application and are designed to pick up any change following treatment.

Table 6.1 gives examples of instruments and response scales that are used to measure health. Few overviews exist of the myriad instruments applied in health measurement. Interested readers could take a look at the books by Spilker (1996) and McDowell (2006).

RESEARCH TRADITIONS

The assessment of health outcomes has developed rapidly. We are therefore confronted with input from three discrete disciplines: psychometrics, health economics, and clinimetrics (Fig. 6.4). Psychometrics is the branch of psychology that deals with the design, administration, and interpretation of quantitative tests or instruments for the measurement of psychological variables such as intelligence, attitudes, and personality traits. Health economics is a branch of economics concerned with issues related to efficiency, effectiveness, values, and behavior in the production and consumption of health and health care. Health researchers are more oriented toward input from clinimetrics. That term denotes an approach to scale development in the area of

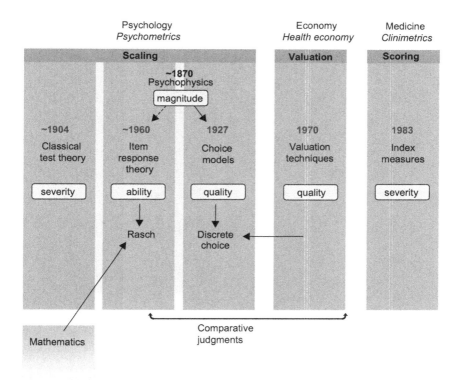

FIGURE 6.4   Different research traditions and different measurement methods.

health, one that ostensibly differs from the more traditional psychometric approach. It has a set of rules to govern the structure of indexes, the choice of component variables, and the evaluation of consistency and validity (Table 6.2).

Much of the progress, but also much of the misunderstanding, in measurement concepts and approaches has taken place among clinimetricians and psychometricians. But progress has also been made through the interaction between health economists and the other research traditions. It is helpful to review the history and development of health measurement from this angle. Research traditions can provide a useful set of categories for examining theories of measurement (Engelhard, 2013).

DISCUSSION

Much attention has been given to understanding the processes that are responsible for producing response shifts. Probably one of the first conceptual

TABLE 6.2 Characteristics of the Five Measurement Traditions

	Classical Test Theory	Item Response Theory	Choice Models	Valuation Techniques	Clinimetrics
Origin	Psychometrics	Psychology/Education	Psychology	Economics	Health science
Measurement level	Interval	Interval	Interval	Interval	Not specified
Responses	Absolute	Absolute	Comparative	Trade-off	Absolute
Unidimensional	Yes, for each domain	Yes	Yes	Yes	Unspecified
Unit of measurement	Unspecified	Worst-best	Worst-best	Dead-full health	Unspecified

papers to address the problems with changes in responses, apart from real intervention effects, originated in the management sciences. Golembiewski et al. (1976) advanced a typology that distinguished observable change (alpha change) from changes in internal standards (beta change) and meaning (gamma change). Parallel in time, but independently, the term "response shift" was coined by Howard et al. (1979) on the basis of research on educational training. They defined response shift in terms of changes in internal standards of measurement.

It is interesting that a large share of the problems and dynamics of health measurement and response shift derive from the type of measurement instruments used. All the studies that investigate response shift, or try to quantify it, have applied simple rating scales or multiitem profile instruments (Chapter 9). Instruments of this type were developed under classical test theory. The way questions are phrased and responses are collected in that tradition allows response—shift mechanisms to emerge and become embedded. We shall return to these matters in Chapters 10 and 11. There, we will look into measurement frameworks that in principle can produce more "objective" measures, as these methods are largely insensitive to the response—shift mechanism.

REFERENCES

Allison, P.J., Locker, D., Feine, J.S., 1997. Quality of life: a dynamic construct. Social Science & Medicine 45 (2), 221—230.

Apgar, V.A., 1953. Proposal for a new method of evaluation of the newborn infant. Current Researches in Anesthesia & Analgesia 32 (4), 260—267.

Barofsky, I., 2012. Quality: Its Definition and Measurement as Applied to the Medically Ill. Springer, New York.

Bollen, K.A., 1989. Structural Equations with Latent Variables. John Wiley & Sons, New York.

Boyer, F., Novella, J.L., Morrone, I., Jolly, D., Blanchard, F., 2004. Agreement between dementia patient report and proxy reports using the Nottingham Health Profile. International Journal of Geriatric Psychiatry 19 (11), 1026—1034.

Boyle, M., Torrance, W., 1984. Developing multiattribute health indexes. Medical Care 22 (11), 1045—1057.

Cappelleri, J.C., Zou, K.H., Bushmakin, A.G., Alvir, J.M.J., Alemayehu, D., Symonds, T., 2014. Patient-reported Outcomes: Measurement, Implementation and Interpretation. CRC Press, Boca Raton.

Carr, A.J., Higginson, I.J., 2001. Are quality of life measures patient centred? BMJ 322 (7298), 1357—1360.

Cella, D.F., Tulsky, D.S., Gray, G., Sarafian, B., Linn, E., Bonomi, A., Silberman, M., Yellen, S.B., Winicour, P., Brannon, J., Eckberg, K., Lloyd, S., Purl, S., Blendowski, C., Goodman, M., Barnicle, M., Stewart, I., McHale, M., Bonomi, P., Kaplan, E., Taylor IV, S., Thomas Jr., C.R., Harris, J., 1993. The functional assessment of cancer therapy scale: development and validation of the general measure. Journal of Clinical Oncology 11 (3), 570—579.

de Vet, H.C.W., Terwee, C.B., Mokkink, L.B., Knol, D.L., 2011. Measurement in Medicine: A Practical Guide. Cambridge University Press, Cambridge.

de Wit, G.A., van Busschbach, J.J.V., de Charro, F.T.H., 2000. Sensitivity and perspective in the valuation of the health status: whose values count? Health Economics 9 (2), 109–126.

Engelhard Jr., G., 2013. Invariant Measurement: Using Rasch Models in Social Behavioral, and Health Sciences. Routledge, New York.

Fayers, P.M., Hand, D.J., 1997. Factor analysis, causal indicators and quality of life. Quality of Life Research 6 (2), 139–150.

Feinstein, A.R., 1983. An additional basic science for clinical medicine: IV. The development of clinimetrics. Annals of Internal Medicine 99, 834–848.

Feinstein, A.R., 1987. Clinimetrics. Yale University Press, New Haven.

Feinstein, A.R., 1999. Multi-item "instruments" vs Virginia Apgar's principles of clinimetrics. Archives of Internal Medicine 159 (2), 125–128.

Gold, M.R., Siegel, J.E., Russell, L.B., Weinstein, M.C., 1996. Cost-effectiveness in Health and Medicine. Oxford University Press, New York.

Golembiewski, R.T., Billingsley, K., Yeager, S., 1976. Measuring change and persistence in human affairs: types of change generated by OD designs. Journal of Applied Behavioural Science 12 (2), 133–157.

Guyatt, G., Feeny, D., Patrick, D., 1991. Issues in quality-of-life measurement in clinical trials. Controlled Clinical Trials 12 (4), 81S–90S.

Hamming, J.F., De Vries, J., 2007. Measuring quality of life. British Journal of Surgery 94 (8), 923–924.

Howard, G.S., Ralph, K.M., Gulanick, N.A., Maxwell, S.E., Nance, D.W., Gerber, S.K., 1979. Internal invalidity in pretest-posttest self-report evaluations and a re-evaluation of retrospective pretests. Applied Psychology Measurement 3 (1), 1–23.

Kaplan, R., Anderson, J., 1990. The general health policy model: an integrated approach. In: Spilker, B. (Ed.), Quality of Life Assessments in Clinical Trials. Raven Press, Ltd, New York, pp. 131–149.

Krueger, A.B., Stone, A.A., 2014. Progress in measuring subjective well-being: moving toward national indicators and policy evaluations. Science 346 (6205), 42–43.

Larson, J.S., 1991. The Measurement of Health: Concepts and Indicators. Greenwood Press, Westport.

Lazarus, R.S., Folkman, S., 1984. Stress, Appraisal and Coping. Springer, New York.

Leidy, N.K., 1994. Functional status and the forward progression of merry-go-rounds: toward a coherent analytical framework. Nursing Research 43 (4), 196–202.

Leplège, A., Hunt, S., 1997. The problem of quality of life in medicine. JAMA 278 (1), 47–50.

McDowell, I., 2006. Measuring Health: A Guide to Rating Scales and Questionnaires, third ed. Oxford University Press, Oxford.

Melzack, R., 1975. The McGill pain questionnaire: major properties and scoring methods. Pain 1 (3), 277–299.

OECD, 2015. Perceived health status. In: OECD (Ed.), Health at a Glance 2015: OECD Indicators. OECD Publishing, Paris.

Reneman, M.F., Brandsema, K.P.D., Schrier, E., Dijkstra, P.U., Krabbe, P.F.M., 2016. Patients first: development of a next-generation, patient-centered, preference-based instrument to measure the impact of chronic pain (Manuscript submitted for publication).

Ridgeway, J.L., Beebe, T.J., Chute, C.G., Eton, D.T., Hart, L.A., Frost, M.H., Jensen, D., Montori, V.M., Smith, J.G., Smith, S.A., Tan, A.D., Yost, K.J., Ziegenfuss, J.Y., Sloan, J.A., 2013. A brief patient-reported outcomes quality of life (PROQOL) instrument to improve patient care. PLoS Medicine 10 (11), e1001548.

Schalock, R.L., 1996. Reconsidering the conceptualization and measurement of quality of life. In: Schalock, R.L. (Ed.), Quality of Life, Volume I: Conceptualization and Measurement. American Association on Mental Retardation, Washington D.C, pp. 123–139.

Schwartz, C.E., Sprangers, M.A.G., 1999. Methodological approaches for assessing response shift in longitudinal health-related quality-of-life research. Social Science & Medicine 48 (11), 1531–1548.

Schwartz, C.E., Ahmed, S., Sawatzky, R., Sajobi, T., Mayo, N., Finkelstein, J., Lix, L., Verdam, M.G.E., Oort, F.J., Sprangers, M.A.G., 2013. Guidelines for secondary analysis in search of response shift. Quality of Life Research 22 (10), 2663–2673.

Spilker, B., 1996. Quality of Life and Pharmacoeconomics in Clinical Trials, second ed. Lippincott-Raven Publishers, Philadelphia.

Sprangers, M.A.G., Schwartz, C.E., 1999. Integrating response shift into health-related quality of life research: a theoretical model. Social Science & Medicine 48 (11), 1507–1515.

Stensman, R., 1985. Severely mobility-disabled people assess the quality of their lives. Scandinavian Journal of Rehabilitation Medicine 17, 87–99.

Streiner, D.L., Norman, G.R., Cairney, J., 2015. Health Measurement Scales: A Practical Guide to Their Development and Use, fifth ed. Oxford University Press, Oxford.

Testa, M.A., Simonson, D.C., 1996. Assessment of quality-of-life outcomes. The New England Journal of Medicine 334 (13), 835–840.

Ubel, P.A., Loewenstein, G., Jepson, C., 2003. Whose quality of life? A commentary exploring discrepancies between health state evaluations of patients and the general public. Quality of Life Research 12 (6), 599–607.

Ubel, P.A., Peeters, Y., Smith, D., 2010. Abandoning the language of "response shift": a plea for conceptual clarity in distinguishing scale recalibration from true changes in quality of life. Quality of Life Research 19 (4), 465–471.

Visser, M.R.M., Oort, F.J., Sprangers, M.A.G., 2005. Methods to detect response shift in quality of life data: a convergent validity study. Quality of Life Research 14 (3), 629–639.

Visser, M., Oort, F., van Lanschot, J., van der Velden, J., Kloek, J., Gouma, D., Schwartz, C., Sprangers, M., 2013. The role of recalibration response shift in explaining bodily pain in cancer patients undergoing invasive surgery: an empirical investigation of the Sprangers and Schwartz model. Psycho-oncology 522 (22), 515–522.

Voigt, K., King, N.B., 2014. Disability weights in the global burden of disease 2010 study: two steps forward, one step back? Bulletin of the World Health Organization 92 (3), 226–228.

Ware, J.E., Brook, R.H., Davies, A.R., Lohr, K.N., 1981. Commentary choosing measures of health status for individuals in general populations. American Journal of Public Health 71 (6), 620–625.

Wilson, I.B., Cleary, P.D., 1995. Linking clinical variables with health-related quality of life: a conceptual model of patient outcomes. JAMA 273 (1), 59–65.

Wilson, I.B., 1999. Clinical understanding and clinical implications of response shift. Social Science & Medicine 48 (11), 1577–1588.

Chapter 7

Validity

Chapter Outline

INTRODUCTION

In science, validity is the extent to which a concept, conclusion, or measurement is well founded and corresponds accurately to the real world. It has also been defined as an overall assessment of the degree to which evidence and theory support the interpretation of the scores entailed by proposed uses of the instrument. To phrase it simply, validity refers to the degree to which evidence and theory support the interpretations of measures. These definitions prompt questions about the "real" meaning and interpretation of scores collected with health measurement instruments. The word "valid" is derived from the Latin "validus," meaning strong. In that sense, the validity of a measurement instrument is the degree to which it measures what it claims to measure.

Many researchers approach the development of health-outcome instruments in two steps. They first design the instrument and make sure it is reliable. Only then do they determine whether it assesses what they wish to measure. Thus, it would make more sense to talk about reliability before introducing validity. However, validity lies at the core of science. Validity is concerned with the nature of "reality" and the nature of the entity being measured. Especially for (partly) subjective phenomena, such as health, the determination of validity involves the incremental accumulation of evidence rather than a one-off comparison. Note the difference between the type of validity we are discussing here and the type known as design validity. In research and experimentation, design validity refers to whether a study is able to provide relevant information and is not biased by certain external factors.

This subject matter is complex and controversial (Borsboom et al., 2004). Validity is particularly problematic in social contexts, where it is notoriously difficult to get an empirical handle on what one really wants to measure. The same is true of health-outcomes research. As opposed to outcomes, such as forced expiratory volume, days of hospital stay, or survival, health is not directly observable.

VALIDITY

Throughout the 1940s, scientists tried to come up with ways to validate measurement instruments on the basis of latent variables. The result was a proliferation of types (intrinsic validity, face validity, logical validity, empirical validity, etc.). Given this variety, it was difficult to tell which ones were actually the same and which were not useful at all. Until the middle of the 1950s, there were very few universally accepted methods to validate scales in the social sciences. Most researchers came to acknowledge that validity has three aspects, each rather broad in scope; content validity, criterion validity, and construct validity (Cronbach and Meehl, 1955).

Over the next four decades, many theorists, including Cronbach himself, voiced dissatisfaction with this three-in-one model. The debate culminated in Messick's article of 1995 that describes validity as a single construct, composed of six "aspects" (Messick, 1995). Modern theory defines construct validity as the overarching concern of validity research, subsuming all other types of validity evidence. In Messick's view, inferences drawn from instrument scores may require different types of evidence, but not different validities. Messick and others have pushed for a unified view of construct validity "... as an integrated evaluative judgment of the degree to which empirical evidence and theoretical rationales support the adequacy and appropriateness of inferences and actions based on instrument scores." Taking a little distance from these modern and elaborated views, we will stick to the traditional three-in-one classification, as for most purposes this framework is sufficient and understandable.

An overall assessment of an instrument's validity is not established once and for all. Instead, it is gradually built up (corroboration process) from cumulative evidence regarding the interrelationships between the content of the instrument and definitions of the construct to be measured, and the interrelationships between scores on the instrument. Relevant intercorrelations and associations should be provided by the developer of an instrument or with each reported use of it. Then, successive users can judge the validity of the instrument in relation to the purpose for which they may wish to use it. Gathering evidence of relationships that could support or refute particular interpretations of measures derived from a health instrument is an ongoing process. The reason is that no single relationship is normally sufficient to conclusively establish the validity of a given interpretation. Especially for a phenomenon such as health, validity is assessed through the incremental accumulation of evidence rather than one conclusive comparison. In short, validation is a process, not a result (Strauss and Smith, 2009).

FACE VALIDITY

Face validity concerns whether an instrument "looks valid" to the respondents who have to fill it up, to the administrative personnel who decide on its use, and to other technically untrained observers. Quite different is content validity (see below), which is assessed by recognized experts. Their expertise in the subject matter is needed to evaluate whether items assess the defined content and to conduct more rigorous statistical tests than those applied for face validity. It has been suggested that the validity of an interpretation or use of a measurement instrument can be judged by those who might use it or those on whom it might be used, on the basis of whether the content of the instrument seems relevant to the construct to be measured. However, such "face validation" may be unreliable and misleading. It is understandable that those examining an instrument would form an impression of whether the items are appropriate to what they perceive the purpose of the measurement to be. Nonetheless, it is impossible to draw legitimate conclusions regarding an instrument's validity or reliability until other information has been gathered and considered (Nunnally and Bernstein, 1994). Face validity is classified as "weak evidence" supporting validity. But that does not mean that it is incorrect, only that caution is necessary.

CONTENT VALIDITY

The concept of content validity is nonanalytical. Its assessment deals with a basic question, "Is the measure really measuring what it is intended to measure?" Crucial to the development of any measure of health is the degree of correspondence between health concepts and specific operational definitions. Typically, claims of content validity rest largely on how comprehensive or detailed the instrument is. A health instrument that encompasses more health domains or attributes than another instrument is assumed to possess a greater content validity. However, this assumption is violated when nonrelevant domains are included, when the instrument does not cover the whole concept to be assessed, or when some domains are redundant. Content validity of a health-outcome instrument is probably the most important type of validity but also the most difficult one to achieve (Kerlinger, 1986c). We say "probably" because this issue is subject to ongoing debate (Froberg and Kane, 1989; Nord, 1992).

Evaluations of content validity tend to be fairly subjective. Ideally, they include systematic comparison of a measure by confronting it with existing standards, widely accepted theoretical definitions, expert opinions, and interviews with individuals to whom the measure is targeted. When a new scale is being developed, a content map is often created so that items can be constructed to cover the domains of interest (Fig. 7.1).

Percentage of responses by chronic pain patients

FIGURE 7.1   Example from a study in which pain patients were given a list of 84 different health attributes related to experienced chronic pain and were asked to select the eight most important ones. *From Reneman, M.F., Brandsema, K.P.D., Schrier, E., Dijkstra, P.U., Krabbe, P.F.M., 2016. Development of a next-generation, patient-centered, preference-based instrument to measure the impact of chronic pain. (Manuscript submitted for publication).*

CRITERION VALIDITY

In contrast to content validity, criterion or predictive validity is determined analytically. The concept is only applicable if another existing instrument can be identified as superior. Correlation coefficients (product-moment, Spearman rank, intraclass) are often estimated between a criterion measure (gold standard) and a competing measure to test for equivalence. By definition, the criterion must be a superior, more accurate measure of the phenomenon if it is to serve as a verifying norm. Widely applied and recognized depression scales such as Hamilton Depression Rating (Hamilton, 1960, 1980) and Beck Depression Inventory (Beck et al., 1961, 1996) are possible gold standards for developing alternative (briefer) depression instruments.

Elevating a single criterion to a "gold standard" is difficult, if not impossible, for assessing health instruments because no such standard exists in this setting. It is difficult even to imagine what would constitute a gold standard, since there is no generally accepted concept of health status, nor of health. If there is no gold standard, criterion validity cannot be proved. Therefore, this type of validity can be ignored in this setting and is not addressed elsewhere in this book.

CONSTRUCT VALIDITY

To evaluate construct validity, we would hypothesize how measures should "behave" and on that basis we could then confirm or reject these hypotheses (Cronbach and Meehl, 1955). Hypotheses are formulated on the direction (and sometimes the strength) of relationships that might be expected, and validity is supported when the associations are consistent with these hypotheses. Construct validation is iterative by nature. The empirical results feed into the revision of measures, retesting, and further revision, if necessary. Over time, and after repeated studies, we become more confident in the validity of the measure and its application in different populations. Many analysts view construct validity as the crucial test of the validity of a health-outcome measure (Brooks, 1995).

Construct validity is now generally viewed as a unifying form of validity for measurements. It subsumes both content and criterion validity, which traditionally had been treated as distinct forms (Landy, 1986). But Messick has argued that even this notion of validity is too narrow. In his view, "Validity is an overall evaluative judgment of the degree to which [multiple forms of] evidence and theoretical rationales support the adequacy and appropriateness of interpretations and actions on the basis of test scores..." (Messick, 1995, p. 741). It should be kept in mind that Messick was writing mainly about educational assessment. So some of his observations and conclusions may not be fully relevant to a health setting.

Convergent/Discriminant

Convergent and discriminant validities are two fundamental aspects of construct validity. Convergent validity refers to how closely the new scale is related to other variables and other measures of the same construct. Not only should the construct correlate with related variables but it should *not* correlate with dissimilar, unrelated ones. A determination along the latter lines is referred to as discriminant validity (de Vet et al., 2011; Streiner et al., 2015). For example, a performance-based measure of walking should be positively correlated with self-reported ability to walk a block. Similarly, performance-based measures of daily activities, such as fastening buttons and preparing and boiling a pot of water, should be associated positively with self-reported activities of daily living.

Multitrait-Multimethod

When two or more constructs (traits) are assessed by more than one method, convergent and discriminant validities can be determined analytically with the multitrait-multimethod (MTMM) approach. In its original form, MTMM was introduced by Campbell and Fiske (1959). Their model examines convergence (evidence that different methods to measure a construct give similar results) and discriminability (ability to differentiate a construct from other related constructs). According to Cook and Campbell (1979, p. 61), assessing construct validity depends on two processes. "First, testing for a convergence across different measures or manipulation of the same 'thing' and, second, testing for a divergence between measures and manipulations of related but conceptually distinct 'things'." The MTMM approach can identify four classes of correlation coefficients (Fig. 7.2). The first class covers monotrait-monomethod reliability correlations (health conditions measured twice for each method separately: test—retest). The second covers heterotrait-monomethod correlations (different health conditions for the same method); the third, heterotrait-heteromethod correlations (different health conditions assessed by different methods). And the last class covers monotrait-heteromethod correlations (same health condition assessed by different methods). Under this methodology, construct validity is supported when correlations among different methods are high for a single trait (convergent validity) but correlations between the same methods measuring different traits are low (discriminant validity). The original developers, Campbell and Fiske, recommended visual inspection of the MTMM matrix for assessment of construct validity. Since then, additional modeling procedures (e.g., confirmatory factor analysis) have been developed that may bring us closer to an unequivocal interpretation of such data (Schmitt and Stults, 1986; Jöreskog and Sörbom, 1989).

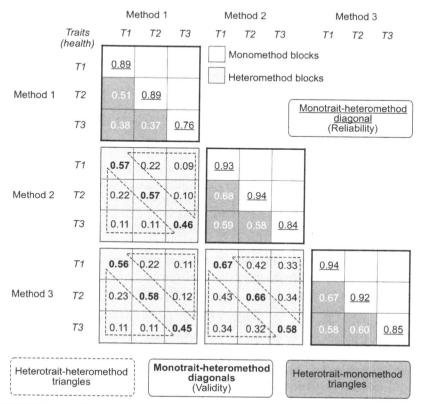

FIGURE 7.2   Multitrait-multimethod matrix to assess the construct validity of a set of measures in a study.

Known Groups

A third variant of construct validity is called the "known groups" strategy. In this approach, the instrument is administered to two groups that are known to have, or that logically should have, different levels of the construct to be confirmed. The assumption is that the hypothesized difference is reflected in the scores of the two groups (Hattie and Cooksey, 1984). This strategy typically involves evaluating an instrument in relation to (clinical) measures of (implicit) disease status (e.g., elderly vs youngsters, ill people vs not ill people).

Statistical Structure

Evidence based on the internal structure of an instrument is another source of information relevant to the interpretation of its scores. Such evidence typically is obtained by techniques such as factor analysis and cluster analysis. The evidence is used to determine whether the items measure a single, unidimensional construct or a multidimensional one. In addition, it is used to ascertain whether internal relationships between the variables is consistent with the definition of the construct to be measured and the intended structure of the instrument. As factor analysis is one of the most prominent statistical tools used in the development of multiitem (multidomain) health instruments (Chapter 9), this methodology will be discussed in detail later.

FACTOR ANALYSIS

Factor analysis is used for many purposes. This statistical tool originated in the early 1900s from attempts to establish intelligence as a unitary or multidimensional construct (Spearman, 1904). It has since developed into a general-purpose data analytical tool with many applications. Perhaps its most frequent use in the modern social sciences is to explore the psychometric properties of an instrument or scale. For instance, it is used to determine whether numerous variables can be explained on the basis of a smaller number of factors. In particular, this technique is used in instrument development to validate the elements (i.e., items, variables) that comprise the construct.

During the century since Charles Spearman's seminal work in this area, few statistical techniques have been so widely used, nor so prone to misperception. Although Spearman is credited with inventing factor analysis, Thurstone is the one who first coined the term. In addition, Thurstone is recognized as the inventor of exploratory factor analysis, a version that is more practical than Spearman's original factor analysis. Exploratory factor analysis determines the number and the nature of latent constructs within a set of observed variables. The goal of Thurstone's model is to determine the number of meaningful common factors in a correlation matrix. Factor analysis differs from most other statistical techniques in one important way. That is, factor analysis makes no distinction between independent and dependent variables; all are treated equally, and the data come from a single group of subjects. In other words, the technique is used to examine the structure of the relationship (correlations) among a set of variables, not to see how they relate to other variables (prediction).

The method has two main types. One, known simply as factor analysis, extracts factors on the basis of their shared variance. The other, known as principal component analysis, extracts factors on the basis of the total variance of the variables. Common factor analysis is used to look for the latent (underlying) factors, whereas principal component analysis is used to find the fewest variables that explain the most variance. That is why factor analysis is more correctly referred to as common factor analysis; it uses only the variance that all of the variables have in common. This has a number of implications for performing the analysis, interpreting the results, and choosing a particular method. Nevertheless, like many other scientists, I will also drop the word "common" and refer to this method simply as factor analysis.

There is considerable confusion about the name factor analysis. Some people (erroneously) think of it as a synonym for principal component analysis. In fact, if we want to run a factor analysis, for example in Statistical Package for the Social Sciences (SPSS), and do not change any of the default options in the menu windows, we are actually running a principal component analysis. The two are not identical, however, and researchers have to make sure that they are performing the analysis they really planned to do.

Principal Component Analysis

The goal of principal component analysis is to decompose a set of data with correlated variables into a new set of uncorrelated (i.e., orthogonal) factors. We would also use principal component analysis to account for the maximum amount of variance in the data with the smallest number of mutually independent underlying factors. Compared to factor analysis, principal component analysis is much more like a data reduction technique in which a researcher reduces a large number of variables to a smaller, more manageable number of factors. This technique maximizes the variance accounted for; it does not find groups of variables that measure the same thing. There are several differences between the two methods; essentially, in principal component analysis there is no unique covariance matrix (ψ) and the components cannot be correlated (Φ) (Box 7.1).

BOX 7.1

It is extremely useful to think in terms of matrices (arrays). The beauty of matrix notation is that it provides an extremely compact way to describe what a particular analytical model is about, and that it reflects in a uniform notation the difference with other statistical or analytical techniques. For example, it reflects the difference between principal component analysis and factor analysis, or between factor analysis and regression analysis, or factor analysis and correspondence analysis. The matrix notation for factor analysis is as follows:

$$\Sigma = \Lambda\Phi\Lambda' + \Psi$$

Σ is population correlation matrix of observed variables; Λ is the matrix of factor loadings (' is transposed); Φ is the matrix of correlations between the factors; Ψ is the matrix of residual variances/covariances.

In Fig. 7.3, the v by v correlation matrix (here 10 variables) of the data is denoted as Σ; the f by f correlation matrix (here three factors) of the underlying factors is denoted as Φ; the v by v correlations of the measurement errors is denoted as Ψ. That is, the correlations among the observed variables can be decomposed into a component attributable to the underlying factors (and the relationships among those factors), and the measurement error variances.

Continued

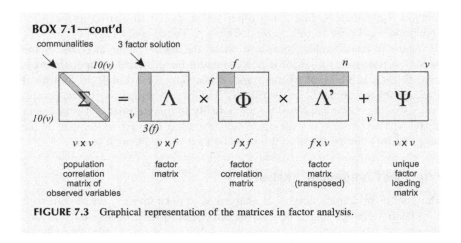

BOX 7.1—cont'd

$v \times v$ $v \times f$ $f \times f$ $f \times v$ $v \times v$

population correlation matrix of observed variables   factor matrix   factor correlation matrix   factor matrix (transposed)   unique factor loading matrix

FIGURE 7.3   Graphical representation of the matrices in factor analysis.

Factor Analysis

Factor analysis recognizes that measurement variance contains both shared and unique variance across variables. At this point, we are already touching on material from Chapter 9, where we will discuss the close connection between factor analysis and the measurement theory under classical test theory (which states that $X = T + E$; where T is the true score). In factor analysis, we are concerned only with the variance that each variable has in common with the other variables, not with the unique variance. This technique uncovers patterns among variables and then clusters highly interrelated variables into factors. Factor analysis is appropriate to create a new scale by eliminating variables (items) that are not associated with other ones or do not load on any factor. Factor analysis has many applications. It is often applied in survey research, where researchers use it to see if a lengthy series of questions can be grouped into shorter sets (Tabachnick and Fidell, 1989).

Multivariate

Multivariate statistical analysis comprises a set of advanced techniques for examining relationships among multiple variables at the same time. Researchers use multivariate procedures in studies that involve more than one dependent variable, more than one independent variable, or both. There are many statistical techniques for conducting multivariate analysis. Which one is the most appropriate will depend on the type of study and the key research questions. Some common multivariate techniques are multiple regression analysis, factor analysis, and multivariate analysis of variance.

Most of the multivariate methods used for exploratory factor analysis are fairly robust against deviations from normality and do not require multivariate normality. This means that five- or seven-point Likert items, often used in health-status measurement, provide data sufficiently robust for factor analysis. Applying factor analysis to dichotomous data (yes/no, true/false, present/absent, and other variables like these) is definitely inappropriate. Nevertheless, examples of performing factor analysis on this type of variable can be found in many journals (Norman and Steiner, 2008).

Correlations

It is important to understand that factor analysis is a statistical technique to analyze the correlations (or covariances) between variables. Here, we choose to start with a correlation matrix mainly because the variables are often measured in different units, so we have to convert all of them to standard scores. If a similar metric is used for all of the variables (such as when we perform factor analysis on Likert items, each using 0 to 6 response categories), under certain circumstances it might be better to begin with a variance—covariance matrix.

What is happening in the background during a factor analysis is the computation of a correlation matrix, followed by estimation procedures involving several mathematical routines. Factor analysis is used in an attempt to identify groups of variables such that there are strong correlations among all the variables within a group of variables, but weak correlations between variables within the group and those outside the group. If factor analysis is expressed in matrix notation, it becomes quite clear what it is all about (Box 7.1).

Communalities

Factor analysis and principal component analysis differ in the communality estimates that are used. Both types of analysis begin with a correlation matrix that has 1.0s along the main diagonal. Those 1.0s mean that we are concerned about all of the variance, from whatever source. In principal component analysis, these 1.0s remain in the analysis until the end. In factor analysis, however, these 1.0s are only used as a starting point to estimate communalities. By using iterative estimation techniques (which require computers), the optimal solution for the communalities is determined, which will always be less than 1.0. As a first step, the communality (common variance) of a variable (how much it has in common with the extracted factors) can be approximated by its multiple correlation (R^2) with all of the other variables.

So, principal component analysis uses all 1.0s for the communalities, whereas in factor analysis the communalities are estimated and lower than 1.0. But the differences between solutions found with principal component analysis and factor analysis tend to be minor. Actually, there are some circumstances for which this statement is untrue. Field (2013) concluded that with 30 or more variables and communalities greater than 0.7 for all variables, different solutions are unlikely. However, with fewer than 20 variables and any low communalities (<0.4), differences can occur. This has mainly to do with the fact that if the number of variables increases, the proportion of communalities (main diagonal) compared to the proportion of other data cells in the data matrix will decrease rapidly.

Estimation

A variety of methods can be used to estimate the factors, and all of these methods would lead to different solutions. Most statistical packages offer at least five or six methods for factor extraction. The only thing these have in common is that they define some measure of fit, which is then maximized. One of the methods ordinarily used is maximum likelihood, which generates estimates that are most likely to produce the observed correlation matrix under assumptions of normal distributions. Maximum-likelihood estimation was recommended, analyzed, and popularized by Sir Ronald Fisher around 1920. Another method, called unweighted least squares, minimizes the sum of the squared differences between the observed and model-predicted correlation matrices. Yet another method, alpha factoring, maximizes the Cronbach's α reliability of the factors so that (for example) the first factor has maximum reliability or internal consistency. Many other methods exist, and each has its proponents.

Unfortunately, different procedures can result in appreciably different solutions unless the underlying structure of the data happens to be particularly clear and simple. Theoretical and empirical studies suggest that when the factor structure of the data is strongly defined, most methods of extraction will yield similar results. However, considerable divergence may arise between the factor solutions, especially when there are small sample sizes, few explanatory variables, and a weak factor structure (Gorsuch, 1983).

Statisticians generally prefer maximum-likelihood to other methods. The main reason is that it is based upon sound mathematical theory that is widely applicable to many situations. Maximum-likelihood estimation also provides foundations for hypothesis testing, including tests for the number of factors. Furthermore, unlike other methods, maximum-likelihood yields the same results whether it is a correlation matrix or a covariance matrix that is factored. It is usually thought to be a disadvantage that maximum-likelihood estimation explicitly assumes that the sample is from a multivariate normal distribution. Yet maximum-likelihood estimation of factor structure is fairly robust against departures from normality (Fayers and Machin, 2000).

Factor Loadings

The relationship of each variable to the underlying factor is expressed by factor loadings. The factor loadings are the correlations (standardized regression coefficients) between the factor and the variables. Typically, a factor loading of 0.4 or higher is required to attribute a specific variable to a factor. The uniqueness for a variable is then simply that portion of a variable that cannot be predicted by (i.e., is unrelated to) the factor(s). The number of factors to retain is one of the most important decisions a factor analyst must make. If too many or too few factors are kept, the results from later steps may be distorted.

Eigenvalues

Not all factors are retained in an analysis, and there is debate over which criterion should be used to decide whether a factor is (statistically) important. The criterion that is still the most frequently used is called the eigenvalue one test, or the Kaiser criterion (1960). It is the default option in most computer packages. In factor analysis, each factor yields an eigenvalue, which is the amount of the total variance explained by that factor. Estimation in factor analysis is done in such a way that the first factor has the largest eigenvalue, the second factor has the second largest eigenvalue, and so on. Only factors with an eigenvalue greater than 1.0 are considered relevant or meaningful, because such a factor explains more variance than one variable in total.

Another test to determine the number of relevant factors is Cattell's scree plot (1966). This is not a real test, as it relies fully on visual inspection. Cattell himself considered it an empirical procedure for reaching a decision. All eigenvalues are plotted for each of the estimated factors. Typically there will be a few factors with quite high eigenvalues and many factors with relatively low eigenvalues, so this graph has a characteristic shape. In that case, there is a sharp descent in the curve followed by a tailing off (Fig. 7.4). Cattell argued that the cut-off point for selecting factors should be at the point of inflexion of

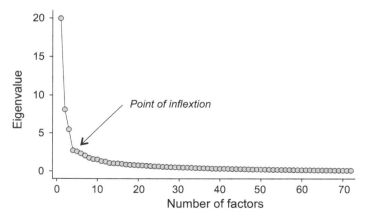

FIGURE 7.4   Scree plot (in this example three factors should be the right choice).

this curve. That is where the slope of the curve changes from negative to nearly zero. The last "real" number of "meaningful" factors is the number found before the scree (the relatively flat portion of the curve) begins. Cattell (1966, p. 250) formulated the issue about the interpretation of the scree plot as "whether the last nontrivial factor is that immediately beyond or at the end of the straight line." And on page 252 he clarifies this: "Moreover, as Thurstone argued, it is always safer to take out one too many rather than the converse, since rotation will reduce it to triviality if it is in excess." In our example, shown in Fig. 7.4, the point of inflexion occurs at the fourth data point (factor). Therefore, we would extract three factors.

No matter which criterion is chosen for determining the number of factors, the results should be interpreted as a suggestion, not as the truth. Although scree plots are very useful, factor selection should not be based on this criterion alone. Typically, factors are extracted as long as the eigenvalues are greater than 1.0. Otherwise, the scree plot visually indicates how many factors to extract.

Rotation

The role of factor estimation is to find an initial solution. Once found, the next step would be to "rotate" the factors. The aim of rotation is to simplify the initial factorization, thereby obtaining a solution that keeps as many variables and factors distinct from one another as possible. Thus, rotation is essential to factor analysis, as the initial factor solution is frequently difficult to interpret. In fact, it will rarely show any interpretable patterns. However, it is easier to interpret the results of a factor analysis if we aim for structural simplicity. This is what rotating the factors is good for; the factors are usually rotated until a solution with a simpler structure is found. Note that if the initial factor analysis solution showed that six factors were detected, the same six would be used for the subsequent rotation.

There are two main types of rotation. Orthogonal rotation is used when the new axes are also mutually orthogonal (the factors are independent from each other). And oblique rotation is used when the new axes are *not* required to be mutually orthogonal (some relationship is believed to exist between the factors). The simplest rotations are orthogonal, which assumes that the underlying factors are not correlated with each other (Fig. 7.5).

Varimax

One of the most commonly used orthogonal rotation methods is "varimax," although many alternatives have been proposed. Compared to the others, varimax (Kaiser, 1958) rotation usually yields sensible solutions, which means that each factor has a small number of large loadings and a large number of zero (or small) loadings. This simplifies the interpretation because, after a varimax rotation, each original variable tends to be associated with one factor,

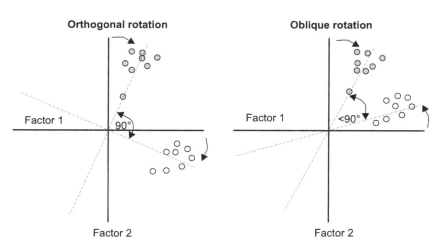

FIGURE 7.5   Schematic representation of two types of factor rotation: orthogonal and oblique.

and each factor represents only a small number of variables. In addition, the factors can often be interpreted from the opposition of few variables with positive loadings to few variables with negative loadings. Formally, varimax searches for a rotation (i.e., a linear combination) of the original factors such that the variance of the loadings is maximized. This implies that after rotation the percentages of explained variance (eigenvalue) for the factors will alter. In particular, the first factor will have a lower eigenvalue after rotation, simply because the estimation criterion is different (less focused on maximization of the largest eigenvalue for the first factor). In practical terms, varimax results in a "simple" factor decomposition because each factor will include the smallest possible number of explanatory variables. For exploratory analysis, the apparent simplicity and the sensible results have led to its near universal implementation in all computer packages.

There are several other methods for orthogonal rotation. These rotate the factors differently. Therefore, the resulting output depends on which method is selected. For example, the quartimax rotation minimizes the number of factors needed to explain each variable. Equamax rotation is a compromise between varimax and quartimax. While other methods exist, none approaches varimax in popularity. For a first analysis, varimax is probably the best choice because it is a good general approach that simplifies the interpretation of factors (Field, 2013).

The rationale for rotating factors comes largely from Thurstone (1947) and Cattell (1978). They advocated this procedure because it simplifies the factor structure, making interpretation easier and more reliable (i.e., easier to replicate with different data samples). Thurstone suggested five criteria to identify a simple structure. According to these criteria, which are still often reported in the literature, a matrix of factor loadings (where the rows correspond to the original variables and the columns to the factors) is simple if: each row contains at least one zero; for each column, there are at least as many zeros as there are columns (i.e., number of factors kept); for any pair of factors, there are some variables with zero loadings on one factor and a large number of loadings on the other factor; for any pair of factors, there is a sizable proportion of zero loadings; for any pair of factors, there is only a small number of large loadings.

Oblique Rotation

The assumption underlying varimax rotation is that the factors are orthogonal and uncorrelated with each other. In many cases, however, that assumption is unrealistic. For example, there is a tendency for seriously ill patients to suffer from both anxiety and depression, and these two factors will be correlated. In statistical terms, we should allow "oblique axes" instead of insisting upon orthogonality. In oblique rotations, the new axes are free to take any position in the factor space, but the degree of correlation allowed among factors is, in general, small because two highly correlated factors are better interpreted as only one factor. Oblique rotations, therefore, relax the orthogonality constraint to gain simplicity in the interpretation. These types of oblique rotations were strongly recommended by Thurstone but are used less than their orthogonal counterparts. Promax, a derivative of varimax, is the most frequently recommended oblique rotation method. Starting from the varimax solution, promax attempts to make the low variable loadings even lower by relaxing the assumption that factors should be uncorrelated with each other. Therefore, it results in an even simpler structure in terms of variable loadings on the factors.

The choice between orthogonal and oblique rotation depends on whether there is a good theoretical reason to suppose that the factors should be related or independent. A downside of applying oblique rotations is that it may become difficult to interpret the factors. One approach is to run the analysis using both types of rotation. If the oblique rotation demonstrates a negligible correlation between the extracted factors, then it is reasonable to use the orthogonally rotated solution. If the oblique rotation reveals a correlated factor structure, then the orthogonally rotated solution might be discarded. In any case, an oblique rotation should be used only if there are good reasons to suppose that the underlying factors could be related.

INTERPRETATION

The first step in factor analysis is to determine the number of factors that are to be extracted. This is one of the most important decisions to be made, since a totally different and erroneous factor structure may be estimated if an incorrect number of factors is used. If too many, or too few, factors are mistakenly entered into the model, the analyses can yield solutions that are extremely difficult to interpret. Often, it is hard to say whether factors are meaningful and which models are likely to be correct. On the other hand, it is often possible to ascribe plausible meanings, and each factor is then interpreted and "named" according to the subset of items having high loadings on the factor.

Some of the most useful "tests" do not involve any statistics at all, other than counting. Tabachnick and Fidell (1989) recommend nothing more sophisticated than an eyeball check of the correlation matrix; if we have only a few correlations higher than 0.30, we may conclude that there is no underlying structure. A slightly more stringent test is to look at a matrix of the partial correlations. If the variables in that matrix correlate with each other because of an underlying factor structure, then the correlation between any two variables should be small after partialing out the effects of the other variables. A related diagnostic test involves looking at communalities. Because they are squared multiple correlations, as opposed to partial correlations, the communalities should be above 0.60 or so, reflecting the fact that the variables are related to each other to some degree.

Among the formal statistical tests, one of the oldest is the Bartlett test of sphericity. This test yields a chi-square statistic. If its value is small, and the associated p-level is over 0.05, then the correlation matrix does not differ significantly from an identity matrix. This means that there is no information. However, Tabachnick and Fidell (1989) state that the Bartlett test is notoriously sensitive, especially with large sample sizes. Another test is the Kaiser-Meyer-Olkin (KMO) measure of sampling adequacy, which is based on squared partial correlations. In the SPSS computer package, the KMO value for each variable is printed along the main diagonal of the antiimage correlation matrix, and a summary value is also given. This allows the user to check the overall adequacy of the matrix.

Although many studies report results that show a factor with only two items loading on it, a factor should consist of at least three variables. A single factor model with two items has two loadings and two error variances to be estimated (four parameters), but there are only three nontrivial entries in the variance–covariance matrix. Therefore, there is not enough information to estimate the four parameters. A single factor model with three items has three loadings and three error variances (six parameters). The variance–covariance matrix has six entries. This means that the model is exactly identified. With more items per single factor, we have an overidentified model (more degrees of freedom than parameters). With more than one factor, a factor model is always identified with three or more items for each factor.

The standard factor analysis model makes assumptions about data. Since the commonly used estimation procedures are based upon either maximum-likelihood or least squares, continuous data from a normal distribution is assumed for a factor analysis. Furthermore, most methods to estimate factors are based on the Pearson product-moment correlation matrix (or, equivalently, the covariance matrix) with a normally distributed error structure. If these distributional assumptions are violated, any test for goodness-of-fit may be compromised.

Many authors have come up with conflicting recommendations and rules of thumb on the basis of simulations and experience. Various recommendations have been made for the threshold number of respondents. It may be as high as 5 or 10 times the number of observed variables. Or it may be set by various functions of the number of factors and observed variables. However, there is no theoretical basis for any of these rules. Moreover, as Gorsuch (1983) has pointed out, if the variables have low reliabilities or the phenomena are weak, then many more individuals will be needed. The best advice is to be conservative and aim for large studies. Health scales are often expected to have five or more factors and perhaps to contain 30 or more items with few items per factor. The items may form highly skewed scales with floor or ceiling effects. Then it would seem likely that a minimum of a few hundred patients would be required, though ideally there should be many hundreds (Fayers and Machin, 2000).

STRUCTURAL EQUATION MODELING

Factor analysis is an exploratory technique. As such, it is suitable for generating hypotheses about the structure of the data. In recognition of this fact, it is often called "exploratory factor analysis." Other, newer, techniques are available for testing whether a postulated model fits the data; among these, one of particular interest is confirmatory factor analysis.

If prior information is available on substantive theory, previous results, or employed research design, it is possible to use confirmatory factor analysis where particular parameters are set to prescribe values, typically zero. Confirmatory factor analysis can be performed where restrictions are placed on specific parameters, error terms, or correlations. This type of factor analysis goes farther than exploratory factor analysis by permitting one to test the viability of different models, not merely the number of factors. For instance, it is often specified that each item loads on one and only one common factor. Researchers can test for the significance of correlated factors, particular factor loadings, or specific correlated disturbances, as well as for the whole factor model. With structural equation modeling (SEM), which is a very general statistical modeling technique, it is possible to "test" whether a specific set of variables belongs to the same factor and to perform confirmatory factor analysis. Factors are sometimes also specified as uncorrelated by setting pertinent off-diagonal elements of Φ to zero. Confirmatory factor analysis is thus a procedure designed to test hypotheses about the relationship between items and factors, whose number and interpretation are determined in advance.

SEM can be seen as a combination of factor analysis (Kerlinger, 1986a; Long, 1983) and multiple regression analysis (Kleinbaum et al., 1988). It is largely confirmatory rather than exploratory. That is, researchers are more likely to use SEM to determine whether a certain model is valid, rather than using SEM to "find" a suitable model, although SEM analyses often involve a certain exploratory element (Skrondal and Rabe-Hesketh, 2004).

SEM incorporates several different approaches or frameworks to represent these models. In SEM, various building blocks (reflective and formative factor structures, regression structures) can be combined and the relations among them can be modeled. The distinction between formative and reflective measures is important here. Proper specification of a measurement model is necessary to assign meaningful relationships in the structural model. This is definitely relevant in SEM, where causality and explanation of connections and processes are more central. The key variables of interest in SEM are usually "latent constructs": abstract concepts such as "intelligence" or "quality of health" (Fayers and Hand, 1997). In one well known framework, popularized by Karl Jöreskog and Dag Sörbom (1979), the general SEMs can be represented by three matrix equations. In applied work, however, structural equation models are usually represented graphically (Fig. 7.6).

A structural equation model implies a structure of the covariance matrix of the measures (hence the alternative name for this field, "analysis of covariance structures") (Kerlinger, 1986b). Once the model's parameters have been estimated, the resulting model implied by the covariance matrix can then be compared to an empirical or data-based covariance matrix. If the two matrices are mutually consistent, then the structural equation model can be considered a plausible explanation for relations between the measures.

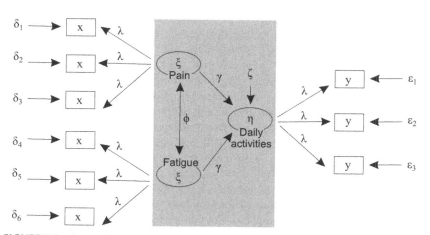

FIGURE 7.6 Structural equation model with factor analysis components and the measurement component (*shaded area*).

DISCUSSION

Unfortunately, the concept of validity has been interpreted and explained quite differently by many users of instruments in health research. Some take it to mean all things that are good about an instrument rather than to specify how well the instrument has met the standard by which it is judged (Nunnally and Bernstein, 1994). Many others think that an instrument is validated if it has been applied in two or three studies that were published in the public domain. Some researchers go so far as to explicitly assert in the title of their publication that part of the research was about validation. But upon close reading, it turns out that only consensus was reached on the final version of the instrument.

Where should we go from here? The awareness of the inadequacy of our measurement technologies should spur us on to improve them. It is inappropriate to speak of a measurement instrument as inherently valid or invalid. It is only meaningful to consider the validity of a specified purpose or interpretation of the resulting measures. Since multiple types of inferences may be entertained for scores from a given instrument, depending upon the situation in which it is to be used, the validity of each inference must be established. Strictly speaking, one validates not a measurement instrument but rather some use to which the instrument is put (Nunnally and Bernstein, 1994, p. 133). An instrument is neither valid nor invalid in and of itself, but only in regard to how it is used and what interpretations are given to the measures for particular groups of people. It is the entire web of types of evidence discussed in this chapter that sheds light on such issues.

Researchers often use principal component analysis when exploratory or even confirmatory factor analysis would be more appropriate (Gorsuch, 1983). Principal component analysis was developed decades ago when most analyses were done by hand. Thus, shortcuts that did not substantially diminish the outcome were valuable. Now, with popular statistical software packages and readily available processing power in even the cheapest laptop computers, principal component analysis is probably not necessary. It is not even considered a true method of factor analysis. Actually, statisticians disagree on when it should be used, if at all. There are some situations where principal component analysis might be an appropriate option, but more often it is not. As mentioned earlier, principal component analysis is still the default procedure in many statistical analysis software packages, although it usually is not the preferred option (Tucker et al., 2013; Taft et al., 2001).

Having arrived at the end of this chapter, I have to offer a piece of advice. Never "try" to perform a complex analysis such as factor analysis by yourself. Even a "simple" exploratory factor analysis can be difficult, especially when it comes to interpreting the results. From my own experience, I know that only after reading at least three handbooks on factor analysis, on top of analyzing a dozen empirical data sets, can a researcher handle this type of analysis. Cattell, who invented the scree plot, performed a hundred or more factor analyses in the course of over 30 years (1966, p. 249). Indeed, as with so many—if not all—things in life and in science, it is all about understanding (and experience). If you need a factor analysis as part of your study or developmental work, seek out collaboration with someone who is dedicated to this analytical type of work. After all, performing a factor analysis has probably more to do with art than with science. Mark my words, checklists and guidelines will definitely not work here.

REFERENCES

Beck, A.T., Ward, C.H., Mendelson, M., Mock, J., Erbaugh, J., 1961. An inventory for measuring depression. Archives of General Psychiatry 4, 53–63.

Beck, A.T., Steer, R.A., Ball, R., Ranieri, W.F., 1996. Comparison of Beck depression inventories -IA and -II in psychiatric outpatients. Journal of Personality Assessment 67 (3), 588–597.

Borsboom, D., Mellenbergh, G.J., van Heerden, J., 2004. The concept of validity. Psychological Review 111 (4), 1061–1071.

Brooks, R.G., 1995. Health Status Measurement: A Perspective on Change. Macmillian Press, London.

Campbell, D.T., Fiske, D.W., 1959. Convergent and discriminant validation by the multitrait-multimethod matrix. Psychological Bulletin 56 (2), 81–105.

Cattell, R.B., 1966. The scree test for the number of factors. Multivariate Behavioral Research 1 (2), 245–276.

Cattell, R.B., 1978. The Scientific Use of Factor Analysis in Behavioral and Life Sciences. Plenum, New York.

Cook, T.D., Campbell, D.T., 1979. Quasi-experimentation: Design & Analysis Issues for Field Settings. Houghton Mifflin, Boston.

Cronbach, L.J., Meehl, P.E., 1955. Construct validity in psychological tests. Psychological Bulletin 52 (4), 281–302.

de Vet, H.C.W., Terwee, C.B., Mokkink, L.B., Knol, D.L., 2011. Measurement in Medicine: A Practical Guide. Cambridge University Press, Cambridge.

Fayers, P.M., Hand, D.J., 1997. Factor analysis, causal indicators and quality of life. Quality of Life Research 6 (2), 139–150.

Fayers, P.M., Machin, D., 2000. Quality of Life: Assessment, Analysis and Interpretation. Wiley, Chichester.

Field, A., 2013. Discovering Statistics Using IBM SPSS Statistics, fourth ed. Sage, London.

Froberg, D.G., Kane, R.L., 1989. Methodology for measuring health-state preferences — II: scaling methods. Journal of Clinical Epidemiology 42 (5), 459–471.

Gorsuch, R.L., 1983. Factor Analysis, second ed. Lawrence Erlbaum Associates, Hillsdale.

Hamilton, M., 1960. A rating scale for depression. Journal of Neurology, Neurosurgery and Psychiatry 23 (56), 56–62.

Hamilton, M., 1980. Rating depressive patients. Journal of Clinical Psychiatry 41 (12), 21–24.

Hattie, J., Cooksey, R.W., 1984. Procedures for assessing the validities of tests using the "known-groups" method. Applied Psychological Measurement 8 (3), 295–305.

Jöreskog, K.G., Sörbom, D., 1979. Advances in Factor Analysis and Structural Equation Models. Abt Books, Cambridge M.A.

Jöreskog, K.G., Sörbom, D., 1989. LISREL 7. A Guide to the Program and Applications, second ed. SPSS, Chicago.

Kaiser, H.F., 1958. The varimax criterion for analytic rotation in factor analysis. Psychometrika 23 (3), 187−200.

Kaiser, H.F., 1960. The Application of Electronic Computers to Factor Analysis. Educational and Psychological Measurement, 20 (1), pp. 141−151.

Kerlinger, F.N., 1986a. Factor analysis. In: Kerlinger, F.N., Lee, H.B. (Eds.), Foundations of Behavioral Research, third ed. CBS Publishing Japan Ltd., New York.

Kerlinger, F.N., 1986b. Analysis of covariance structures. In: Kerlinger, F.N., Lee, H.B. (Eds.), Foundations of Behavioral Research, fourth ed. CBS Publishing Japan Ltd., New York.

Kerlinger, F.N., 1986c. Foundations of Behavioral Research, third ed. CBS Publishing Japan Ltd, New York.

Kleinbaum, D.G., Kupper, L.L., Muller, K.E., 1988. Applied Regression Analysis and Other Multivariable Methods. PWS-KENT Publishing Company, Boston M.A.

Landy, F.J., 1986. Stamp collecting versus science: validation as hypothesis testing. American Psychology 41 (11), 1183−1192.

Long, S.J., 1983. Confirmatory Factor Analysis. Series: Quantitative Applications in the Social Sciences (07-001). Sage Publications, Beverly Hills.

Messick, S., 1995. Validity of psychological assessment: validation of inferences from persons' responses and performances as scientific inquiry into score meaning. American Psychologist 50 (9), 741−749.

Nord, E., 1992. An alternative to QALYs: the saved young life equivalent (SAVE). BMJ 305 (6858), 875−877.

Norman, G.R., Streiner, D.L., 2008. Biostatistics: The Bare Essentials, third ed. BC Decker Inc., Hamilton.

Nunnally, J.C., Bernstein, I.H., 1994. Psychometric Theory, third ed. McGraw-Hill, New York.

Reneman, M.F., Brandsema, K.P.D., Schrier, E., Dijkstra, P.U., Krabbe, P.F.M., 2016. Development of a next-generation, patient-centered, preference-based instrument to measure the impact of chronic pain. (Manuscript submitted for publication).

Schmitt, N., Stults, D.M., 1986. Methodology review: analysis of multitrait-multimethod matrices. Applied Psychological Measurement 10 (1), 1−22.

Skrondal, A., Rabe-Hesketh, S., 2004. Generalized Latent Variable Modeling: Multilevel, Longitudinal, and Structural Equation Models. Chapman and Hall/CRC, Boca Raton.

Spearman, C., 1904. General intelligence, objectively determined and measured. The American Journal of Psychology 15 (2), 201−293.

Strauss, M.E., Smith, G.T., 2009. Construct validity: advances in theory and methodology. Annual Review of Clinical Psychology 5, 1−25.

Streiner, D.L., Norman, G.R., Cairney, J., 2015. Health Measurement Scales: A Practical Guide to Their Development and Use, fifth ed. Oxford University Press, Oxford.

Tabachnick, B.G., Fidell, L.S., 1989. Using Multivariate Statistics, second ed. Harper Collins Publishers, New York.

Taft, C., Karlsson, J., Sullivan, M., 2001. Do SF-36 summary component scores accurately summarize subscale scores? Quality of Life Research 10 (5), 395−404.

Thurstone, L.L., 1947. Multiple-Factor Analysis. University of Chicago Press, Chicago.

Tucker, G., Adams, R., Wilson, D., 2013. Observed agreement problems between sub-scales and summary components of the SF-36 version 2-An alternative scoring method can correct the problem. PLoS One 8 (4), e61191.

Chapter 8

Reliability

Chapter Outline

INTRODUCTION

Reliability refers to the consistency between an instrument and the measures it produces. For a seemingly simple concept, the estimation of the level of reliability and interpretation of the resulting numbers are surprisingly confusing in the literature. In part, this lack of clarity stems from the fact that reliability can be assessed in a variety of contexts. The psychometric literature has provided considerable guidance on the evaluation of the reliability of measurement instruments that are designed for subjective phenomena (Guilford, 1954; Crocker and Algina, 1986; Streiner et al., 2015).

The reliability of a measurement instrument is the extent to which it yields consistent, reproducible estimates of what is assumed to be an underlying true score (see Chapter 9). Reliability concerns the stability of the measurements and the congruence between raters. In particular, it concerns the assessment of one person by another person or the assessment of internal body structures by diagnostic images (e.g., medical imaging modalities such as ultrasonography, CT/PET/MRI scans). Various sources of measurement error are traditionally evaluated in terms of reliability coefficients. There are three basic categories of reliability estimation, reflecting the ways by which random error of measurement is estimated: test—retest, internal consistency, and interrater reliability. Test—retest of a measure is carried out by repeated examinations of an unchanged characteristic. Internal consistency refers to the degree of association among items addressing equivalent concepts. Interrater reliability is about the congruency in assessment between two or more raters or observers.

Some of the confusion stems from the jargon, e.g., consistency, precision, repeatability, and agreement. Intuitively, these terms seem to describe the same concept, but in practice some are operationalized differently. Precision and accuracy have definite meanings of their own. They come from work in the clinical laboratory, where measurements are usually made by machines, not people, and there is often a gold standard. The problem arises when these terms are applied to instruments that are administered by humans and in situations where no gold standard exists (Streiner and Norman, 2006).

Reliability Versus Validity

Validity and reliability are usually treated as two distinct concepts, although the boundary is blurred, both mathematically and conceptually. Achieving reliability is a technical matter (e.g., using larger sample sizes, conducting repeated measurements, and increasing the number of items to be included). Much more insight is needed to achieve validity, however, as it resembles the scientific method itself. A question about validity is largely an inquiry into the nature of reality and thus concerns the nature of the properties being measured (Kerlinger, 1986). The relationships between reliability and validity are shown in Fig. 8.1. A large set of measurements could be valid in the aggregate but not

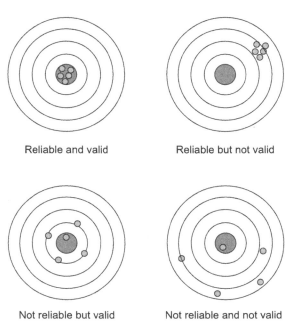

FIGURE 8.1 Graphic illustration of the four possible combinations based on the fact that reliability and validity can vary independently.

reliable if the measures obtained are scattered widely around the true value. On the other hand, an instrument can be very reliable but systematically fall wide of the mark. A single measurement with poor reliability has a low validity because it is likely to be off target by mere chance. The magnitude of the validity coefficient (typically, obtained by correlating the measure with a credible indicator of the variable of interest) cannot exceed the square root of the reliability coefficient.

Factors Affecting Reliability

Intuitively, it seems obvious that taking the average of multiple items (that measure the same concept) should yield a higher reliability than the value for any single item, since the errors are random and those errors associated with each observation are averaged out (Streiner et al., 2015). Another strategy to increase reliability is to perform repeated measures. But the latter course is only possible if we assume that the object of measurement is stable (no changes). At the group level, we can increase the reliability of a mean score by increasing the sample size.

Several factors may influence the reliability of a measure derived from a health instrument. For instance, it may be affected by the conditions under which the instrument was administered; by learning effects; by specific factors affecting the participants in their daily lives; or by the length of time between administrations.

THEORY

Much, if not all, of the theory about the reliability of health measurement instruments has its roots in classical test theory (Chapter 9). That theory deals with measurement error, which is why reliability lies at its core. (There are comprehensive and complicated ways to estimate the level of measurement error for various sources, but this topic lies outside the scope of this book.) Therefore, some overlap may arise between the present chapter and Chapter 9, which is about classical test theory. In fact, much what is explained here about reliability could also be explained in Chapter 9. Classical test theory has been described variously as a theory and a model. Conceived as a model of how the world of (subjective) measurement operates, it postulates that each person has a true score (T) that would be obtained if there were no errors in measurement. It could be envisaged (Eq. (8.1)) as the composite of two hypothetical components: a true score (T) and an undifferentiated random-error component (E).

$$X = T + E \tag{8.1}$$

The correlation coefficient that expresses the degree of relationship between true and observed scores is known as the reliability index: T/X. These relations can say something about the reliability of measures. The reliability of the observed measure X, which is denoted as ρ_{XT}^2, is defined as the ratio of true score variance σ_T^2 to the observed measure variance σ_X^2:

$$\text{Reliability} = \rho_{XT}^2 = \frac{\sigma_T^2}{\sigma_X^2} = \frac{\sigma_T^2}{\sigma_T^2 + \sigma_E^2} \tag{8.2}$$

Expressed in words, reliability is the ratio of the variance of the true scores to the observed variance (Eq. (8.2)). Specifically, we may define an index of reliability in terms of the proportion of true score variability that is captured across subjects (respondents, patients), relative to the total observed variability (Eq. (8.3)). In the form of an equation, it is the ratio between subjects s to total variability e (subject variability + measurement error).

$$\text{Reliability} = \frac{\sigma_s^2}{\sigma_s^2 + \sigma_e^2} \tag{8.3}$$

Variance in the observed score includes both systematic variance and error variance. Accordingly, the variance of the observed scores will always be greater than the systematic variance in the scores themselves. Therefore, reliability can be thought of as systematic variance divided by total variance.

Most existing health instruments were developed within the measurement framework of classical test theory. Eq. (8.2), which formulates a signal-to-noise ratio, has intuitive appeal. The reliability of scores becomes higher as the proportion of error variance in the test scores becomes lower and vice versa. More details about the theoretical underpinnings of reliability as worked out under classical test theory can be found in (classical) textbooks such as Crocker and Algina (1986), Nunnally and Bernstein (1994), and Streiner et al. (2015).

TEST–RETEST

This type of reliability assessment captures the extent to which an instrument consistently yields the same scores on two successive occasions if the health of the individual to whom it is applied has not changed in the intervening period. Thus, reliability concerns the reproducibility of an instrument. Failure to give the same results on different occasions, when identical outcomes would be expected, suggests that some source of error variance is affecting the scores.

The test–retest design yields relevant information on instrument reliability when there is reasonable assurance that the person being assessed has not changed. Test–retest is not always appropriate for evaluating instrument reliability in certain situations. For instance, it is not used when the process of making the first measurement might have affected the second measurement significantly (e.g., a recall effect); when the characteristic being measured may have been subjected to some intervening influence or temporal fluctuation; or when there are subjective elements in the scoring process (e.g., observer ratings). The most appropriate way of testing an instrument's reliability is by computing the intraclass correlation (ICC) (discussed at the end of this chapter) between the test and retest.

INTERNAL CONSISTENCY

Internal consistency can only be estimated if an instrument consists of several related items that together measure a hypothetical phenomenon (construct or domain). Increasing the reliability (reducing error) of sum scores is the main reason why there are so many health profile instruments (Chapter 9) with multiple items for each domain.

Interitem Correlation

Interitem correlations indicate whether or not the item is part of the scale. The interitem correlations found for items within one domain (or one dimension, based on factor analysis) should be neither too low nor too high. For example, if the correlation of two items is higher than 0.9, these items are said to measure almost the same thing, so one of them could be deleted. If the correlation between two items is only 0.2, one of the items is probably not part of the same construct.

Item—Total Correlation

This is the correlation of one item with the total set of items on the scale after omitting that particular item. Accordingly, another name for this type of correlation is item-partial total correlation. Many statistical software packages make computations for this purpose very easy. In fact, a simple estimation of the item—total correlation can be quite informative. A small item correlation provides empirical evidence that the item is not measuring the same construct measured by the other items included and may thus be dropped. If an item shows an item—total correlation of less than 0.3 (Nunnally and Bernstein, 1994), it is a candidate for deletion.

Split Halves

In the old days, split-halves reliability was commonly used to evaluate the reliability of a measurement instrument. The items are randomly assigned to one of the "split halves." Then the intercorrelation of scores derived from each half is calculated. A high level of correlation is taken as evidence that the items are consistently measuring the same underlying construct. The split-halves design was Likert's choice for a reliability estimate. He preferred this method because of the minimal amount of computational effort it took to generate it. Scores on both halves are correlated to obtain estimates of the reliability of each half. The total set of scale items, however, is twice as long as the set of items for each half. The reliability of the total scale is computed using the Spearman—Brown prophecy formula:

$$\rho_{XX}^* = \frac{k\rho_{XX}}{1 + (k-1)\rho_{XX}} \qquad (8.4)$$

where ρ_{XX} is the reliability of the current scale and ρ_{XX}^* is the reliability after changing the length (number of items) of the scale. In Eq. (8.4), k is the increase in length of the instrument (e.g., $k = 3$ would mean the instrument is $3\times$ longer). For example, if the reliability of a 20-item instrument is 0.70 and 40 items are added to the instrument (increasing the length of the instrument by $3\times$), the estimated reliability of the new 60-item instrument will be 0.88 (Eq. (8.5)).

$$\rho_{XX} = \frac{(3)(0.7)}{1 + (3-1)(0.7)} = 0.88 \qquad (8.5)$$

While computationally simple, the split-halves method is not without its drawbacks. Chief among these is that different reliability coefficients may be obtained with each of the possible ways of splitting the set of scale items into two equal groups. Faced with this indeterminacy, Lee Cronbach (1951) developed coefficient α (alpha), which provides a unique estimate of reliability based upon the interitem correlation. With its introduction, the split-halves method became outmoded. In the setting of health measurement, nobody uses this procedure anymore. Today, Cronbach's α is preferred as an estimate of scale reliability.

Cronbach's Alpha

The internal consistency of a scale is a function of the number of items on it and their covariation. Random error due to item selection is modeled by using Cronbach's α to estimate the reliability of health instruments based on internal consistency (also called homogeneity or scalability). In the field of psychometrics, this type of analysis has been used to construct hundreds of tests (e.g., intelligence, personality, and psychopathology). In the field of health outcomes, health instruments are constructed in the same manner, though in that context we mostly speak of domains or scales instead of constructs or tests.

Cronbach's α is related conceptually to the Spearman—Brown prophecy formula. Both arise from classical test theory. The insight that Cronbach's α can offer is how well the composite of all the items associated with the underlying domain or scale is measuring the construct. Substantial intercorrelation of these elements is interpreted to mean that the items are measuring the same or closely related constructs in a reliable manner. Low levels of correlation among items would suggest that the construct is not being measured reliably; there are sources of unexplained error in the measurement. Cronbach's α will generally increase as the intercorrelations among test items increase. Thus, it is known as an internal consistency estimate of the reliability of test scales. It can be easily computed as follows:

$$\alpha = \frac{n}{n-1}\left(1 - \frac{\sum_i \sigma_i^2}{\sigma_t^2}\right) \tag{8.6}$$

Here n is equal to the number of items, σ_i^2 is the variance of scores on each individual item, and σ_t^2 equals the variance of the overall total score on the entire set of items. Cronbach α does not reveal the unidimensionality of a scale. Nor does it detect whether the items in the analysis are reflective or formative in their relationship to the underlying construct to be measured (Streiner, 2003).

In the field of psychometrics, the Kuder–Richardson Formula 20 (KR-20), first published in 1937, is a measure of internal consistency reliability for measures with dichotomous choices (Kuder and Richardson, 1937). Ever since Cronbach's α was published in 1951, KR-20 has not been considered as having any advantage over Cronbach's alternative. Rather, KR-20 is seen as a derivative of the Cronbach formula, whereby the advantage of α is that it can handle both dichotomous and continuous variables.

Because coefficient α is determined both by the number of items and the strength of the correlations (or covariances, which are merely unstandardized correlations) among those items, increasing either of those influences will typically increase reliability. The exception is when the items added to a scale are so poorly correlated with the original items that their negative impact on the overall correlations among items outweighs their positive impact on the total number of items. It is often easier to improve a scale by using more items than to improve it through better (i.e., more highly correlated) items. An unfortunate consequence of this fact is that measures developed under classical test theory often contain many items.

BOX 8.1

Lee Cronbach was an American educational psychologist who made contributions to psychological testing and measurement. The work of Thurstone intrigued Cronbach, motivating him to complete and receive his doctorate in educational psychology from the University of Chicago in 1940. His research can be clustered into three main areas: measurement theory, program evaluation, and instruction. In 1948, he moved on to the University of Illinois, Urbana, where he produced many of his key works: the "Alpha" paper (Cronbach, 1951), as well as an essay on "The Two Disciplines of Scientific Psychology" (Cronbach, 1957), in which he discussed the divergence between the fields of experimental psychology and correlational psychology (to which he himself belonged). Cronbach then took a faculty position at Stanford University in 1964 (Fig. 8.2).

Continued

BOX 8.1—cont'd

FIGURE 8.2   Lee Cronbach (1916–2001).

From 0.70 to 0.95

A reliability level of 0.90 is considered the minimum for any measurement that is designed to interpret scores at the individual level (Nunnally and Bernstein, 1994). In practice, the 0.90 threshold may even be too stringent. To achieve this level of reliability requires including several items per scale, which many highly regarded instruments fail to do. For example, none of the eight scales in the SF-36 satisfied the 0.90 reliability criterion (Ware and Sherbourne, 1992; Hays and Morales, 2001).

A reliability coefficient of 0.80 is considered sufficient in the initial stages of developing a measure (Nunnally and Bernstein, 1994). Some researchers have stressed the need to set higher standards for reliability. Others have stated that in the case of Cronbach's α, coefficients between 0.7 and 0.9 are usually acceptable, >0.9 implies some possible redundancy, and values <0.7 indicate that the items do not correlate very well with one another.

The principle of attaining an α coefficient larger than 0.7 can be used to reduce the number of items. With a high Cronbach's α, we can afford to delete items to make the instrument more efficient (Fig. 8.3). Conversely, with a low Cronbach's α, we can construct new items that are manifestations of the same construct and add these to the scale. This principle implies that with a large number of items in a scale, Cronbach's α may show a high coefficient despite rather low interitem correlations.

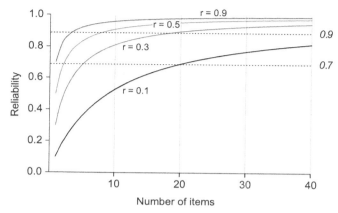

FIGURE 8.3   Relation between number of items and reliability.

Sample Dependent

Like all other reliability coefficients, Cronbach's α depends on the variation in the population. Responses from heterogeneous populations will produce a higher Cronbach α than homogeneous populations. As such, Cronbach's α is sample dependent, just like validity statistics and test–retest reliability. The estimated statistics reflect a characteristic of an instrument used in a specific population, not a characteristic of a measurement instrument in general.

INTERRATER

Often we want to assess how consistently subjects perform their task of rating symptoms, signs, health conditions, health descriptions, or any other observable phenomena that are attributable to other persons other than themselves. Such situations draw upon raters. When the measurement process involves the judgment of one person by another, a third source of measurement error must be considered: interrater (or interobserver) variability. That is because an additional source of subjectivity is introduced.

Test–retest and Cronbach's α deal with two sources of variance (measurement error): differences between patients and differences between the characteristic of interest (items) being scored. Instruments designed for patient-reported measurements do not make use of raters, even if subjective matter is reported. Thus, a concern about interrater error is irrelevant. But it is relevant when a radiologist is rating bone scans, or when therapists are scoring psychiatric patients. Interrater reliability refers to a comparison of scores assigned to the same target (a patient or whatever is being judged) by two or more raters. In the analysis, the target is the attribute that has to be judged rather than the characteristics of the judges. All we need to take into account is the variability of the raters as revealed by their scores. Unreliability of the measurements due to variation between raters is evaluated by considering the average association between scores obtained from different raters rating the same patients. If reliability is determined using a research design in which each patient is assessed by only one rater (hence, measurement error attributable to the rater is ignored), the estimate of measurement reliability will be overly optimistic. In most instances, the degree of overestimation is substantial.

Kappa

The kappa statistic is useful when we want to estimate exact agreement between raters for scores commonly measured on a nominal or ordinal scale. Kappa is known as a reliability index because it compares observed agreement with agreement expected by chance (Cohen, 1968). The general formula for kappa is:

$$K = \frac{\text{observed agreement} - \text{change agreement}}{1 - \text{change agreement}} = \frac{P_o - P_e}{1 - P_e} \qquad (8.7)$$

where P_o denotes the observed percentage of agreement, and P_e denotes the expected percentage of agreement due to chance (Hallgren, 2012). Rules of thumb for interpreting the magnitude of kappa have been provided by Landis and Koch (1977) and Fleiss (1981). For nominal data, kappa is most likely the best option. For ordinal categorical and continuous data, weighted kappa and the ICC (see below) are equivalent under certain conditions (Fleiss and Cohen, 1973).

From Kappa to Alpha

Internal consistency, which refers to the degree of association among items, can be estimated by applying Cronbach's α. Consistency between raters is estimated by computing interrater reliability. Although these two tests are used to address different research questions, they are based on equivalent concepts. Formally, the interrater reliability coefficient is a simple adaptation of the conventional Cronbach's α; but instead of examining multiple items, we are now investigating multiple raters (Appendix: Krabbe et al., 1996; Kottner and Streiner, 2011).

In some situations, however, α is not an appropriate indicator of interrater reliability. For example, it is not used when the goal is to ascertain precise agreement rather than mere consistency. Nor is it used when raters are considered as individuals chosen from a population of potential raters rather than as the only raters of interest (DeVellis, 2006). To deal with situations such as these, adapted statistical routines named

ICC statistics are required. Not surprisingly, Cronbach's α is a special case of the ICC.

Despite their differences, the methods for quantifying reliability share a definition: reliability is the proportion of observed variation that can be ascribed to the characteristic being assessed. Types of reliability may be computed differently. That is because variation representing the true score and variation representing error are dependent on the context of the research. However, the reliability coefficient always compares the proportion of variance in obtained scores that can be attributed to true scores (DeVellis, 2006).

INTRACLASS CORRELATIONS

The main reason to apply the ICC method is that it is far more flexible than other approaches. All sources of variability in a single study can be systematically combined by utilizing all of the data to estimate the variance between subjects and to estimate the components of error (Shavelson et al., 1989).

The ICC is one of the statistics ordinarily used to assess test−retest; internal consistency; and interrater reliability for ordinal, interval, and ratio data. ICCs are suitable for studies with two or more raters. An ICC may be used when all subjects in a study are rated by multiple raters, or when only a subset of the subjects is rated by multiple raters and the rest are rated by one rater. ICCs are suitable for fully crossed designs or when a new set of raters is randomly selected for each participant.

Beginning with Fisher (1925), the ICC has been continuously refined within the statistical framework of analysis of variance (ANOVA). Recently, it has been expanded in the framework of random-effect models. A number of ICCs have been proposed.

Regarding reliability of responses, which are given by respondents, not raters, the ICC coefficient is based on between-subjects variability (see also Eq. (8.3)). This is the variance due to how subjects differ from each other and the variability how subjects differ in their own scores (e.g., repeated measures). In this context, reliability is formally defined as follows:

$$\text{ICC} = \frac{\sigma_b^2}{\sigma_b^2 + \sigma_w^2} \tag{8.8}$$

where σ_b^2 is the variance between subjects, and σ_w^2 is the pooled variance within subjects.

The ICC is a relative measure of reliability. In that sense, it is conceptually akin to R^2 from regression analysis and less like the Pearson correlation coefficient. Theoretically, the ICC can vary between 0.0 and 1.0. An ICC of 0.0 indicates no reliability, whereas an ICC of 1.0 indicates perfect reliability. The relative nature of the ICC is reflected in the fact that the magnitude of an ICC depends on between-subjects variability. That is, if subjects differ little from each other, ICC values are small even if trial-to-trial variability is small. If subjects differ from each other a lot, ICCs can be large even if trial-to-trial variability is large. Thus, just as for all reliability coefficients, the ICC is context specific. Further, one intuitively assumes that small differences between individuals are more difficult to detect than large ones, and the ICC is reflective of this (Weir, 2005). At one level, interpreting the ICC is fairly straightforward; it represents the proportion of variance in a set of scores that is attributable to the object of interest. An ICC of 0.95 means that an estimated 95% of the observed variance in the scores is due to the variability in the reported scores among responders.

Different Versions

Unfortunately, there is considerable confusion about both the estimation and interpretation of the ICC. Indeed, there are several versions of the ICC and it is not intuitively obvious which version one should choose. Shrout and Fleiss have presented six forms of the ICC (Shrout and Fleiss, 1979). In their nomenclature, there are three general models labeled 1, 2, and 3, and each can be calculated in one of two ways. If the scores used in the analysis are single scores assigned by each subject for each study (or rater, when assessing interrater reliability), then the ICC is given a second designation of 1. If the scores entered into the analysis represent the average of the k scores assigned by each subject (i.e., the average across the studies), then the ICC is given a second designation of k. An ICC with a model designation of (2,1) indicates an ICC calculated using model 2 with single scores. The one-way ANOVA models coincide with the model $(1,k)$ for situations where scores are averaged and with model (1,1) for single scores for a given study (or rater).

More recently, McGraw and Wong (1996) expanded the system of Shrout and Fleiss to include two more general forms, again each with a single-score or average-score version, resulting in 10 ICCs. These ICCs have now been incorporated into SPSS statistical software. Given the 6 ICC versions of Shrout and Fleiss and the 10 versions presented by McGraw and Wong, the choice among ICCs has become perplexing. Broadly speaking, there are four issues to be addressed when choosing an ICC:

- one- or two-way model,
- fixed- or random-effect model,
- include or exclude systematic error (absolute or relative consistency), and
- single or mean score.

In case raters are involved in the study and each rater is expected to exactly rate all patients, variability among the raters is generally treated as a second source of systematic variability. Raters then become the second factor in a two-way model.

TABLE 8.1 Intraclass Correlation Coefficient Nomenclature (Shrout & Fleiss, 1979)

Type	Raters	Subjects	Assessment
ICC (1,1)	Raters are randomly chosen from a larger population of raters.	Each subject is assessed by a different set of k raters.	The researcher is interested in assessing the reliability among raters on a single measure.
ICC (1,k)			The researcher assesses the mean of several recordings.
ICC (2,1)	Raters are randomly chosen from a larger population of raters.	Each subject is assessed by the same raters.	The researcher is interested in assessing the reliability among raters on a single measure.
ICC (2,k)			The researcher assesses the mean of several recordings.
ICC (3,1)	Raters only represent the raters of interest for the reliability study.	Each subject is assessed by the same raters.	The researcher is interested in assessing the reliability among raters on a single measure.
ICC (3,k)			The researcher assesses the mean of several recordings.

Regarding fixed versus random effects, a fixed factor is one in which all levels of the factor of interest (e.g., all academic hospitals in the Netherlands, all radiologists on the ward) are included in the analysis and no attempt at generalization of the reliability data is expected beyond the confines of the study. On the other hand, if patients in a study are randomly selected from a larger population and their ratings are meant to generalize to that population, then the researcher may use a random-effect model. These models are called random because patients are considered to be randomly selected.

Absolute consistency concerns the consistency among the scores of individuals. Relative consistency, in contrast, concerns the consistency of the position or rank of individuals in the group with respect to others.

In case of multiple raters or repeated measures, the ICC can estimate reliability for a single score or for the mean or sum of scores. Which measure would be most appropriate in any given study depends on whether the researchers plan to rely in practice on a single score or a combination of scores. Of course, combining multiple scores generally produces more reliable measurements (Table 8.1).

Generalizability Theory

Generalizability theory, introduced by Cronbach et al. (1972), has found only limited acceptance in the medical sciences. But despite its infrequent application, this approach has some potential that is worth noting. Generalizability theory involves the identification and measurement of multiple sources of error variance. Generalizability theory may be regarded as a generalization of the ICC. It also provides a conceptual framework to construct dedicated study designs in combination with fit-for-purpose statistical analyses (ANOVA) to explore components that (may) contribute to the reduction of the reliability of the measured object of interest. Interested readers who wish to learn more about Generalizability theory can consult published sources (Shavelson et al., 1989; Streiner et al., 2015).

DISCUSSION

There are many misconceptions about interpreting scores and using measurement instruments. One of these is the mistaken assumption that the interpretation remains the same when the instrument is used for purposes other than those for which it was developed. Its developer may have given evidence regarding an instrument's score distributions, reliability, and the validity of particular interpretations of its scores. Nonetheless, these properties are not fixed. Nor can they be presumed relevant to all subsequent measurement situations, regardless of differences in the purposes, populations, or settings for which the instrument is adopted.

Cronbach's α is seen as a statistical procedure to show the reliability of a bundle of items that together represent a general domain. If all of the items in a domain are related, at least to some extent, then there is a single latent variable (factor) that is common to all of the items in that domain. Alpha will generally increase as the intercorrelations among scale items increase. Accordingly, α is known as an internal consistency estimate of the reliability of scale scores. Cronbach's α is highly dependent on the number of items in the scale. It is easy to demonstrate, however, that scales with the same length and variance but different underlying factorial structures can result in the same values for Cronbach's α. Indeed, several investigators have shown that α can take on quite high values even when the set of items measures several unrelated latent constructs. Many publications report study results in which a factor analysis clearly identified several factors. In a subsequent analysis the researchers did not perform an internal consistency analysis (Cronbach's α) for the items of each factor separately but for the complete set of items. Cronbach's α is believed to indirectly indicate the degree to which a set of items measures a single unidimensional latent construct. However, Cronbach's α shows whether the items in a set are on average (highly) correlated and therefore are tapping into the same construct. But this result provides no evidence as to whether or not the items measure the construct that they claim to measure.

Another confusing aspect of reliability calculations is that many different procedures have been used to determine reliability. In the past, the Pearson product-moment correlation coefficient (r) was often applied to quantify reliability, but its use is typically discouraged for assessing test–retest reliability. The primary, though not the only, weakness of the Pearson r is that it cannot detect systematic error. More recently, a procedure based on the limits of agreement (LOA), as described by Bland and Altman (1986), has come into vogue in the biomedical literature. The procedure was developed to examine agreement between two different techniques of quantifying some variable (so-called method comparison studies; e.g., one could compare testosterone concentration using two different bioassays), not to determine reliability. The use of LOA as an index of reliability has been criticized, as set forth in detail elsewhere (Streiner et al., 2015). Under most circumstances, all these different techniques can be replaced by the ICC coefficient.

It has been argued that the ability to check the internal structure (validity) and the internal consistency (reliability) of multiitem scales is essential. It may be clear from the presentation in this chapter and the preceding one that such tests cannot be employed on a "scale" that contains only a single item. We will see that the same restriction holds for measurement approaches that are based on "holistic" judgments of a set of distinct health attributes (Chapters 11 and 12). This restriction may partially explain why so many health instruments are developed on the basis of multiple items. Almost all of these multiitem instruments are worked out under the assumptions of classical test theory. The next chapter will expand on this framework.

REFERENCES

Bland, J.M., Altman, D.G., 1986. Statistical methods for assessing agreement between two methods of clinical measurement. Lancet 327 (8476), 307–310.

Cohen, J., 1968. Weighted kappa: nominal scale agreement with provision for scaled disagreement or partial credit. Psychological Bulletin 70 (4), 213–230.

Crocker, L., Algina, J., 1986. Introduction to Classical and Modern Test Theory. Holt, Rinehart and Winston, New York.

Cronbach, L.J., 1951. Coefficient alpha and the internal structure of tests. Psychometrika 16 (3), 297–334.

Cronbach, L.J., 1957. The two disciplines of scientific psychology. American Psychologist 12 (11), 671–684.

Cronbach, L.J., Gleser, G.C., Nanda, H., Rajaratnam, N., 1972. The Dependability of Behavioral Measurements: Theory of Generalizability for Scores and Profiles. Wiley, New York.

DeVellis, R.F., 2006. Classical test theory. Medical Care 44 (11), S50–S59.

Fischer, R.A., 1925. Intraclass correlations and the analysis of variance. In: Fischer, R.A. (Ed.), Statistical Methods for Research Workers. Oliver and Boyd, Edinburgh.

Fleiss, J.L., 1981. Statistical Methods for Rates and Proportions, second ed. Wiley, New York.

Fleiss, J.L., Cohen, J., 1973. The equivalence of weighted kappa and the intraclass correlation coefficient as measures of reliability. Educational and Psychological Measurement 33 (3), 613–619.

Guilford, J.P., 1954. Psychometric Methods. McGraw-Hill, New York.

Hallgren, K.A., 2012. Computing inter-rater reliability for observational data: an overview and tutorial. Tutorials in Quantitative Methods for Psychology 8 (1), 23–34.

Hays, R., Morales, L., 2001. The RAND-36 measure of health-related quality of life. Annals of Medicine 33 (5), 350–357.

Kerlinger, F.N., 1986. Foundations of Behavioral Research, third ed. CBS Publishing Japan Ltd., New York.

Kottner, J., Streiner, D.L., 2011. The difference between reliability and agreement. Journal of Clinical Epidemiology 64 (6), 701–702.

Krabbe, P.F.M., Essink-Bot, M.L., Bonsel, G.J., 1996. On the equivalence of collectively and individually collected responses: standard-gamble and time-tradeoff judgments of health states. Medical Decision Making 16 (2), 120–132.

Kuder, G.F., Richardson, M.W., 1937. The theory of the estimation of test reliability. Psychometrika 2 (3), 151–160.

Landis, J.R., Koch, G.C., 1977. The measurement of observer agreement for categorical data. Biometrics 33 (1), 159–174.

McGraw, K.O., Wong, S.P., 1996. Forming inferences about some intraclass correlation coefficients. Psychological Methods 1 (1), 30–46.

Nunnally, J.C., Bernstein, I.H., 1994. Psychometric Theory, third ed. McGraw-Hill, New York.

Shavelson, R.J., Webb, N.M., Rowley, G.L., 1989. Generalizability theory. American Psychology 44 (6), 922–932.

Shrout, P.E., Fleiss, J.L., 1979. Intraclass correlations: uses in assessing rater reliability. Psychological Bulletin 82 (2), 420–428.

Streiner, D.L., 2003. Starting at the beginning: an introduction to coefficient alpha and internal consistency. Journal of Personality Assessment 80 (1), 99–103.

Streiner, D.L., Norman, G.R., 2006. "Precision" and "accuracy": two terms that are neither. Journal of Clinical Epidemiology 59, 327–330.

Streiner, D.L., Norman, G.R., Cairney, J., 2015. Health Measurement Scales: A Practical Guide to Their Development and Use, fifth ed. Oxford University Press, Oxford.

Ware, J.E., Sherbourne, C.D., 1992. The MOS 36-item short-form health survey (SF-36): I. Conceptual framework and item selection. Medical Care 30 (6), 473–483.

Weir, J.P., 2005. Quantifying test-retest reliability using the intraclass correlation coefficient and the SEM. Journal of Strength and Conditioning Research 19 (1), 231–240.

Part III

Chapter 9

Classical Test Theory

Chapter Outline

INTRODUCTION

Classical test theory is a body of related psychometric theory that was developed to predict the outcomes of psychological testing. Originally, the aim was to understand and improve the reliability of psychological tests. Later, classical test theory provided grounds for the development and evaluation of all kinds of tests in areas such as social attitudes, personality traits, and health. In this book, these are not called "tests" but "instruments."

The term "classical" refers to the more recent psychometric theories, collectively known as item-response theory (Chapter 10), which may bear the appellation "modern," as in "modern latent trait theory."

Classical test theory is based on several principles. The first is the recognition that measurements always contain errors. Second, as with many other measurement models that are devised within a statistical structure, those errors are assumed to be random. That is, factors other than the true score of the variable of interest are as likely to increase as to lower the observed score for any item. Third, the errors for items are assumed to be independent of one another. That is, the error associated with each item is unique to that item. The fourth is that variables measuring more or less the same "thing" or concept are associated or correlated.

Classical test theory was founded upon the notion of parallel testing. This implies that each item, be it a question or statement, reflects the underlying construct. For example, when evaluating anxiety, each item (e.g., insecurity, panic attacks) should reflect the underlying level of a patient's anxiety. Each item should be distinct from the others, yet similar and comparable in all important respects. The responses should therefore differ only as a consequence of random error. Such items are then said to be parallel. The notion of parallel testing underpins the construction of simple summated scales, whereby the scale score is computed by simply adding up the item scores. This explains why analytical techniques such as factor analysis (Chapter 7) and Cronbach's α (Chapter 8) play a prominent role in the development of most (multidomain) multiitem health instruments. In fact, most health status and other health-outcome instruments have been developed under classical test theory.

TRUE SCORE

Classical test theory is a theory or model of how (subjective) measurement operates in the real world. It is roughly synonymous with true score theory, which, like many powerful models, is very simple. True score theory regards observed responses as consisting of the sum of the true score and any error. It assumes that for each person there is a true score (T) that would be obtained if there were no errors in measurement. If no random error is present, the reliability is 1.0. Reliability approaches zero as the relative amount of random error increases. However, what we actually measure is the observed score. The latter may be envisaged as the composite of two hypothetical components: a true score (T) and an undifferentiated random-error component (E).

$$X = T + E \tag{9.1}$$

Now that we have introduced the concept of true score, let us pause to clarify it. The true score stands for a specific relationship between validity coefficients (typically, the correlation coefficient obtained by correlating the measure with a credible indicator of the variable of interest or by factor analysis) and reliability coefficients. The fact is, the validity coefficient for a specific instrument cannot exceed the size of its reliability coefficient. That is because no indicator can give a better estimate of the true score than the true score itself. No validity coefficient can exceed the correlation between the scale and the true score (Kerlinger, 1986; DeVellis, 2006). More details about classical test theory can be found in classical textbooks, notably in those written by Lord and Novick (1968), Crocker and Algina (1986), and Nunnally and Bernstein (1994).

INSTRUMENTS

Literally thousands of instruments have been developed in the framework of classical test theory. These are loosely referred to as "questionnaires." Basically, classical test theory is used to derive scale measures by bundling a set of items that tap into the same construct. Many health-status instruments and other health-outcome instruments, whose purpose is to measure specific health aspects as experienced by patients or observed by others, have been constructed on the principles of classical test theory.

Many health instruments are disease specific or intended to tap a specific aspect of health (such as pain or depression). For such instruments, a single scale may be sufficient to measure the aspect of interest. Such a scale is based on a couple of related items that are scored and summed up. In case the aim is to measure health in a broader and generic sense, so-called profile instruments may be more pertinent. These instruments measure each one of the multiple health domains on a distinct scale. Each scale comprises a select set of items.

The next section will expand on the origin of the SF-36 profile instrument. The SF-36 was developed almost three decades ago and is probably still the most widely used health-status instrument. The following description will clarify its ancestry as well as the lengthy process of item selection. In addition, some other "classical" health instruments will be presented to contextualize it among the various ideas and methodologies that were used in the development of these instruments.

Short Form-36

Perhaps the most common outcome measure in use today is the Medical Outcomes Study (MOS) Short Form-36 (SF-36). It is one of the products of the RAND Corporation's Health Insurance Experiment (HIE) and subsequent MOS (Ware and Gandek, 1998). Both were based on long-form general health surveys, the precursors of short-form instruments. A novel feature was the decision to collect patient-assessed outcome measures as well as traditional clinical and laboratory measures of health and illness.

Health Insurance Experiment

The RAND HIE was a social experiment funded by the US federal government. For that procedure, representative samples of different communities were assigned to several different health insurance plans through a nonbiased selection process. The intention was to assess the effects that variation in the cost of health services to the patient and of provision of services in either the fee-for-service system or a prepaid group practice would have on the use of services, quality of care, patient satisfaction, and health status (Brook et al., 1979). To determine whether those receiving "free" care were healthier as a result, measures were developed to evaluate the effect of cost sharing on health status. Among these was an instrument to assess subjective health; the HIE health instrument contained 108 items and was administered at entry and upon leaving the study.

The HIE health instrument provided the background for the MOS of patient-assessed health measures. That baseline HIE study, conducted on a sample of 7708 people, began in 1974 and ended in 1986. It measured physical, mental, and social health and general health perceptions (Jenkinson and McGee, 1998).

Medical Outcomes Study

The MOS was a 2-year prospective study with two major aims. The first was to determine whether variations in patient outcomes could be explained by variations in the system of care, a clinician's specialty, and a clinician's technical and interpersonal style. The second was to develop instruments for routine monitoring of patient outcomes in medical practice; specifically the aim was to develop self-administered instruments and generic scales. The Short Form 20 (SF-20) general health survey was part of the quest for a generic health-status measure that could satisfy a number of potentially conflicting criteria. Ideally, the researchers wanted a measure that was short enough to be completed quickly yet was also comprehensive (covering as many domains of health as possible) with psychometrically sound, multiitem scales. The HIE provided the basis for the 20-item short-form instrument. While the results indicated that the SF-20 had high levels of internal reliability consistency on multiitem domains, the aspects of social functioning and pain were single-item measures and could therefore not be evaluated for internal reliability consistency. The decision to include domains measured with single items prompted the development of an improved short-form health survey instrument (Jenkinson and McGee, 1998). However, enthusiasm for the SF-20 waned when a study of hospitalized patients showed that further decrements in health went undetected for persons who scored at the floor (floor effect) on some of the SF-20 subscales.

Medical Outcomes Study 36-Item Short-Form Health Survey

The MOS 36-item Short-Form Health Survey (SF-36) became available in a "standard" form in 1990. Besides eliminating more than one-fourth of the words contained in the original versions, it also made improvements in item wording, format, and scoring (Ware et al., 1993). The SF-36 contains eight domains and a single item requesting information on perceived change in health over the past year. The eight domains are measured by scores that are computed from the responses on two or more items. These domains are presented in Table 9.1, along with the number of items per domain in the instrument.

TABLE 9.1 Short Form-36 Characteristics of the Set of Eight Domains

Domain	Number of Items	Number of Categories	Response Labels
Physical functioning	10	3	Limited...Not limited
Role limitations due to physical problems	4	5	All...None of the time
Role limitations due to emotional problems	3	5	All...None of the time
Social functioning	2	5	Different labels
Mental health	5	5	All...None of the time
Energy/vitality	4	5	All...None of the time
Pain	2	5	Different labels
General health perception	5	5	Definitely True...False

The eight health domains of the SF-36 are physical functioning, role-physical, role-emotional, social functioning, mental health, vitality, bodily pain, and general health perceptions (Kosinski et al., 1999). Physical health was defined as functional status and the ability to perform a variety of activities: mobility, self-care, household, physical, role, and leisure. Mental health measures refer to symptoms of affective (mood) disorders and of anxiety disorders, positive well-being, and self-control. These measures emphasize psychological states (rather than somatic or physiological manifestations of these states). Social health was defined in terms of interpersonal interactions and activities indicating social participation. Measures of general health perceptions require a self-rating of overall health.

The eight health concepts were selected from the 40 concepts included in the MOS (Stewart et al., 1988). These eight domains cover the aspects most frequently measured in widely used health surveys and those aspects most affected by disease and treatment (Ware et al., 1993; Ware, 1995). Most SF-36 items have their roots in instruments that have been in use since the 1970s and 1980s (Ware and Sherbourne, 1992). Their derivation is quite diverse: various physical and role-functioning measures (Patrick et al., 1973; Hulka and Cassel, 1973; Reynolds et al., 1974; Stewart et al., 1981); the Health Perceptions

Questionnaire (Ware, 1976); and other measures that proved to be useful during the HIE. The items selected represent multiple elements of health, e.g., behavioral function and dysfunction; distress and well-being; objective reports and subjective ratings; and both favorable and unfavorable self-evaluations of general health status (Ware et al., 1993).

The SF-36 was constructed to broaden the range of health concepts being measured and improve measurement precision for each concept over the precision achieved with the SF-20. To that end, the number of health domains was increased to eight and the number of items was almost doubled to 36. Another reason was to satisfy the minimum psychometric standards for group comparisons. Noteworthy improvements include the addition of items tied to vitality, better representation of general health perceptions, distinguishing physical from mental causes of role limitations, and more precise measurement on the physical, role, social, and bodily pain scales (Stewart et al., 1988).

The SF-36 can be presented by a trained interviewer or be self-administered. It has many advantages: besides being brief, its reliability and validity are well documented; moreover, it can be machine-scored and has been evaluated in large population studies (Stewart and Ware, 1992; Kosinski et al., 1999).

Two Versions

The items included in the SF-36 were developed at RAND as part of the MOS, but a replica of the survey, called MOS SF-36, can be obtained elsewhere. The scoring method in the replica is slightly different, so the original version goes by its proprietary name: The RAND 36-Item Health Survey. Because the MOS SF-36 uses slightly different scoring algorithms for two of the eight scales (bodily pain, general health), the results for those scales are not comparable with results from the RAND SF-36. Strict adherence to item wording and scoring rules is required in order to use the original RAND SF-36 trademark, which may explain the small adjustments made in the MOS SF-36. The differences are mainly in the bodily pain scale. The RAND SF-36 seems to be used less frequently. Much confusion is caused by having two almost identical instruments with different ways of scoring. We often see that one version of the SF-36 is used by researchers and the other version for scoring, clearly an undesirable situation. To make it even less transparent, the MOS SF-36 is a commercial product, whereas use of the RAND SF-36 is free. Measures developed by RAND, such as those in the MOS study, have traditionally been placed in the public domain. The differences in scoring were summarized by Hays et al. (1993).

Sickness Impact Profile

The Sickness Impact Profile (SIP) was created in 1975 by Gilson et al. and revised in 1981 by Bergner et al. Along with the Katz ADL index, for many years it was one of the most widely used scales. The story of the SIP is interesting, illustrating as it does some of the discernible trends in health-status measurement. This instrument provides a descriptive profile of the changes in a person's behavior due to sickness. The SIP is used in measuring the outcomes of care, in health surveys, in program planning and policy making, and in monitoring patient progress (Larson, 1991). It grew out of the belief that health care should reduce sickness and its effects on daily living. The objective was to construct a measure that was less sensitive to cultural variation and demographic limitations. From that angle, work began in 1972 on a measure that could be applied to many conditions and across many locations. Following a series of field trials, a prototype was constructed comprising 312 items (yes/no) grouped into 14 domains (Brooks, 1995). It was finally reduced to a more manageable format of 136 items under 12 domains covering such areas as work, recreation, emotional behavior, and social interaction. Respondents were asked to give an affirmative answer only when they were sure that an item described them and was related to their health. Three of the twelve domains could be used to form a physical score and five to form a psychosocial score.

The resulting SIP is composed of statements such as "I have difficulty reasoning and problem solving" or "I don't walk at all." Respondents are asked to check items that describe them on a given day and are related to their health. The items have been weighted in accordance with the judgments of over 100 raters, who valued each item on an equal-interval 11-point scale ranging from minimal to severe dysfunction. The SIP score is derived by adding up the scale values for each item, checking these values across all domains, dividing the sums by the maximum possible dysfunction score for all domains, and then multiplying by 100 to obtain the overall SIP value. Bergner et al. (1981) noted some advantages of the SIP: it can be either self-administered or presented by an interviewer and can be completed in 20–30 min. The SIP is particularly valuable when assessing the impact of illness on the chronically ill and measuring the effect of noncurative interventions. On the other hand, as Hunt et al. (1985) mentioned, it has two major limitations: its length; and the fact that it can be used only with people who are regarded, or who regard themselves, as ill. The SIP has been widely adopted. It has been translated into several languages and has been used in a number of countries.

Indeed, the SIP is very lengthy, which has raised doubts about its application in clinical and care settings (Brooks, 1995). As described earlier, the SIP presents 136 questions and takes an average of 30 min. To its credit, however, independent observers, such as myself, give the measure rather high marks on methodological grounds. Subsequent developments with other profiles have usually reduced the number of items assessed in instruments, partly to make the measures more applicable in a clinical setting. Yet when reducing the number of items, care should be taken to retain the sound methodological features of instruments (Essink-Bot et al., 1996). Note that the SIP consists of binary response data. Therefore, it is not possible to apply psychometric methods such as factor analysis and Cronbach's α to confirm the internal structure of the instrument.

Nottingham Health Profile

A profile that has achieved some popularity in the British context is the Nottingham Health Profile (NHP) (Hunt and McEwen, 1980; Hunt, 1988). Its developers came up initially with 2200 health-state descriptors. Over several years of experimentation, the profile was reduced to 45 items. This corresponds with a similar process in the construction of the SIP. Its relative simplicity probably accounts for some of its popularity in the clinical environment.

After preliminary work outlining the sort of structure and contents of the NHP, the instrument was published in the mid-1980s. The first of its two sections, containing 38 items, is intended to measure perceptions of subjective health on six domains that ill health may adversely affect. The items are distributed over the areas of pain, physical mobility, emotional reactions, sleep disturbance, social isolation, and energy. In the second section, the questions broaden out into the domains of employment, social life, household work, home life, sex life, interests and hobbies, and holidays (Hunt et al., 1985).

Items were generated for the instrument by interviewing lay people and clinicians. Each item carries a specific weight to reflect the relative seriousness of items on the instrument. Weights were assigned by Thurstone's method of paired comparisons (Chapter 11) (McKenna et al., 1981). Respondents can only answer "yes" or "no" to the questions. They can affirm any number of the items on the NHP that they believe reflect their health status "at the moment." Scores range from 0 (indicating good health) to 100 (indicating poor health).

The developers of the NHP decided it should cover only the severe end of ill health. Therefore, items reflecting minor health problems were discarded. The restriction to items that represent severe problems was apparently necessary to avoid picking up large numbers of false positives. This focus is in keeping with the original intention, which was to design an instrument for use in population surveys, an instrument that could be employed in planning health service provision.

Substantial effort has gone into validating the NHP and ensuring its reliability over time. It has been tested for face, content, and criterion validity. The measure seems to be good at identifying people with chronic illnesses and distinguishing between conditions. Although researchers also initiated studies to transform the scores for the domains into single-index figures, that vein has not been fruitful (Hunt and McEwen, 1980). Some translated versions of the instrument permit its use in cross-cultural trials. The instrument is short and easy to complete. However, as with all measurement tools, it has limitations. The decision to tap only the severe end of ill health means that data obtained from the instrument tend to be highly skewed. Most respondents came out with zero or low scores on many or, indeed, all of the domains in the profile. Due to this "floor effect," the NHP may not always be an appropriate tool. Other problems with the NHP have been mentioned elsewhere. For example, it has been noted that the domains of pain and mobility are confounded and that the method of weighting the severity of items may lead to illogical and incoherent results (McKenna et al., 1993; Jenkinson and McGee, 1998). This instrument is no longer in vogue and has been replaced by seemingly better ones.

EORTC QLQ C-30

The EORTC QLQ C-30 is a disease-specific health instrument. The EORTC (European Organization for Research and Treatment of Cancer) QLQ (Quality of Life Questionnaire) C-30 (Core 30 items) has been used widely in the field of cancer research. It takes a modular approach to quality-of-life assessment. The intent was to generate a core instrument incorporating a range of physical, emotional, and social health issues relevant to a broad range of cancer patients irrespective of the specific diagnosis, supplemented by diagnosis-specific (e.g., lung cancer, breast cancer), and/or treatment-specific instrument modules. Adoption of a modular approach was supposed to reconcile the two main requirements of a quality-of-life measure, which are as follows: (1) a sufficient degree of generalizability to allow for cross-study comparisons; and (2) a level of specificity adequate for addressing those research questions of particular relevance in a given clinical trial. A first-generation core instrument, the EORTC QLQ-C36, was developed in 1987 (Aaronson et al., 1993).

This instrument consists of 30 items comprising five functional scales (physical, role, cognitive, emotional, and social), three symptom scales (fatigue, pain, and nausea/vomiting), a global health scale with two items (overall health and quality of life during the past week), a number of single items assessing symptoms (dyspnea, loss of appetite, sleep disturbances, constipation, and diarrhea), and perceived financial impact of the disease. In its current form (version 3), all items except those referring to global health have four response levels, while the global health items have seven. Transformed raw scores for the functional scales range from 0 (worst) to 100 (best), whereas scores for the symptom scales range from 0 (no symptoms) to 100 (severe symptoms).

The EORTC QLQ C-30 has proven to be a valid and reliable instrument (Aaronson et al., 1993). It has been used in a wide variety of clinical cancer settings, including studies on responsiveness, for which it was used as a comparative instrument.

APPLICATION

The study that will be presented in this section as an example is about the development of a symptom-specific, multidomain (profile), multiitem health-related quality-of-life instrument (Hinderink et al., 2000). We will explain how the items were selected for the various domains and walk through the first steps in the evaluation of its construct validity.

Cochlear implantation (CI) is a widely accepted treatment. It is routinely used for postlingually (after the acquisition of speech and language, usually after the age of 6 years) profound deafness. Objective speech perception results have improved during the past decades thanks to this advanced technology. Today most patients achieve a certain degree of open-set speech understanding with or without lip reading. Apart from better hearing, important improvements may also be expected in other respects. In the case of CI, this means that the treatment not only affects hearing and speech production, but may also have an impact on self-esteem, daily activities, and social functioning.

Construct

The development of the Nijmegen Cochlear Implant Questionnaire (NCIQ) began by formulating the health domains relevant to CI users. Taking a conventional approach, the developers distinguished three general domains: physical, psychological, and social functioning. They then specified subdomains: basic sound perception, advanced sound perception, and speech production in the physical subdomain; and activity and social functioning in the social domain. Psychological functioning has only one subdomain: self-esteem (Fig. 9.1).

Specific criteria that should be satisfied by the NCIQ were formulated beforehand: (1) each subdomain should consist of an equal number of items; (2) all items should be phrased in a similar way; (3) responses to the items should be uniform; and (4) the items should be suitable for constructing Likert scales. Additionally, the NCIQ should be a self-report instrument. The next step consisted of the selection and construction of sets of items for each subdomain. An empirical psychometric approach normally has three phases: (1) generation of a large pool of items; (2) completion of all items by the target population; and (3) reduction of the number of items by psychometric techniques (e.g., factor analysis). In this study, however, the items were selected by intuitive judgment. Numerous items were taken from other published instruments and then adapted. Several other items were formulated on the basis of interviews and previous instruments that had been used over the years to monitor the rehabilitation of CI users within the Nijmegen team. The goal was to include 10 items for each subdomain to assure a sufficient level of reliability for the scales of these subdomains. Each item was formulated as a statement

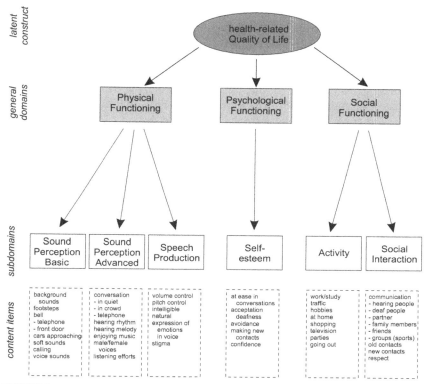

FIGURE 9.1   Diagram of the construction of the Nijmegen Cochlear Implant Questionnaire (NCIQ). *From Hinderink, J.B., Krabbe, P.F.M., van den Broek, P., 2000. Development and application of a health-related quality of life instrument for adults with cochlear implants: the Nijmegen cochlear implant questionnaire. Otolaryngology-Head and Neck Surgery 123 (6), 756–765.*

APPENDIX: NIJMEGEN COCHLEAR IMPLANT QUESTIONNAIRE*

*A native English speaker using forward-backward translation translated the NCIQ questionnaire.

Please answer the following 60 questions regarding the CI situation (use "not applicable" [N/A] only if none of the possibilities is applicable).

	Never	Sometimes	Regularly	Usually	Always	N/A
1. Can you hear background noises (toilet flushing, vacuum cleaner)?	☐	☐	☐	☐	☐	☐
2. Does your hearing impairment present a serious obstacle in your contact with persons with normal hearing?	☐	☐	☐	☐	☐	☐
3. Are you able to whisper if you have to?	☐	☐	☐	☐	☐	☐
4. Do you feel at ease in company despite your hearing impairment?	☐	☐	☐	☐	☐	☐
5. Can you hold a conversation in a quiet environment (with or without lip-reading) with one person?	☐	☐	☐	☐	☐	☐

FIGURE 9.2   Example of the first five questions based on Likert items with five level categories. *From Hinderink, J.B., Krabbe, P.F.M., van den Broek, P., 2000. Development and application of a health-related quality of life instrument for adults with cochlear implants: the Nijmegen cochlear implant questionnaire. Otolaryngology-Head and Neck Surgery 123 (6), 756–765.*

with a five-point response scale to indicate the degree to which the statement was true. These five response categories were as follows: never (1), sometimes (2), regularly (3), usually (4), and always (5) for 55 of the total of 60 items (Fig. 9.2). The other five items were answered according to the CI user's ability to perform the action in question. Response categories for these five remaining items (not shown) were as follows: no (1); poorly (2); moderate (3); adequate (4); and good (5). Respondents were also offered a sixth response category to cover items that were not relevant to them.

Study

The NCIQ was sent to 60 adults who subsequently received a CI during the period 1989—97 under the supervision of the Nijmegen/St. Michielsgestel CI team. All were using oral—aural communication. 13 of the 60 subjects were excluded from the study: 10 were prelingually deaf; and 3 were postlingually deaf but had been fitted with a single-channel implant. The remaining 47 postlingually deaf adult participants had been fitted with a multichannel implant. They had all been using the implant for at least 1 year. The NCIQ was administered twice in a crossover design: once in the past tense to obtain retrospective information and once in the present tense to evaluate their current health. Half of the CI users filled out the retrospective version first (CI-pre), and the other half filled out the standard version (CI-post). Two weeks after the subjects completed and returned the first instrument, they were sent the other version. The retrospective answers of the CI users were compared with the answers from the control group (baseline) of 53 postlingually deaf candidates for CI on the waiting list at the institute. To study the effect of CI and the construct validity of the NCIQ in more detail, two generic health-status instruments also formed part of this study.

Coding

Before computing the six subdomains of the NCIQ, the scores for 27 items of the instrument that were phrased in the opposite form were recoded (6—score). Next, the answer categories (1—5) for all items were transformed: $1 = 0$; $2 = 25$; $3 = 50$; $4 = 75$; and $5 = 100$. Scores for the subdomains were computed by adding up the 10-item scores of each subdomain and dividing by the number of completed items. Missing values and the response category "not applicable" were both treated as "not completed." The maximum number of incomplete answers for a specific subdomain was set at three items per subject; above this number, the scores were excluded.

Reliability and Validity Tests

The internal consistency of the six subdomains for the NCIQ was assessed by using Cronbach's α. An α coefficient of 0.70 or higher was considered sufficient for the purpose of group comparisons. Test—retest reliability was evaluated by readministration of the standard version of the NCIQ to the CI users $(n = 43)$ 2 months later. Scores for the six subdomains of the NCIQ were related to the answers on the two auditory performance tests to evaluate convergent validity. Unfortunately, due to the small number of CI users in relation to the large number of items, the confirmatory factor analysis to test the construct validity (relationship of 60 items vs 6 subdomains) of the NCIQ could not be carried out.

Table 9.2 shows the correlation coefficients for the scores on the six scales of the NCIQ and the scores on the Spondee and Environmental Sounds Identification Tests. There did not appear to be any prominent positive correlation between the health scores of the NCIQ subdomains and objective measures of performance as measured by the two auditory perception tests. Especially on the sound perception sub-domains, higher correlations could be expected, which would confirm the convergent validity for at least two subdomains of the NCIQ. This low association might be explained by the small range in outcomes on the sound perception tests, which would inflate the differentiation among the scores of the subjects.

TABLE 9.2 Correlations of the Six Subdomains of the NCIQ From the CI Users $(n = 45)$ With the Spondee Identification Test and the Environmental Sounds Identification Test

Subdomain	Spondee Identification Test	Environmental Sounds Identification Test
Sound perception—basic	0.23	0.38
Sound perception—advanced	0.49	0.54
Speech production	0.36	0.59
Self-esteem	0.38	0.23
Activity	0.40	0.43
Social interactions	0.32	0.36

From Hinderink, J.B., Krabbe, P.F.M., van den Broek, P., 2000. Development and application of a health-related quality of life instrument for adults with cochlear implants: the Nijmegen cochlear implant questionnaire. Otolaryngology-Head and Neck Surgery 123 (6), 756—765.

TABLE 9.3 Internal Consistency (Cronbach's α), Test–Retest Reliability (Intraclass Correlation Coefficient) of the Assessment of the Six Subdomains of the NCIQ by the Cochlear implantation Users

Subdomain	Internal Consistency ($n = 45$)	Test–Retest Reliability ($n = 35$)
Sound perception –basic	0.81	0.83
Sound perception –advanced	0.84	0.85
Speech production	0.73	0.78
Self-esteem	0.75	0.64
Activity	0.89	0.82
Social interactions	0.84	0.81

From Hinderink, J.B., Krabbe, P.F.M., van den Broek, P., 2000. Development and application of a health-related quality of life instrument for adults with cochlear implants: the Nijmegen cochlear implant questionnaire. Otolaryngology-Head and Neck Surgery 123 (6), 756–765.

Table 9.3 presents internal consistency statistics and test–retest coefficients for the six subdomains. Internal consistency was high (>0.80) for the two related subdomains of basic and advanced sound perception but also for the subdomains of activity and social interaction. This suggests that the items in these subdomains are closely related to each other, producing reliable scores for these subdomains. Moderate Cronbach's α values (>0.70) were found for the subdomains of speech production and self-esteem. Test–retest reliability of the NCIQ also proved to be satisfactory. All subdomains had coefficients that exceeded 0.60.

Cochlear Implantation Effect

Before discussing the NCIQ results, it is important to note that strong agreement was observed between the retrospective answers of the CI users regarding their preimplant health and the health presently perceived by the deaf candidates on the waiting list for CI. This strong resemblance between the CI users and the controls in this study provides support for the validity of interpreting the retrospective information.

Fig. 9.3 presents the descriptive statistics of the six subdomains for the two groups. Scores from the CI-pre and the control assessments were similar, whereas scores from the CI-post assessment were statistically significantly better on all subdomains. The largest differences between CI-post and CI-pre/controls were for basic and advanced sound perception. The differences for the

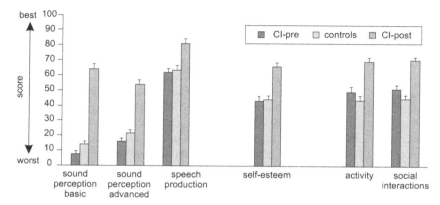

FIGURE 9.3   Mean scores (with standard errors of the mean) for the CI users (standard version: post; retrospective version: pre) and the control group on the six subdomains of the NCIQ. *From Hinderink, J.B., Krabbe, P.F.M., van den Broek, P., 2000. Development and application of a health-related quality of life instrument for adults with cochlear implants: the Nijmegen cochlear implant questionnaire. Otolaryngology-Head and Neck Surgery 123 (6), 756–765.*

other four subdomains were smaller, although the overall improvement due to a CI was still greater than 30% on all four of those subdomains. Variation (SD) in the answers to the instruments was about equal for the six subdomains, for the two groups, and for the CI-pre and CI-post assessments.

The NCIQ measured considerable improvement in all six subdomains between the CI-pre and CI-post assessments. Great improvement in the perceived health was measured not only for the physical subdomains but also for the psychological and social subdomains. Most improvement was recorded for the basic and advanced sound perception subdomains, which is understandable because deaf subjects tend to score almost nothing on these subdomains. Improvements were found for all items in these two subdomains, even for listening to music, even though the implant systems are specially designed for processing speech. Being able to enjoy music can contribute considerably to a person's quality of life. Consequently, music should not be categorized at the same level as environmental sounds but rather on a more cognitive level. In that regard, it was decided to integrate music items with items for speech discrimination in the advanced sound perception subdomain.

DISCUSSION

Even though the multiitem instruments developed within the classical test theory framework have their limitations, multiitem scales are still superior to single-item scales (Nunnally and Bernstein, 1994, p. 66). Let us consider some reasons why. First, it is very unlikely that a single item can fully represent a complex theoretical concept, or any specific attribute for that matter. For example, no single item in a clinical rating evaluation will allow a physician to accurately and validly measure a person's level of anxiety. Second, single-item measures lack precision because they cannot discriminate among fine degrees of an attribute. In fact, commonly used dichotomous items can only distinguish between two levels of the attribute. Third, single-item measures are usually less reliable than multiitem scales. That is, they are more prone to random error. However, as Nunnally and Bernstein (1994, p. 67) observed, "All of these difficulties are diminished by the use of multi-item measures. The tendency of items to relate to incidental factors usually averages out when they are combined because these different incidental factors apply to the various items. (...) reliability increases (measurement error decreases) as the number of items increases."

Most health instruments do not usually provide a single score but instead generate a set of profiles for each patient. The profile approach has been extensively applied in the clinical environment. It clearly has its merits as a provider of information on health status and health-related quality of life. The problem with single-item measures is not simply that they tend to be less valid and less reliable than their multiitem equivalents. Rather, because they provide only a single measurement, the scientist rarely has sufficient information to estimate their measurement properties. Thus, the fundamental problem is that their degree of validity and reliability may be unknowable. Without this vital information, we might overlook the serious deficiencies of single-item measures.

One well-known shortcoming of classical test theory is that respondents' characteristics and item characteristics cannot be separated. Each of them can only be interpreted in the context of the other. Another shortcoming lies in its definition of reliability. Various reliability coefficients provide either lower bound estimates of reliability or reliability estimates with unknown biases. Yet another one involves the standard error of measurement. The problem is that the standard error of measurement is assumed to be the same for all examinees. Furthermore, in classical test theory, the level for each respondent is relative to the levels for the rest of the sample. As such, all conventional multiitem health instruments provide nonstandardized measures on (multiple) health domains about the relative level of complaints.

In all the measurement instruments developed under classical test theory, the scales of these instruments are fully dependent on the phrasing of the lowest and highest categories of the items used and on the phrasing of the items itself. In fact, the choice for a highest category determines the end of the scale. Moreover, the choice for the phrasing of the items or statements determines how people will respond on these items. If we mainly include items that have to do with simple activities that most people can easily perform, then the scores will be high (good health). But if we mainly include items about difficult and demanding activities, the scores will surely be lower. Therefore, the interpretation of scores on such scales is often only modestly meaningful.

Measures from profile instruments are particularly helpful to compare results from different treatments or different subgroups. However, as they lack a clear interpretation of their highest and lowest scores, these instruments do not provide numbers that are easily compared. In the next chapter, methods will be introduced that are able to provide more informative measures.

The reliability of an instrument is normally not examined in the empirical (clinical) studies that use the instrument. Nevertheless, it is the researchers' responsibility to review the available information and determine whether adequate reliability has already been established for the instrument in relation to their intended purpose. If it has not, they bear the responsibility of obtaining and making available any relevant information supporting their interpretation of scores in the new situation.

REFERENCES

Aaronson, N.K., Ahmedzai, S., Bergman, B., Bullinger, M., Cull, A., Duez, N.J., Filiberti, A., Flechtner, H., Fleishman, S.B., de Haes, J.C., et al., 1993. The European Organization for Research and Treatment of Cancer QLQ-C30: a quality-of-life instrument for use in international clinical trials in oncology. Journal of the National Cancer Institute 85 (5), 365−376.

Bergner, M., Bobbitt, R.A., Carter, W.B., Gilson, B.S., 1981. The Sickness Impact Profile: development and final revision of a health status measure. Medical Care 19 (8), 787−805.

Brook, R.H., Ware Jr., J.E., Davies-Avery, A., Stewart, A.L., Donald, C.A., Rogers, W.H., Williams, K.N., Johnston, S.A., 1979. Overview of adult health status measures fielded in Rand's health insurance study. Medical Care 17 (7), 1−131, iii−x.

Brooks, R.G., 1995. Health Status Measurement: A Perspective on Change. Macmillian Press, London.

Crocker, L., Algina, J., 1986. Introduction to Classical and Modern Test Theory. Holt, Rinehart and Winston, New York.

DeVellis, R.F., 2006. Classical test theory. Medical Care 44 (11), S50−S59.

Essink-Bot, M.L., Krabbe, P.F.M., van Agt, H.M.E., Bonsel, G.J., 1996. NHP or SIP − a comparative study in renal insufficiency associated anemia. Quality of Life Research 5 (1), 91−100.

Gilson, B., Gilson, J., Bergner, M., Bobbitt, R., Kressel, S., Pollard, W., Vesselago, M., 1975. The sickness impact profile: development of an outcome measure of health care. American Journal of Public Health 65 (12), 1304−1310.

Hays, R.D., Sherbourne, C.D., Mazel, R.M., 1993. The RAND 36-item health survey 1.0. Health Economics 2 (3), 217−227.

Hinderink, J.B., Krabbe, P.F.M., van den Broek, P., 2000. Development and application of a health-related quality of life instrument for adults with cochlear implants: the Nijmegen cochlear implant questionnaire. Otolaryngology-Head and Neck Surgery 123 (6), 756−765.

Hulka, B.S., Cassel, J.C., 1973. The AAFP-UNC study of the organization, utilization, and assessment of primary medical care. American Journal of Public Health 63 (6), 494−501.

Hunt, S.M., McEwen, J., 1980. The development of a subjective health indicator. Sociology of Health & Illness 2, 231−246.

Hunt, S.M., McEwen, J., McKenna, S.P., 1985. Measuring health status: a new tool for clinicians and epidemiologists. The Journal of the Royal College of General Practitioners 35 (273), 185−188.

Hunt, S.M., 1988. Subjective health indicators and health promotion. Health Promotion 3 (1), 23—24.

Jenkinson, C., McGee, H.M., 1998. Health Status Measurement: A Brief But Critical Introduction. Radcliffe Medical Press Ltd., Abingdon, Oxon.

Kerlinger, F.N., 1986. Foundations of Behavioral Research, third ed. CBS Publishing Japan Ltd., New York.

Kosinski, M., Keller, S.D., Hatoum, H.T., Kong, S.X., Ware Jr., J.E., 1999. The SF-36 Health Survey as a generic outcome measure in clinical trials of patients with osteoarthritis and rheumatoid arthritis: tests of data quality, scaling assumptions and score reliability. Medical Care 37 (Suppl. 5), MS10—MS22.

Larson, J.S., 1991. The Measurement of Health: Concepts and Indicators. Greenwood Press, Westport, CT.

Lord, F.M., Novick, M.R., 1968. Statistical Theories of Mental Test Scores. Reading. Addison-Wesley.

McKenna, S., Hunt, S., McEwen, J., 1981. Weighting the seriousness of perceived health problems using Thurstone's method of paired comparisons. International Journal of Epidemiology 10 (1), 93—97.

McKenna, S.P., Hunt, S.M., Tennant, A., 1993. The development of a patient completed index of distress from the Nottingham Health Profile: a new measure for use in cost-utility studies. British Journal of Medical Economics 6, 13—24.

Nunnally, J.C., Bernstein, I.H., 1994. Psychometric Theory, third ed. McGraw-Hill, New York.

Patrick, D.L., Bush, J.W., Chen, M.M., 1973. Methods for measuring levels of well-being for a health status index. Health Services Research 8 (3), 228—245.

Reynolds, J.W., Rushing, W.A., Miles, D.L., 1974. The validation of a function status index. Journal of Health and Social Behavior 15 (4), 271—288.

Stewart, A.L., Hays, R.D., Ware Jr., J.E., 1988. The MOS short-form general health survey: reliability and validity in a patient population. Medical Care 26 (7), 724—735.

Stewart, A.L., Ware Jr., J.E., Brook, R.H., 1981. Advances in the measurement of functional status: construction of aggregate indexes. Medical Care 19 (5), 473—488.

Stewart, A.L., Ware Jr., J.E., 1992. Measuring Functioning and Well-being: The Medical Outcomes Study Approach. Duke University Press, Durham.

Ware Jr., J.E., 1976. Scales for measuring general health perceptions. Health Services Research 11 (4), 396—415.

Ware Jr., J.E., Sherbourne, C.D., 1992. The MOS 36-item short-form health survey (SF-36): I. Conceptual framework and item selection. Medical Care 30 (6), 473—483.

Ware Jr., J.E., Snow, K.K., Kosinski, M., Gandek, B., 1993. SF-36 Health Survey Manual and Interpretation Guide. New England Medical Center, The Health Institute, Boston, MA.

Ware Jr., J.E., 1995. The status of health assessment 1994. Annual Review of Public Health 16 (1), 327—354.

Ware Jr., J.E., Gandek, B., 1998. Overview of the SF-36 health survey and the international quality of life assessment (IQOLA) project. Journal of Clinical Epidemiology 51 (11), 903—912.

Chapter 10

Item Response Theory

Chapter Outline

INTRODUCTION

In the past, most health and other clinical scales were based on traditional psychometric theory, with summated scales composed of multiple items being particularly common. Newer instruments are based on so-called modern measurement theory: item response theory (IRT) models. IRT use to evaluate item characteristics and to scale both items and respondents (patients) on a unidimensional construct. Essentially, IRT allows a comprehensive understanding of item responses and their relationships to the construct to be measured. IRT resembles classical test theory (Chapter 9) in that it can be applied when the underlying measurement model is reflective (Chapter 5). In such a model, the correlation among the observed variables is caused by, and therefore reflects, variation in the latent (unobservable) variable. Any changes in the latent variable should cause comparable changes in all the manifest (observable) variables.

IRT models are also referred to as latent trait models. Because a latent trait is not manifested directly, it can only be assessed indirectly by the various items. IRT, as its name suggests, explicitly models a person's responses at the item level, whereas classical test theory pertains to the scale as a whole.

Briefly, IRT offers several advantages over classical test theory: (1) it provides more in-depth insight at the item level; (2) it facilitates the development of shorter measures (by applying computerized adaptive routines); (3) it detects cross-group variations in item performance (called differential item functioning or DIF); (4) and it permits linking scores from one measure to another.

In IRT-based models, items vary in "difficulty." It is assumed that patients will have different probabilities of responding positively to an item, according to their level of functioning and symptoms. The procedure for designing scales using IRT methods differs markedly from traditional methods that fall under classical test theory. Under classical test theory, Likert summated scales assume that the items are of broadly similar difficulty, with each item having response categories to reflect the severity or degree of the response level. In contrast, IRT scales are based on items of varying difficulty; moreover, the steps between response categories are not necessarily of equal distance.

Unlike classical test theory, IRT can measure a respondent's scale score from an arbitrary subset of items specific to that respondent while preserving the origin and unit of the measurement scale of the full instrument (invariance principle; Chapter 4 and below). The arbitrary sum-of-item scores (Likert scales) as worked out under classical test theory do not have that capability.

Typically, IRT models are used to measure a patient's ability, for example, cognitive ability. That is why many handbooks speak of "difficulty" and "tests" to describe them. The most common application of IRT is in education, particularly for developing and designing exams, maintaining banks of items for exams, and equating the difficulty of items for successive versions of exams (for example, to allow comparison between results over time). It is only since the turn of the century that health-outcome instruments using IRT have appeared. The use of IRT methods in health settings is one of the most important developments of recent years (Holman et al., 2005; Schultz-Larsen et al., 2007; Young et al., 2009, 2011).

HISTORY OF ITEM RESPONSE THEORY

The item response function was around before 1950, but only as a concept. Pioneering theoretical work on IRT was done during the 1950s and 1960s: by the Educational Testing Service statistician Allan Birnbaum (Lord and Novick's book from 1968 contains four chapters on IRT, written by Birnbaum); by the Danish mathematician Georg Rasch; and by the Austrian sociologist Paul Lazarsfeld, who pursued parallel research independently. Further advances in IRT, in particular the elementary Rasch model, were made by Benjamin

Wright (Chicago) and David Andrich. IRT did not become widely used until the late 1970s, when practitioners were told about the "usefulness" and "advantages" of IRT. Followed in the 1980s by personal computers that gave researchers access to the computing power necessary for IRT.

As pointed out by Bock (1997), a glimpse of what would eventually become IRT could be seen in an article by Thurstone (1925), "A method of Scaling Psychological and Educational Tests." Apparently, Thurstone never recognized that both items and persons could be simultaneously calibrated onto the same underlying scales (Engelhard, 2013).

Most early applications of IRT were in education. Educational assessment has been particularly amenable to IRT modeling for a number of reasons. First, one usually gives an exam to assess a single aspect, such as mathematical ability, that can be easily placed on a unidimensional scale. Second, in education it is natural to think of binary test items that are marked "right" or "wrong." And third, there is an infinite pool of potential test items, so one can usually select items that fit the IRT model well and reject those that do not.

IRT models differ in the functional form of the item characteristic curve to handle different item response formats. Models for binary or dichotomously (e.g., right/wrong, true/false, yes/no) scored items were the first to be developed. But models for ratings as well as for graded or multicategory (Likert-type) item responses (e.g., strongly disagree, disagree, neutral, agree, strongly agree) are now available to handle the more complex instrument designs and scoring procedures seen in recent health-outcomes applications (van der Linden and Hambleton, 2010).

RESPONSE STRUCTURE

IRT models are mathematical functions that specify the probability of a discrete outcome, such as a correct response to an item, in terms of both item and person parameters. The parameters of an item include its difficulty (for health: severity) and its discrimination (for health: the agreement among patients on the severity or quality of health aspects). The parameters of a person may represent a student's ability or a patient's quality of health. The items may be questions that have incorrect and correct responses, or they may be statements that allow the respondents to indicate their level of agreement.

The most common assumption is unidimensionality, which means that only one latent trait (e.g., physical functioning) is measured by the items in a scale. In practice, that assumption has to be supported by a "dominant" factor that overly influences an individual's responses to scale items. The unidimensionality assumption underpins the Rasch and other unidimensional IRT models. According to Hattie (1985), however, no satisfactory method for assessing dimensionality had been put forward. At present, various techniques are being explored but there is still no single method for assessing unidimensionality.

As we shall see, early models dealt with the same response mechanisms as in IRT. Crucially, all of these generate the same specific data structure that is necessary for the IRT models that have since been developed.

Guttman Scaling

A positive response to a more "difficult" item on a Guttman (1950) scale predicts that all less-difficult items will result in positive responses. In a perfect Guttman scale, a subject's score indicates a positive result on all items whose difficulty is at or below the score and a negative result on all items whose difficulty exceeds the score. Showing that a set of items meets the assumptions of a Guttman scale provides considerable information about the meaning of particular scores. The Guttman model works best for constructs that are hierarchical and highly structured, such as ADL (activities of daily living) and severity of health conditions.

This basic mechanism, which underlies all IRT models, is clarified in the following example. Fig. 10.1 (top) shows the responses of seven patients on eight health conditions (A−H). Subsequently, this matrix is sorted. The patients who agree that all health conditions are preferred over their own health condition are listed at the top, and those agreeing with fewer statements are at the bottom. For patients with an equal number of agreements, the health conditions are sorted from left to right, starting with the conditions that most had agreed on and ending with the conditions that fewest had agreed on. After this sorting process, a perfect Guttman structure would look like a diagonal filled with check marks. Items (here, health conditions) and subjects (here, patients) can be scaled on a unifying scale if there are not too many misfits.

Mokken Scaling

Mokken analysis is a method of Guttman scaling that was developed in the Netherlands. Although originally intended for dichotomous items, it has been extended to handle ordinal data as well (Molenaar and Sijtsma, 1988). Mokken analysis employs several criteria to assess whether a set of items forms a Guttman scale. Of particular importance is the H coefficient, which indicates the extent to which the scale resembles a perfect Guttman scale. The standards for evaluating goodness of fit are as follows: an H coefficient of 0.30 is considered an acceptable but weak fit; an H coefficient of 0.40 is considered a medium or good fit; and an H coefficient of 0.50 or greater is considered a strong fit of the Guttman model to the data. Unlike deterministic approaches such as Guttman scaling, the Mokken model tests whether scalability (as measured by the H coefficient) is greater than would be expected by chance.

The Guttman scale and the Mokken scaling approach are both deterministic (nonparametric and therefore nonprobabilistic). The next section introduces the Rasch model, the most basic of all IRT models. There, we clarify

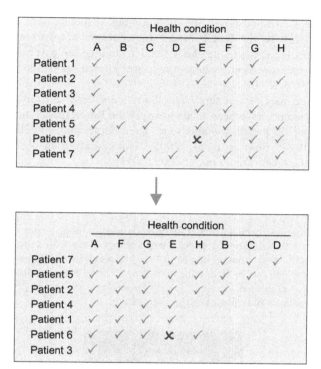

FIGURE 10.1 Representation of the raw data (top) and after sorting (bottom) the columns (health conditions) and the rows (patients) to arrive at the hierarchical Guttman/item response theory data structures (a check indicates that this health condition is preferred over the next health state, whereas a cross indicates a misfit). *From Arons, A.M.M., Krabbe, P.F.M., 2013. Probabilistic choice models in health-state valuation research: background, theories, assumptions and applications. Expert Review of Pharmacoeconomics and Outcomes Research 13 (1), 93–108.*

the close connection between Guttman and Rasch scaling with regard to the required data structure. Whereas the Guttmann model is deterministic, the Rasch model is probabilistic.

RASCH MODEL FOR BINARY ITEMS

The Rasch model—named after the Danish mathematician, statistician, and psychometrician Georg Rasch (1901–80)—is a probabilistic measurement or scaling model. Rasch had studied for 1 year (in 1935) under the famous statistician and biologist Sir Ronald Fisher. He went on to develop a measurement model with unique mathematical properties. Though commonly referred to as the Rasch model, it is also called the one-parameter logistic (1-PL) model. While primarily employed to assess attainment, it is increasingly used in other areas (Rasch, 1980). Its original setting was the field of reading skills, where it was designed for dichotomous response data (e.g., right/ wrong). Versions of the Rasch model are common in psychometrics, the discipline concerned with the theory and technique of psychological and educational measurement, where they are known as response models (Chapter 5).

Health-outcomes researchers have shown considerable interest in Rasch modeling. Attempts have been made to apply the Rasch model and other IRT models to specific health domains (e.g., pain, depression, mobility; Revicki and Cella, 1997; Holman et al., 2005; Schultz-Larsen et al., 2007; Ten Klooster et al., 2008; Young et al., 2009, 2011).

BOX 10.1

Georg Rasch was a Danish mathematician, statistician, and psychometrician, most famous for the development of a class of measurement models known as Rasch models. In 1919, Rasch began studying mathematics at the University of Copenhagen. He completed a masters degree in 1925 and received a doctorate in science in 1930. Unable to find work as a mathematician in the 1930s, he turned to work as a statistical consultant. In this capacity he worked on a range of problems, including problems of biological growth (Fig. 10.2).

FIGURE 10.2 Georg Rasch (1901–80).

Rasch Mechanism

Rasch developed measurement theory and his measurement model independently of research by others. According to Wright (a major proponent of the mechanism), "The Rasch model is so simple that its immediate relevance to contemporary measurement practice and its extensive possibilities for solving measurement problems may not be fully apparent" (Wright, 1977).

Rasch's starting point was not real data (practice) but an axiomatic definition of measurement (theory). He formulated a "model," actually an equation, fixing the "ideal" relationship between an observation and the "thing" to be measured. Let us consider three features of this relationship. First, the observed response (e.g., pass/yes/agree/right = 1, rather than fail/no/disagree/wrong = 0) depends on the difference between only two parameters: the "ability" of the individual and the "difficulty" of the item. No extraneous factors should bias this linear relationship. Second, the parameters "ability" and "difficulty" are independent of each other. (It should be pointed out that the invariance principle is also a theoretical requirement for measurement in the field of physics.) In his "separability theorem," Rasch showed that his measurement model is the only one that satisfies this requirement. Third, the model is probabilistic: uncertainty surrounds the expected response, a condition that is consistent with the real-world situation. The Rasch model is a mathematical function that relates the probability of a (correct) response on an item to characteristics of the person (one's ability) and to characteristics of the item (its difficulty).

Rasch developed the model for dichotomous data. He applied it to response data derived from intelligence and attainment tests, including data collected by the Danish military (Rasch, 1966). It does not confront the respondents with a paired comparison task or a ranking task (Chapter 11). Instead, the responses are collected separately (monadic) for a set of items. The most important claim of the Rasch model is that due to the operation of collecting response data, in combination with the conditional estimation procedure, the derived measures comply with the three most important measurement principles: interval level, unidimensionality, and invariance. Because it uses a specific mechanism that integrates a specific mode of response collection with a specific statistical analysis, the application of the Rasch model is sometimes referred to as fundamental or objective measurement.

This all means that the Rasch model makes stronger assumptions about the data than the other IRT models and may therefore not be appropriate for any given set of items. However, when the Rasch model is considered appropriate, it allows one to draw "specific objective comparisons."

Mathematics

The Rasch model, like other IRT models, assumes that if a more difficult item is affirmed, then there is a high probability that easier items will also be affirmed (Rasch, 1980). Furthermore, it assumes that if a person is more able (for example, to perform ADL activities), he or she has a higher probability of affirming any item compared to a less able person.

Like other IRT models, the Rasch model is based on statistical models. Central to its development is the idea that the probability of a correct response to an item is modeled as a logistic function of the difference between the difficulty of an item (parameterized by β) and certain characteristics of a person (e.g., ability or health; parameterized by θ). For a dichotomous item, the item characteristic function is the probability $P(\theta)$ of a correct response to the item.

$$P(\theta) = \frac{e^{(\theta-\beta_i)}}{1 + e^{(\theta-\beta_i)}} \qquad (10.1)$$

which can also be written as follows:

$$P(\theta) = \frac{1}{1 + e^{(\theta-\beta_i)}} \qquad (10.2)$$

Ideally, the difficulty of the items is evenly distributed along a broad continuum of disability. As the difficulty of items and ability of patients are combined in a common metric, we can accurately differentiate and measure disability among those who are severely disabled and those who are less so.

By an iterative conditional (maximum likelihood) estimation approach, a scale estimation is obtained without involvement of the person parameter, which is specific to the Rasch model. Therefore, Rasch models have a specific measurement property, namely invariance, which is a critical criterion of fundamental measurement. Obviously, this demands a strong specification of the structure (Guttman structure, Fig. 10.1) of the response data, a requirement that is not often satisfied. However, the Rasch and other IRT models are robust to minor violations, and no real data will ever correspond with the assumptions perfectly.

When the model only involves the estimation of the difficulty parameter, it is a Rasch model or one-parametric IRT model. (Of course, there are two parameters, θ and β, but because the person parameter θ is standard in all models, this parameter is not "counted.") The central assumption is that the discrimination parameter a (see in the following) is constant. The mathematical expression for the item response is called an "item response function," an "item characteristic curve," or a "trace line." The item characteristic curve for the Rasch model and the other IRT models is usually S-shaped (Fig. 10.3). A characteristic of the item characteristic curves for models with only one parameter (β) is that the lines are assumed to be parallel.

ORDERED ITEM MODELS (1-PL)

The dichotomous Rasch model has been extended to make it suitable for polytomous items (items that do not have binary response categories but do have more than two response categories) with the introduction of the threshold parameter τ for the kth response category (Eq. (10.3)). This threshold

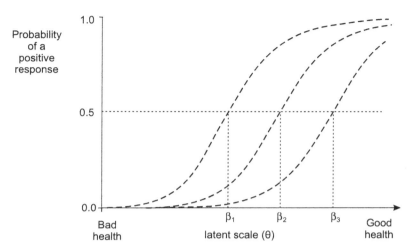

FIGURE 10.3 *Trace lines* Rasch model (location at 0.5 probability level determines the β parameter).

represents the probabilistic midpoint between two adjacent response categories (50% chance of choosing one over the other):

$$P_i(\theta) = \frac{e^{(\theta - \beta_i - \tau_k)}}{1 + e^{(\theta - \beta_i - \tau_k)}} \qquad (10.3)$$

Polytomously scored items are often highly advantageous because of the extra information (e.g., varying intensity or severity) provided about respondents. For that reason they are frequently used in patient-reported outcome measures. Polytomous response data (e.g., "not at all," "a little bit," "somewhat," "quite a bit," and "very much") would require polytomous (multiple category) IRT models to represent the nonlinear relationship between a respondent's trait level and the probability of responding in a particular response category.

Rating Scale Model

Andrich's rating scale model (RSM) is an extension of the dichotomous Rasch model. It is intended for polytomous items and assumes that the distances between each pair of ordinal response categories are the same across all items (Andrich, 1978a,b,c). Response categories are assigned threshold parameters that are equal across the items, and an item position along the measured continuum or location is described by a single-scale difficulty or location parameter (the average difficulty for a particular item relative to the category intersections).

Partial Credit Model

In contrast, distances between ordinal response categories are not assumed to be equal across all items in the partial credit model (PCM) developed by Masters in 1982. For example, the distance between the threshold (i.e., probabilistic midpoint) separating response categories 0 and 1 can differ from the threshold separating response categories 1 and 2 (Tennant and Conaghan, 2007). Masters originally developed his model to analyze test items that require multiple steps and for which partial credit should be assigned to complete several steps in the solution process (Masters, 1982). The PCM is also appropriate for analyzing attitude or personality scale responses where subjects rate their beliefs or respond to statements on a categorical scale. Because the PCM is an extension of the Rasch model, it assumes that all items are equally good at discriminating among persons. The PCM and RSM differ in that the PCM makes no assumptions about the relative difficulty of the steps (the categories) within any item. That is, the spread of the category threshold parameters or step values can differ across the items.

Invariance

Given the theoretical assumptions on which the Rasch model was conceptualized, it may be said to embody a fundamental aspect of measurement theory, namely invariance (Engelhard, 1992). Invariance of measurement has two implications: (1) estimates of individual characteristics (person parameter; e.g., ability or health status) as measured by the instrument are comparable regardless of which items are included in the instrument; and (2) estimates of the position (e.g., difficulty, severity) of the items on the scale of the instrument are comparable regardless of how the sample of respondents was selected. The potential for invariant parameters is one of the chief purported advantages of the Rasch model over classical test theory. Some researchers state that such potential is present in all IRT models, not just the Rasch model (McHorney and Monahan, 2004). Several other scaling models and their extensions (e.g., discrete choice models) are strongly related to the Rasch model, but in the absence of the person parameter they also lack the invariance characteristics.

Advantages

It should be noted that the Rasch model makes relatively strong assumptions. Nonetheless, if the assumptions hold sufficiently, this measurement model can produce scales offering a number of advantages over those derived by standard measurement techniques or even contemporary choice models (Chapter 11).

The Rasch model is unique among IRT models in that it provides what Rasch called "specific objectivity." This property is related to the fact that the summed score is a sufficient statistic in the Rasch models. Some theorists believe that specific objectivity is very important, while others believe it is less important than other aspects of IRT models. Specific objectivity and the claim of an interval scale of measurement are related to what Rasch called "fundamental measurement." However, some measurement theorists believe that neither the Rasch nor the other IRT models provide interval measurement of latent psychological constructs (McHorney and Monahan, 2004).

Another advantage of IRT is that parameter estimates are not sample dependent, as is the case with instruments developed under classical test theory. Properties of some items and scales (e.g., test–retest, Cronbach's α) that were developed under classical test theory are based on correlations computed on the sample. Different samples with different variances will not yield equivalent responses or data that can easily be compared across samples.

ITEM RESPONSE THEORY MODELS WITH MORE THAN ONE PARAMETER

Extensions of the Rasch model (i.e., other IRT models) relax somewhat the strong requirements posed on the response data for the Rasch or one-parameter IRT models. The extensions differ from the original model in the sense that they have an additional parameter to express the discrimination of an item (the degree to which the item discriminates between persons in different regions on the continuum of the scale). These IRT models relax to some extent the strict requirements for responses (e.g., data) posed by the Rasch model. But many researchers state that IRT models do not possess the specific fundamental measurement property of the Rasch model and therefore do not necessarily produce interval measures (Rasch, 1966; Engelhard, 1992).

A major distinction among the most popular dichotomous IRT models is the number of parameters used to describe the items. The three most popular IRT models are the one-, two-, and three-parameter logistic (1-PL, 2-PL, and 3-PL) models. The names denote the number of item parameters that each incorporates to characterize an item's functioning and, thus, that need to be estimated. All three models have an item difficulty, or threshold, parameter (β), which is the point on the curve where there is 50% probability that people with a specific value of θ will endorse the item (Fig. 10.3). The 2- and 3-PL models also specify a slope, or discrimination, parameter (a) that allows items to differentially discriminate among respondents (Fig. 10.4). As mentioned earlier, the 1-PL (Rasch) model assumes a scale in which all items have the same discriminating power (only the item threshold parameter may vary).

$$P(\theta) = \frac{1}{1 + e^{a_i(\theta - \beta_i)}} \qquad (10.4)$$

A commonly applied IRT model for binary items is the two-parameter logistic (2-PL) model (Eq. (10.4)). It is called a "two"-parameter model

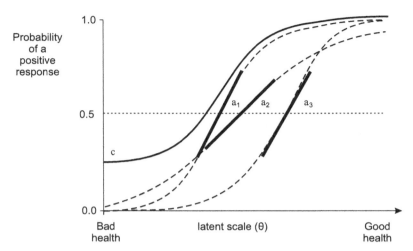

FIGURE 10.4 *Trace lines (broken lines)* for the two-parametric item response theory models (slope at 0.5 probability level determines the a-parameter; β parameters are not depicted but are the same as in Fig. 10.3), the *unbroken line* is an example of a trace line for a three-parametric model (*c* parameter for "guessing").

because items are allowed to differ in two ways: by discrimination (slope) and location (severity).

The 3-PL model has an additional parameter (*c*), also referred to as the pseudo-chance (guessing) parameter, that allows for respondents' random guessing of correct responses to multiple-choice items (Fig. 10.4). Although the guessing parameter can be challenging to interpret in health-outcomes measurement, it is potentially useful to understand atypical response patterns (Chang, 2005).

Graded Response Model

A popular and flexible 2-PL polytomous IRT model is the graded response model (GRM) (Samejima, 1969). The item characteristics that are estimated with the GRM are discrimination and threshold (location) parameters. The threshold parameters provide information on an item's difficulty or severity and locate it along the measured construct. The GRM is a direct extension of the two-parameter dichotomous model, and it assumes that item discrimination (parameter *a*) is not equal across all items. It also assumes that differences between each of the response categories are not the same across all items. The discrimination parameter indicates the strength of the relationship between an item and the measured construct.

Samejima's GRM and its extensions were proposed for the analysis of multiple-item tests when responses are graded on an ordinal scale or when they can be characterized as ordered categorical responses, such as exist in Likert-type rating scales (Samejima, 1969, 1972).

ITEM RESPONSE THEORY STATISTICS

Because IRT uses mathematical models to relate health levels to item responses, it is possible to evaluate model fit at the scale, item, or person level. This idea is represented in a sizable number of person-fit indices (scalability, appropriateness) that assess the degree to which an individual's item response pattern is consistent with an IRT model (Meijer and Sijtsma, 2001; Reise and Waller, 2009).

IRT measurement models are considered more sophisticated than classical test models in that IRT analyses allow the following: (1) to estimate performance on each item for any respondent who is located at a particular point on the underlying health scale; (2) to make more precise measures by estimating the maximum discrimination of an item at a particular value of the health outcome; (3) to estimate error of measurement at each health level; and (4) to make estimates that are not sample dependent (Crocker and Algina, 1986).

As with any use of mathematical models, it is important to assess how well the data fit the model. If item misfit is diagnosed as due to poor item quality, then the unsuitable items may be removed or replaced. If, however, a large number of misfitting items occur for no apparent reason, the construct validity of the scale will need to be reconsidered. Thus, misfit provides invaluable diagnostic clues for developers, allowing the hypotheses upon which specifications are based to be empirically tested against data.

Items should not be removed simply because they misfit the model. A relevant reason must have been diagnosed, such as a nonnative speaker of English grappling with items from a health-outcome instrument written in highbrow British English.

Information

One of the major contributions of IRT is the extension of the concept of reliability. Traditionally, reliability refers to the reproducibility of measurement (i.e., the degree to which measurement is free of error). Usually, it is measured using a single coefficient defined in various ways, such as the ratio of true and observed score variance. This coefficient is helpful in characterizing the average reliability of a scale. IRT makes it clear that precision is not uniform across the entire scale. Scores at the edges of the range, for example, generally have more errors associated with them than scores closer to the middle of the scale.

In IRT, scale reliability is replaced by item and scale information. These concepts are expressed as the information function that indicates the range over the construct being measured for which an item or scale is most useful for discriminating among individuals. The information function characterizes the precision of measurement (i.e., reliability) for persons at different levels of the underlying health outcome, with higher information denoting more precision. It should be noted that the information functions for the items (i.e., item function) on the same scale sum to produce the information function for the whole scale (i.e., information). Unlike classical expressions of reliability (e.g., test−retest, Cronbach's α), IRT information functions are not single co-efficients but functions showing how "reliably" an instrument measures persons at each possible value of the scale.

Information is also a function of the model parameters. For example, according to Fisher information theory, the item information supplied in the case of the 1-PL for dichotomous response data is simply the probability of a correct response multiplied by the probability of an incorrect response, or as follows:

$$I(\theta) = P_i(\theta)q_i(\theta). \tag{10.5}$$

The standard error of estimation is the reciprocal of the instrument information at a given scale level, or as follows:

$$SE(\theta) = \frac{1}{\sqrt{I(\theta)}}. \tag{10.6}$$

Differential Item Functioning

Another major improvement of IRT models is DIF. Its contribution is particularly evident when developing scoring algorithms to be used across different subgroups: for instance, cultures, age groups, or residential settings (Lutomski et al., 2016). We speak of DIF when an item function differs for two different groups of respondents (Camilli and Shepard, 1994). In other words, respondents who show similar levels on a specific health outcome but belong to different identifiable subgroups (e.g., male vs female; Hispanic vs non-Hispanic; ill vs healthy) have a different probability of responding to an item. Items exhibiting DIF across groups pose a serious threat to the validity of a measure and may have reduced validity for between-group comparisons. The reason is that their scores may be indicative of attributes other than those the scale is intended to measure. Such items can result in one group showing higher scores than another, merely because of their group membership.

For example, to make quantitative and valid comparisons cross-culturally, it is vital to establish the cross-cultural equivalence of the translated instruments. Knowing the extent to which items in a measure perform similarly across linguistic, racial, or cultural lines is critical when determining whether data collected with translated instruments can be used as unbiased grounds for comparing clinically relevant groups.

To give another example, an item about pain may not evoke comparable information from different categories of people. A variety of culture-based (e.g., beliefs about the responsiveness of health providers to one's complaints), age-based (e.g., expectations about the inevitability and normalcy of aches and pains), or gender-based (e.g., perceptions of how acceptable it is to acknowledge pain) factors might evoke different scale scores from people with equivalent subjective experiences.

Item-parameter invariance (or equivalence) makes IRT modeling a powerful tool for detecting DIF because the psychometric properties of items (e.g., threshold and discrimination parameters) do not change as a function of the sample. Under the IRT framework, DIF analysis comes down to comparing the results derived from fitting the IRT model to each group separately versus fitting it to both groups jointly. In fact, DIF is nothing more than testing whether the invariance principle holds. As such, it is a very important check. Differences between item-parameter estimates suggest the possibility of finding DIF items. Results exhibiting DIF should be reviewed by content experts who can choose between removing the items or revising their wording to make them cross-culturally sensitive. DIF analyses also can help explain the physiological or psychological processes underlying group differences in responding to particular items.

CONTROVERSY

It seems to be a standard practice in science to choose a particular statistical model if it accounts better than another one for the data. In general, the more parameters a model contains, the better it accounts for the data. But if available statistical checks show that it does not, a model with fewer parameters is favored. This choice pattern is part of a widespread research paradigm. In the traditional paradigm, it is simply taken for granted that the model of choice is the one that fits the data ("give me the data and I will model it"). This may be an appropriate and defensible approach in many research situations where the goal is to predict, but perhaps not when we want to measure. The Rasch paradigm takes the opposite perspective. In it, the model provides an operational criterion for fundamental measurement, so the data must fit the model. Indeed, the goal of the Rasch model is to construct procedures or operations that provide data that meet the relevant criteria (Andrich, 2004). Therefore, according to some measurement theorists, the Rasch model allows for truly objective (fundamental) measurement. When misfit is identified, Rasch proponents will not look for another model. Many other researchers believe that it is essential to use realistic models unless it can be shown that a simple (one-parameter) model, such as the Rasch model, will provide adequate fit. In short, proponents of IRT prioritize the data, whereas proponents of Rasch measurement prioritize the model.

INSTRUMENTS

In the area of health-outcome research, there is considerable interest in Rasch modeling (Revicky and Cella, 1997; Andrich, 2004). Health researchers have started to apply the Rasch model or one of the two-parameter models as an alternative to conventional descriptive measurement (classical test theory) of health. Recent attempts have been made to apply the Rasch model to measure specific health domains (e.g., pain, depression, mobility; Noerholm et al., 2004; Lindeboom et al., 2004).

An interesting example is found in a study by Thompson et al. (2013). Building on and extending the analysis of preexisting scales, they developed a scale for use in prion disease, a group of rare neurodegenerative conditions for which no proven disease-modifying treatment is available. Prions are misfolded proteins that replicate by converting their properly folded counterparts, in their host, to the same misfolded structure they possess. A case in point is Creutzfeldt–Jakob disease (CJD). CJD causes brain tissue to degenerate rapidly, and as the disease destroys the brain, the brain develops holes and the texture changes.

In their approach, the partial credit variant of the polytomous Rasch model was chosen to reflect multiple successively ordered response categories of varying difficulty. The analysis included 437 participants (311 patients died during the study). The final scale consisted of five domains and in total 11 items (with response categories from 2 through 5). Fig. 10.5 illustrates the typical progression of disease in a patient with sporadic CJD through the functional/cognitive milestones of the scale. It shows the spread of difficulty of items/(categories) in the five measured domains. This diagram illustrates the validity of the progression construct, as the ordering of the items is consistent with clinical experience and there is a reasonable spread of item difficulty in different functional domains.

APPLICATION

Rheumatoid Arthritis (RA) is a systemic autoimmune disease characterized by chronic joint inflammation, pain, fatigue, disability, and a risk for joint damage. Its course ranges from mild to very severe and disabling. Worldwide, RA has been estimated to affect about 1% of the population. It can be disabling and interfere with participation in daily activities. Therefore, the therapeutic aims are to inhibit disease activity, pain, and stiffness; to prevent joint damage; to preserve muscle strength and condition; and to maintain daily activities and social participation. In recent years, the diagnosis and therapy for RA has developed toward early diagnosis followed by early and aggressive drug therapy, which has led to a more favorable course of the disease and less impact on functioning.

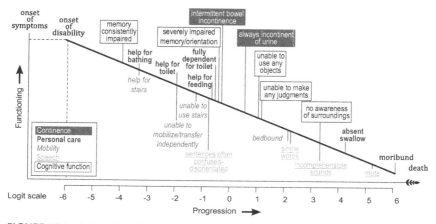

FIGURE 10.5   Schematic of the pattern of decline in a patient with sporadic Creutzfeldt–Jakob disease that would be most consistent with the Rasch (one-parametric IRT) model. Progression is represented by the logit scale used in the Rasch analysis reflecting the relative difficulties of the thresholds that comprise the scale. *By permission of Oxford University Press from Thompson, A.G.B., Lowe, J., Fox, Z., Lukic, A., Porter, M.C., Ford, L., Gorham, M., Gopalakrishnan, G.S., Rudge, P., Walker, A.S., Collinge, J., Mead, S., 2013. The Medical Research Council prion disease rating scale: a new outcome measure for prion disease therapeutic trials developed and validated using systematic observational studies. Brain 136 (4), 1116–1127.*

Health Assessment Questionnaire Disability Index

Effectiveness of treatment strategies can be measured using various instruments, one of which is the Health Assessment Questionnaire Disability Index (HAQ or HAQ-DI). This instrument was first published in 1980 by Fries et al. as a self-reported, patient-oriented outcome measure for use in RA. The HAQ-DI can be used independently in various disabling health conditions (including RA) and is valued for its easy application and high correlation to relevant physiological and biochemical measures. Nowadays, the HAQ-DI is one of the most frequently applied instruments in clinical trials and observational studies in rheumatology for assessment of functional status (Bruce and Fries, 2003a,b).

The HAQ-DI comprises 20 items about a patient's ability to complete everyday tasks, with higher scores representing higher levels of disability. The items are divided over eight categories related to daily activities. For every category, two or three questions are posed in the form of "are you able to…?" There are four response options: without any difficulty (0 points); with some difficulty (1 point); with much difficulty (2 points); or unable to do (3 points). A stepwise calculation method is used to derive an overall score. This calculation, however, does not seem optimal, as concluded by Tennant et al. (1996). And from a measurement perspective, the scoring algorithm of the HAQ-DI has some limitations.

First, the items are divided into eight categories (i.e., domains). The highest component score in each category determines the score for the category, unless aids or devices are required. Dependence on equipment or physical assistance raises a lower score to a score of 2, thereby more accurately representing the underlying disability. The eight category scores are averaged into an overall HAQ-DI score on a scale from 0 (no disability) to 3 (completely disabled) (Bruce and Fries, 2003a,b). However, this division into eight categories may be suboptimal, in view of the well-documented fact that the 20 items of the HAQ-DI represent a unidimensional structure (Tennant et al., 1996; Cole et al., 2005; Ten Klooster et al., 2008).

Second, the overall score of the HAQ-DI and the scoring for the response options of each item separately are considered to lie on a continuous scale (indicating that a score of 2 represents twice as much disability as a score of 1). It would be more appropriate to consider the response options as ordinal (with a score of 2 merely representing "more" disability than a score of 1).

The third limitation—an aspect often neglected in clinical measurement instruments—is that items may vary in their difficulty. Given this variance, some of the items should be assigned more weight when calculating the overall score.

Item Response Theory

Rasch analysis has been performed on HAQ-DI data in some other studies (Tennant et al., 1996; Wolfe, 2001; Wolfe et al., 2004; Ten Klooster et al., 2008) in an effort to improve the construct validity. Although these models offer more sophisticated approaches to measurement, the application of the HAD-DI in practice remains the same. Although the analytical part might be more complex, current developments in terms of computer applications will be able to resolve any problems with this. One of the first papers that reported fitting HAQ data to the Rasch model was performed by Tennant (1996). About two decades later, the researchers of the current study investigated whether measurement approaches based on IRT improve the scoring precision of the HAQ-DI as compared to standard practice (Vermeulen et al., 2015). Given the improvement in therapeutic regimens, compared to the 1980s when the instrument was first published, special attention is now given to improving the reliability of measurement for patients with lower levels of disability.

Patients

The Department of Rheumatology at the University Medical Center St Radboud and the Maartenskliniek Nijmegen maintain a long-term observational cohort study of early RA. This inception cohort continuously includes all patients with newly diagnosed RA who have attended the rheumatology clinic at Radboud University Nijmegen Medical Center since 1985. All patients who meet the criteria of RA according to the American College of Rheumatology, whose disease has had a duration of less than 1 year, and who have had no prior use of disease-modifying antirheumatic drugs are asked to take part in the cohort study. One of the purposes of the project is to establish a database for validation of outcome measures (Welsing and van Riel, 2004). For this study, data have been analyzed from the first available measurement of all included RA patients with an available HAQ-DI ($n = 768$) in the period 2000–06 of the RA cohort.

The standard calculation method for the HAQ-DI score (HAQ-DI$_{standard}$) was compared to two alternatives. The first alternative, summing the individual item scores and taking the mean, is based on classical test theory (HAQ-DI$_{CTT}$). The second one is based on the IRT/Rasch (HAQ-DI$_{IRT}$).

Support for the assumption of unidimensionality was sought with factor analysis (maximum likelihood extraction). Overall fit of the data to the Rasch model was assessed with the inlier-pattern-sensitive fit statistic (infit) and outlier-sensitive fit statistic (outfit). These fit statistics are described by a mean square with an expected value of 1.0 (Bond and Fox, 2001). The data were accepted if values were between 0.7 and 1.3.

Reliability of the data was assessed with person separation reliability and item separation reliability, which are bounded by 0 and 1 (Wright and Masters, 1982; Bond and Fox, 2001). These estimates verify whether the models provide the same measures for the patients' ability and level of disability represented in items when based on a subset of items, for example, a subset of patients.

Results

Patients had a mean age of 57.9 years; 261 were male (34%) and 507 female (66%). The mean score as calculated with the standard method (HAQ-DI$_{standard}$) differed from the score based on classical test theory (HAQ-DI$_{CTT}$) by 0.25 points. The highest score for HAQ-DI$_{standard}$ was 3.0 versus 2.8 for HAQ-DI$_{CTT}$. For the HAQ-DI$_{standard}$, the frequencies of scores from 0 to 1 were rather evenly distributed. For the HAQ-DI$_{CTT}$, in contrast, the frequencies showed a downward slope from the beginning (Fig. 10.6). Overall, the distributions were skewed and showed large standard deviations.

The HAQ-DI$_{IRT}$ clearly showed a unidimensional structure. For the overall model, the infit was 1.01 and the outfit was 0.98. For the separate items, only item 11 (take tub bath) did not fit the model; the infit for this item was 1.41 and the outfit was 1.51. The person separation reliability and the item separation reliability were 0.91 and 0.99, respectively. The distribution of the disability levels (Fig. 10.6) showed a high number of patients that scored 0.0 on all items and consequently were considered to have no disability at all ($n = 118$). Except for this extraordinarily high number of patients without disability, the Rasch model approximates a normal distribution, as opposed to the

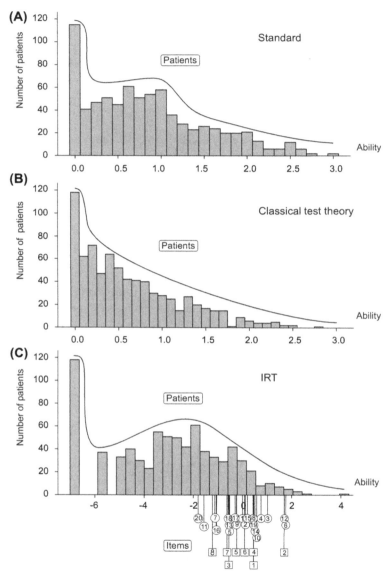

FIGURE 10.6 Disability of the patients measured with the HAQ-DI according to (A) standard scoring; (B) classical test theory; (C) IRT. For the IRT solution, besides the disability of the patients, the difficulty of the items is also estimated (circles stand for the 20 items, squares for the 8 HAQ-DI categories). *HAQ-DI*, Health Assessment Questionnaire Disability Index; *IRT*, item response theory.

HAQ-DI$_{standard}$ and HAQ-DI$_{CTT}$. This implies that more patients were classified as having a mild disability, whereas they would be classified as having (close to) no disability with the HAQ-DI$_{standard}$ and HAQ-DI$_{CTT}$. The HAQ-DI$_{standard}$ is confined to an established set of scores due to its calculation method (25 steps in the range of 0–3). The distribution of bars of the HAQ-DI$_{CTT}$ was arbitrarily set to steps of 0.10. Since the broader range of the Rasch models means there are more possible scores, there is also more discrimination in scores and therefore more differentiation in disability of patients.

The IRT model showed excellent fit and differentiated patients' abilities better than the standard HAQ-DI scoring. Given the position that patients' ability takes in the IRT model (mostly on the negative left-hand side of the scale), the conclusion is that the HAQ-DI is not aimed at patients with little disability. From a clinical perspective, the differences compared to standard HAQ-DI scoring were relatively small in terms of additional relevant information.

DISCUSSION

Most health-status instruments encompass multiple health domains and are therefore commonly used to produce "profile" measures. Most, if not all, of these instruments have been developed under the classical test theory measurement framework (Chapter 9). In this chapter, we showed that if the main interest is in the measurement of specific single and unidimensional health-status concepts, then IRT has much to offer. Of course, an IRT instrument can also be devised analogous to a multidomain profile instrument. In that case, a separate IRT scale has to be developed for each domain.

If IRT provides so many useful tools for evaluating and developing instruments, why is it not used more widely? There are several obstacles. First, most researchers have been trained in classical test theory, are comfortable interpreting these statistics, and can easily apply readily available software to generate familiar summary statistics such as Cronbach's α. In contrast, IRT models require advanced knowledge of measurement theory to understand their mathematical complexity, to check that assumptions are met, and to choose the appropriate model (Reeve and Fayers, 2005). On the other hand, the IRT approach has the advantage that certain measurement conditions and response structures can be evaluated. All too often, classical methods, such as the classical test theory approach, ignore these problems and uncritically use sum-scoring for all scales.

IRT has shown its potential for the development and characterization of outcomes for clinical trials. Notably, it provides a statistical model of how/why individuals respond as they do to an item. Characterizations derived from classical test theory pertain only to total scores and are specific to the sample from which they are derived. In contrast, IRT-derived characterizations, their constituent items, and individual scores are general for the entire population of items or individuals. This is another feature of IRT methods that makes them highly attractive in clinical settings. Furthermore, under IRT, the reliability of an outcome measure has a different, but more informative meaning than under classical test theory.

Ordinal models (polytomous items with ordered categories) appear to be less robust than simpler models. This is especially noticeable when the ordinal models are two-parameter models in which the item slopes (discrimination) and item difficulties (category thresholds) must be estimated. Ordinal models seem less robust in the sense that the estimation of the larger number of parameters implies greater dependence on the assumptions of the model. Accordingly, the computational algorithms may provide unstable estimates or even fail to converge. It has been commented that the large number of parameters and more complicated structure may make models more realistic. But this benefit comes at a price: with less available data per parameter, the parameter estimates may show serious instability. Thus, much larger sample sizes may be required when estimating IRT models with multiple response categories. In that light, it is important to ensure that the subjects in the data set adequately cover the full set of response options. Many IRT handbooks provide crude rules of thumb based on practical experience. Most authors recommend including hundreds of patients for simple models, and even more for models that estimate numerous parameters. Certainly the two-parameter IRT models require a relatively large number of respondents. A large sample is all the more necessary if the number of categories for the items is expanded to, let us say, five or more. By increasing the sample size and number of items, errors of estimation apparently can be made small enough to solve many practical problems in IRT (Hulin et al., 1983). This exposition on sample size makes clear that it is only feasible to develop instruments based on IRT in settings where a substantial number of patients can be approached. Unfortunately, in many cases it would therefore not be an option to develop disease-specific IRT scales for rare diseases.

REFERENCES

Andrich, D., 1978a. A rating formulation for ordered response categories. Psychometrika 43 (4), 561–573.

Andrich, D., 1978b. Scaling attitude items constructed and scored in the Likert tradition. Educational and Psychological Measurement 38, 665–680.

Andrich, D., 1978c. Application of a psychometric rating model to ordered categories which are scored with successive integers. Applied Psychological Measurement 2 (4), 581–594.

Andrich, D., 2004. Controversy and the Rasch model: a characteristic of incompatible paradigms? Medical Care 42 (Suppl. 1.), I7–I16.

Arons, A.M.M., Krabbe, P.F.M., 2013. Probabilistic choice models in health-state valuation research: background, theories, assumptions and applications. Expert Review of Pharmacoeconomics and Outcomes Research 13 (1), 93−108.

Birnbaum, A., 1968. Part 5: some latent trait models and their use in inferring an examinee's ability. In: Lord, F.M., Novick, M.R. (Eds.), Statistical Theories of Mental Test Scores. Addison-Wesley, Reading M.A., pp. 397−479

Bock, R.D., 1997. A brief history of item response theory. Educational Measurement: Issues and Practice 16 (4), 21−33.

Bond, T.G., Fox, C.M., 2001. Applying the Rasch Model: Fundamental Measurement in the Human Sciences. Lawrence Erlbaum Associates, Mahwah, NJ.

Bruce, B., Fries, J.F., 2003a. The Stanford Health Assessment Questionnaire: a review of its history, issues, progress, and documentation. Journal of Rheumatology 30 (1), 167−178.

Bruce, B., Fries, J.F., 2003b. The Stanford Health Assessment Questionnaire: dimensions and practical applications. Health and Quality of Life Outcomes 1 (1), 20.

Camilli, G., Shephard, L., 1994. Methods of Identifying Biased Test Items. SAGE, Thousand Oaks C.A.

Chang, C.H., 2005. Item response theory. In: Lenderking, W.R., Revicki, D.A. (Eds.), Advancing Health Outcomes Research Methods and Clinical Applications. International Society for Quality of Life Research. McLean V.A., pp. 37−55

Cole, J.C., Motivala, S.J., Khanna, D., Lee, J.Y., Paulus, H.E., Irwin, M.R., 2005. Validation of single-factor structure and scoring protocol for the Health Assessment Questionnaire-Disability Index. Arthritis and Rheumatism 53 (4), 536−542.

Crocker, L., Algina, J., 1986. Introduction to Classical and Modern Test Theory. Holt, Rinehart and Winston, New York.

Engelhard Jr., G., 1992. Historical views of invariance: evidence from the measurement theories of Thorndike, Thurstone, and Rasch. Educational and Psychological Measurement 52 (2), 275−291.

Engelhard Jr., G., 2013. Invariant Measurement: Using Rasch Models in Social, Behavioral, and Health Sciences. Routledge, New York.

Fries, J.F., Spitz, P., Kraines, R.G., Holman, H.R., 1980. Measurement of patient outcome in arthritis. Arthritis & Rheumatism 23 (2), 137−145.

Guttman, L., 1950. The basis for scalogram analysis. In: Stouffer, S.A., Guttman, L., Suchman, E.A., Lazarsfeld, P.F., Star, S.A., Clausen, J.L. (Eds.), The American Soldier: Measurement and Prediction, vol. IV. Princeton University Press, Princeton, pp. 60−90.

Hattie, J., 1985. Methodology review: assessing unidimensionality of tests and items. Applied Psychological Measurement 9 (2), 139−164.

Holman, R., Weisscher, N., Glas, C.A., Dijkgraaf, M.G.W., Vermeulen, M., de Haan, R.J., Lindeboom, R., 2005. The Academic Medical Center Linear Disability Score (ALDS) item bank: item response theory analysis in a mixed patient population. Health and Quality of Life Outcomes 3 (1), 83.

Hulin, C.L., Drasgow, F., Parsons, C.K., 1983. Item Response Theory: Application to Psychological Measurement. Dow Jones-Erwin, Homewood.

Lindeboom, R., Holman, R., Dijkgraaf, M.G., Sprangers, M.A., Buskens, E., Diederiks, J.P., de Haan, R.J., 2004. Scaling the sickness impact profile using item response theory: an exploration of linearity, adaptive use, and patient driven item weights. Journal of Clinical Epidemiology 57 (1), 66−74.

Lutomski, J.E., Krabbe, P.F.M., den Elzen, W.P.J., Olde-Rikkert, M.G.M., Steyerberg, E.W., Muntinga, M.E., Blijenberg, N., Kempen, G.I.J.M., Melis, R.J.F., on behalf of TOPICS Consortium, 2016. Rasch analysis reveals comparative analyses of ADL/IADL summary scores from different residential settings is inappropriate. Journal of Clinical Epidemiology 74, 207–217.

Masters, G.N., 1982. A Rasch model for partial credit scoring. Psychometrika 47 (2), 149–174.

McHorney, C.A., Monahan, P.O., 2004. Postscript: applications of Rasch analysis in health care. Medical Care 42 (1), I73–I78.

Meijer, R.R., Sijtsma, K., 2001. Methodology review: evaluating person fit. Applied Psychological Measurement 25 (2), 107–135.

Molenaar, I.W., Sijtsma, K., 1988. Mokken's approach to reliability estimation extended to multicategory items. Kwantitatieve Methoden 9 (28), 115–126.

Noerholm, V., Groenvold, M., Watt, T., Bjorner, J.B., Rasmussen, N.A., Bech, P., 2004. Quality of life in the Danish general population-normative data and validity of WHOQOL-BREF using Rasch and item response theory models. Quality of Life Research 13 (2), 531–540.

Rasch, G., 1966. An item analysis which takes individual differences into account. British Journal of Mathematical and Statistical Psychology 19 (1), 49–57.

Rasch, G., 1980. Probabilistic Models for Some Intelligence and Attainment Tests (Copenhagen, Danish Institute for Educational Research), expanded edition (1980) with foreword and afterword by B.D. Wright. University of Chicago Press, Chicago.

Reeve, B.B., Fayers, P., 2005. Applying item response theory modelling for evaluating questionnaire item and scale properties. In: Fayers, P., Hays, R. (Eds.), Assessing Quality of Life in Clinical Trials, second ed. Oxford University Press, New York, pp. 55–73.

Reise, S.P., Waller, N.G., 2009. Item response theory and clinical measurement. Annual Review of Clinical Psychology 5, 27–48.

Revicki, D.A., Cella, D.F., 1997. Health status assessment for the twenty-first century: item response theory, item banking and computer adaptive testing. Quality of Life Research 6 (6), 595–600.

Samejima, F., 1969. Estimation of latent ability using a response pattern of graded scores. Psychometrika 17, 1–100.

Samejima, F., 1972. A general model for free-response data. Psychometrika 37, 1–68.

Schultz-Larsen, K., Kreiner, S., Lomholt, R.K., 2007. Mini-Mental Status Examination: mixed Rasch model item analysis derived two different cognitive dimensions of the MMSE. Journal of Clinical Epidemiology 60 (3), 268–279.

Ten Klooster, P.M., Taal, E., van de Laar, M.A., 2008. Rasch analysis of the Dutch Health Assessment Questionnaire disability index and the Health Assessment Questionnaire II in patients with rheumatoid arthritis. Arthritis and Rheumatism 59 (12), 1721–1728.

Tennant, A., Conaghan, P.G., 2007. The Rasch measurement model in rheumatology: what is it and why use it? when should it be applied, and what should one look for in a Rasch paper? Arthritis Care & Research 57 (8), 1358–1362.

Tennant, A., Hillman, M., Fear, J., Pickering, A., Chamberlain, M.A., 1996. Are we making the most of the Stanford Health Assessment Questionnaire? British Journal of Rheumatology 35 (6), 574–578.

Thompson, A.G.B., Lowe, J., Fox, Z., Lukic, A., Porter, M.C., Ford, L., Gorham, M., Gopalakrishnan, G.S., Rudge, P., Walker, A.S., Collinge, J., Mead, S., 2013. The Medical Research Council prion disease rating scale: a new outcome measure for prion disease therapeutic trials developed and validated using systematic observational studies. Brain 136 (4), 1116–1127.

Thurstone, L.L., 1925. A method of scaling psychological and educational tests. Journal of Educational Psychology 16 (7), 433–451.

van der Linden, W.J., Hambleton, R.K., 2010. Handbook of Modern Item Response Theory. Springer, New York.

Vermeulen, K.M., Mulder, B., Fransen, J., den Broeder, A.A., Krabbe, P.F.M., 2015. Use of the HAQ-DI in Patients With Rheumatoid Arthritis and Low Levels of Disability: The Value of Alternative Measurement Approaches (Report).

Welsing, P.M., van Riel, P.L., 2004. The Nijmegen inception cohort study of early rheumatoid arthritis. The Journal of Rheumatology 69 (Supplement), 14—21.

Wolfe, F., 2001. Which HAQ is best? A comparison of the HAQ, MHAQ and RA-HAQ, a difficult 8 item HAQ (DHAQ), and a rescored 20 item HAQ (HAQ20): analyses in 2,491 rheumatoid arthritis patients following leflunomide initiation. Journal of Rheumatology 28 (5), 982—989.

Wolfe, F., Michaud, K., Pincus, T., 2004. Development and validation of the Health Assessment Questionnaire II: a revised version of the Health Assessment Questionnaire. Arthritis and Rheumatism 50 (10), 3296—3305.

Wright, B.D., 1977. Solving measurement problems with the Rasch model. Journal of Educational Measurement 14 (2), 97—116.

Wright, B.D., Masters, G.N., 1982. Rating Scale Analysis: Rasch Measurement. MESA Press, Chicago, IL.

Young, T., Yang, Y., Brazier, J.E., Tsuchiya, A., Coyne, K., 2009. The first stage of developing preference-based measures: constructing a health-state classification using Rasch analysis. Quality of Life Research 18 (2), 253—265.

Young, T.A., Yang, Y., Brazier, J.E., Tsuchiya, A., 2011. The use of Rasch analysis in reducing a large condition-specific instrument for preference valuation: the case of moving from AQLQ to AQL-5D. Medical Decision Making 31 (1), 195—210.

Chapter 11

Choice Models

Chapter Outline

INTRODUCTION

Several of the measurement models or scaling methods that will be discussed in this chapter were developed in the field of psychometrics. The main reason to treat them here, and not along with classical test theory (covered in Chapter 9), is that they are based on procedures that cannot be found under classical test theory (part of psychometrics), item response theory (Chapter 10), or clinimetrics (Chapter 13). Another reason to devote a separate chapter to these models/methods is that the field of psychological measurement has diverged. With growing specialization, many researchers who were trained in psychology are unfamiliar with psychometrics; although some are acquainted with it, they are not necessarily familiar with special areas in psychometrics. In particular, Thurstonian scaling, multidimensional scaling, and generalizability theory are not widely known; specialization arises even within a specialized field. These circumstances may explain why researchers who are deeply involved in developing and exploring conventional multiitem profile instruments are largely unaware of forceful measurement methods such as Thurstone's paired comparison model. Yet another reason to discuss choice measurement methods in a separate chapter is that extensions of the original psychometric methods have been introduced by researchers working outside psychometrics. Particularly relevant is the work of the econometrician McFadden (2001), who has elaborated the classical Thurstone model.

In the field of health economics, various measurement methods serve largely one and the same purpose. However, because economic methods are based on different arguments and theories, these will be discussed elsewhere (Chapter 12). Health economists tend to call their methods "valuation techniques." Conventional methods for valuing health states stemming from economics (e.g., standard gamble, time trade-off) are complex, require abstract reasoning skills, and are prone to various biases (e.g., adaptation, time preference, risk aversion; Froberg and Kane, 1989; Gafni, 1992; Attema et al., 2013; Versteegh et al., 2013).

PREFERENCE BASED

This chapter and Chapter 12 are about a class of measurement methods with a common goal, namely to express a person's overall health condition or health status in a single number. Apart from their theoretical and methodological differences (this chapter and Chapter 12), preference-based methods share an underlying assumption. They are all based on the general notion that individuals possess implicit preferences for health conditions that range from good to bad and that in principle it should be possible to reveal these preferences and express them quantitatively as numbers.

The term "preference" generally denotes the (relative) "desirability" of something or someone (Chapter 5). Within a preference-based measurement framework, distinct health attributes are assigned weights. These are produced by specific measurement strategies that elucidate the relative importance of attributes. Unlike the conventional instruments that have been developed under classical test theory (Chapter 9), preference-based measurement does not concern the frequency or intensity of complaints in any given health domain.

Preference-based measurement can be highly relevant to many applications, where it is used to express multiple health attributes in a single metric. The methodology associated with translating the various attributes and components of health into a number that indicates the quality of health is pretty complex, which may explain why preference-based measurement is not applied as often as other methods.

Comparative Element

The basic element of subjective measurement in a preference-based framework is a simple and straightforward response task comprising a comparison between at least two objects (e.g., descriptions, photos, scenarios). The preference-based methods (the choice models discussed in this chapter and the valuation techniques in the next one) require a comparative element in the judgmental task of the measurement procedure. In a well-grounded measurement model, the comparison is made in such a way that it yields data that contain compelling information.

Compound Objects

Another typical feature of preference-based measurement is that all relevant characteristics of the object of study have to be evaluated together. In the case of health, this implies evaluating all relevant health attributes in combination. A health-state description is presented as a set of attributes, and together this compound description has to be evaluated by the respondent. In this way, the evaluation is different from evaluation under classical test theory (Chapter 9), which deals with monadic measurement. There, items are evaluated and scored one by one.

Another type of evaluation in choice models requires respondents to judge holistic objects such as people's faces or skin in photos (Fig. 11.1). In general,

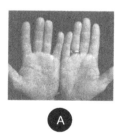
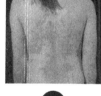

- Severe problems in walking about
- Moderate problems washing or dressing myself
- Unable to do my usual activities
- Moderate pain or discomfort
- Severely anxious or depressed

- Moderate problems in walking about
- Slight problems washing or dressing myself
- No problems doing my usual activities
- Slight pain or discomfort
- Extremely anxious or depressed

FIGURE 11.1   Examples of pairs of holistic (top) and multiattribute compound (below) objects in paired comparisons. The multiattribute example is based on the EQ-5D-5L health classification (see also Chapter 12).

holistic objects such as these are extremely easy to compare and judge. However, features (attributes) by which to describe the object specifically are often absent, so such objects cannot be analyzed in detail. The original paired comparison model of Thurstone was based on simple statements or holistic objects (Krabbe, 2008). In such a model, the objects that are part of the paired comparison tasks are scaled, but no weights are derived for specific features (attributes) of the objects (because such attributes are not specified or are unknown).

The need to evaluate whole set of attributes is one of the key impediments to preference-based measurement, as it may make the tasks more difficult. Another constraint is that there is a limit to the number of attributes that a description can contain. Typically, people can discriminate seven (plus or minus two) pieces of information at a time, so the number of attributes in most preference-based studies is usually no more than nine (Miller, 1956). Miller's study was based on simple unidimensional stimuli (sound, colors) that differed from one another in only one respect. Therefore, to extrapolate these findings to the more complex setting of health may not be entirely appropriate. Nevertheless, Miller's study demonstrates that the span of immediate memory poses severe limitations on the amount of information that people are able to receive, process, and remember. By extension, there is a practical constraint on the number of attributes that may be included within a health description: it is unlikely that respondents would be able to handle a very large number of attributes when undertaking compound judgmental tasks.

Another typical feature of preference-based measurement is that respondents are not asked to score items or attributes. The attributes are presented to the respondents, who are expected to read and mentally process these. Subsequently, they are asked to compare the whole set of attributes covered in a description to those couched in another description. Next, a preference for one of the descriptions is evoked.

There are approaches that circumvent the need to make compound judgments. Thus, it is not always necessary to judge each health attribute separately and then mentally add the distinct weights of the attributes to arrive at the overall (aggregated) value. However, these methods (most of them adopted from decision science and operation research) make strong assumptions, have long administration times, and their purpose is to deal with small group of experts (Keeney and Raiffa, 1976; Roy, 1996; Craig et al., 2014).

Item response theory (Chapter 10) is also used to measure health, particularly the impact of deprived health, but is directed at only one specific domain or attribute (e.g., mobility), not a patient's overall condition. In that sense, it falls into the category of monadic measurement, which means that items are assessed separately. Most of its applications use simple statements about ability or functioning. Furthermore, only specific behaviors that are attributable to the respondent (patient) are assessed. One example is, "Can you go shopping without assistance?" Item response theory differs from choice models, where the response task is more of an overall judgment than an assessment. Moreover, it differs from preference-based models in that people are not asked to decide between two or more presented objects but to compare a presented object or statement with their own belief, experience, or attitude (see also Chapter 14).

VALUES

The overall impact of a health condition can be measured with preference-based methods, which quantify the quality of a patient's health or constructed hypothetical health descriptions. Instead of measuring the level of the reported complaints (i.e., their frequency and intensity), these methods express the quality of health (or of specific health conditions) numerically. The respondents are asked to make a value judgment about a specific phenomenon, condition, or outcome. Therefore, we do not speak of measures but of "values" in preference-based methods.

The values (variously called utilities, strengths of preferences, indices, or weights) that these methods generate are assumed to have a specific measurement property. Any differences between values for health are assumed to correspond to the increments of difference in quality of these states, which implies that the values should be interval-level or cardinal data. Thus, the measures should lie on a continuous scale so that the differences between values indicate true differences (e.g., if a patient's value increases from 0.40 to 0.60, this increase is the same as from 0.70 to 0.90). Preference-based measurement can be very convenient because it produces one overall numerical score, which makes analyzing and interpreting results a straightforward procedure.

It can also be very convenient for another reason. With this type of measurement, we recognize that patients may care little about apparently high levels of problems on specific health attributes, whereas these same patients may care a great deal about apparently minor problems. For example, patients may experience "moderate pain" as more serious (and thus giving it a lower value) than "serious walking problems."

In short, preference-based methods come down to teasing out or discovering the "true" values that people assign to health. This task is not inherently straightforward: health is a complex matter, and a highly emotional one at that.

Health values are frequently used in health services research and economic evaluations of health interventions. Typically, rescaled health-state values (Chapter 12) are combined with the amount of time spent in health states to compute quality-adjusted life years (Chapter 3). Other research applications that require health values for a wide range of health conditions include monitoring of disease and health in populations or evaluating the burden of disease using units of disability-adjusted life years. Increasingly, health values are also used as additional outcome measures in clinical studies. Access to accurate values for a wide variety of health conditions is also advantageous for monitoring programs in the context of public health.

WEIGHING OF THE ATTRIBUTES AND THEIR LEVELS

Surprisingly, preference-based measurement is strongly related to an approach used in clinimetrics, though many, if not all, researchers in this field seem to be unaware of this connection. As explained in Chapter 13, preference-based measurement differs from the index approach used in clinimetrics whereby weights are seldom given to the different domains or attributes, nor are different weights given to the levels or categories of the domains or attributes. In the event that weights are attached to attributes and/or their levels, this is almost always based on expert opinion. In clinical settings the weights are not worked out in a formal measurement (statistical) framework. As such, the index instruments developed in the setting of clinimetrics cannot be considered genuine preference-based instruments.

CHOICE-BASED MODELS

As discussed earlier (Chapter 5), discrimination is a basic operation of judgment and a means of generating knowledge. Most judgments in daily life consist of making choices between competing alternatives and are thus inherently comparative. Therefore, the core activity in many choice methods is to compare two or more health aspects in such a way that the responses will provide relevant information.

Basically, choice models offer people two standard judgmental tasks. One is to order a set of objects, for example, descriptions of different health conditions. The other, being based on paired comparisons, asks respondents to choose their preferred health state from a pair of health states. Paired comparisons between health outcomes are generally feasible, even when the phenomena presented are hard to differentiate or when the feature being measured is not easy to comprehend.

Paired Comparisons

Choice models have a long tradition, starting in 1927 with Louis Thurstone, who began his career as an electrical engineer and worked for many years as a psychometrician at the University of Chicago. Thurstone built upon work by his predecessors, in particular Gustav Fechner (1801–87), a German experimental psychologist. As a pioneer in experimental psychology and the founder of psychophysics, Fechner inspired many scientists and philosophers, including Thurstone, during the 20th century.

Early psychophysical work was built upon precise but simple experiments. A typical example might be worded thus: consider the following two objects with weights w_1 and w_2; which one is heavier? Such experiments would demonstrate that the greater the difference in object weight, the greater the probability of choosing correctly. This measurement approach is based on making comparative judgments. Another example; when presented with pairs of noises, respondents were asked to say which was louder.

Later, the method was extended to obtain value judgments on, say, the pleasantness of different colors. During the late 1920s, Thurstone extended this general approach to a representation of judgments made in regard to nonphysical continua such as the seriousness of crime (Thurstone, 1927), nationalities to be judged for desirability (Peterson and Thurstone, 1932; Fig. 11.2), handwriting samples to be judged for excellence, or the beauty of forest scenes, seriousness of environmental losses (Brown et al., 2002, 2009), and the quality of health (Krabbe, 2008).

In Thurstone's model, the separation of objects (otherwise called items, scenarios, descriptions, profiles, alternatives, stimuli, or graphics) is represented on an underlying unidimensional continuum. There, the distance between pairs of objects is a function of the probability of one object being selected as having a degree of intensity greater than that of the other object in the pair. Although psychologists typically refer to stimuli, for our purposes we will use the term objects. In the context of this book, these objects will be mostly related to health. In that sense, it would be natural to refer to objects as "health conditions" or "health states" and to characterize as "quality of health" the underlying attribute of these health outcomes that we want to measure.

Probabilistic Choice Model

Thurstone postulated that each object in a set would possess some attributes in varying but unknown degrees and that a preference will exist for each object and among all subjects. Furthermore, he postulated that for each object the overall preference will be distributed normally around the most frequent (modal) response. To measure such overall preferences, each person's preference for each object versus every other object has to be obtained. The more people who select one object of a pair over the other object, the greater the preference for that object, and thus the greater is its scale weight.

Accordingly, the basic element of subjective measurement in the framework of preference-based measurement is a simple and straightforward response task based on a comparison between two compound objects (e.g., descriptions, photos) made in such a way that the data yield compelling information.

Thurstone formulated a mathematical model, which he called the Law of Comparative Judgment (LCJ), that could be used to estimate scale values on the basis of binary choices between objects (e.g., health descriptions; Thurstone, 1994, 1927). He proposed that perceived physical phenomena (e.g., brightness, weight) or subjective concepts (e.g., seriousness of crimes, taste) could be expressed as a true weight and a random component.

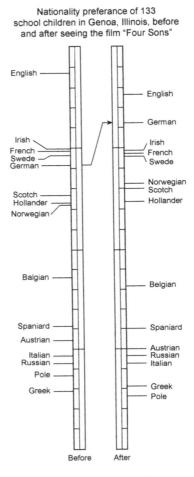

Nationality preferance of 133
school children in Genoa, Illinois, before
and after seeing the film "Four Sons"

FIGURE 11.2 On April 29, 1930, the motion picture "Four Sons" was shown at a special matinee performance in a local theater. The experimental group consisted of 133 children of grades 7–12 inclusive. The figure shows the scale values, before (17 April) and after seeing the film, for each of the nationalities that appeared in the paired comparison schedule. It is evident that the largest shift in scale value concerns the Germans. It shows clearly that the film made the children more friendly toward the Germans. *From Peterson, R., Thurstone, L.L., 1932. The effect of a motion picture film on children's attitudes toward Germans. Journal of Educational Psychology 23 (4), 241–246.*

This approach is indirect; underlying it is a theory allowing raw individual data to be transformed into aggregate data. For that reason, psychometricians regard it as scaling. In terms of modern psychometric theory, however, it is more aptly regarded as a measurement model. Because Thurstone's model derives group scale values from imprecise individual data, it is also regarded as a probabilistic choice model. The LCJ can only be used to model paired (two objects) comparisons. Elaborations of probabilistic choice models have their roots in mathematical psychology. Since all of these choice models are probabilistic, there is no need to keep mentioning the term probabilistic in this chapter.

Discrimination

The LCJ model presumes that the assessment of any object with respect to a specific attribute (magnitude) can be represented by a theoretical distribution of points located along an underlying continuum. In Thurstone's terminology, choices are mediated by a "discriminable process," which he defined as the process by which an organism identifies, distinguishes, or reacts to objects.

Thurstone proposed that perceived physical phenomena or subjective concepts (e.g., health outcomes) can be expressed as follows:

$$\theta_i = \alpha_i + \varepsilon_i \tag{11.1}$$

where θ_i is the true weight of an object i, α is the measurable component of that weight for the object i, and ε is a random error term. The assumption in the model is that ε is normally distributed, which is consistent with the binomial probit model. Consider the theoretical distributions of the discriminable process for any two objects: for instance, two different descriptions of health conditions i and j. In the LCJ model, the standard deviation of the distribution associated with a given health condition is called the discriminable dispersion (now called variance) of that health condition. Discriminable dispersions may be different for different health conditions.

Let θ_i and θ_j correspond to the scale values of the two health conditions. The difference $\theta_i-\theta_j$ is measured in units of discriminable differences. This discriminal (difference) process, $\theta_i-\theta_j = (\alpha_i-\alpha_j) + (\varepsilon_i-\varepsilon_j)$, is normally distributed with the mean $\theta_i-\theta_j$ and variance σ_{ij}^2 corresponding to the following:

$$\sigma_{ij}^2 = \sigma_i^2 + \sigma_j^2 - 2\rho_{ij}\sigma_i^2\sigma_j^2 \tag{11.2}$$

Thurstone (Fig. 11.3) stated that the relation between the difference in the means, $\theta_i-\theta_j$, the z score of the probability of selecting the one object as larger (better) than the other, and the variance and correlations of the random variables θ_i and θ_j can be modeled. This is known as the law of comparative judgment:

$$\theta_i - \theta_j = z_{ij}\sqrt{\sigma_i^2 + \sigma_j^2 - 2\rho_{ij}\sigma_i^2\sigma_j^2} \tag{11.3}$$

where σ_i and σ_j denote the standard deviations of the two objects (health conditions) i and j, ρ_{ij} denotes the correlation between the pairs of discriminal processes i and j, and z_{ij} is the unit normal deviate (z value) corresponding to the theoretical proportion of times that object j is judged better than object *i*. In its most basic form (known as Case V), the model can be represented as, $\theta_i - \theta_j = z_{ij}$, for which the probability that object j is judged to have more of an attribute than object i is:

$$P_{ij} = \phi\left(\frac{\theta_i - \theta_j}{\sqrt{\sigma_i^2 + \sigma_j^2}}\right) = \phi\left(\frac{\theta_i - \theta_j}{\sigma_{ij}}\right) \tag{11.4}$$

where ϕ is the cumulative normal distribution with a mean of zero and a variance of unity.

BOX 11.1 Louis Leon Thurstone

Thurstone was born in Chicago, Illinois, to Swedish immigrants. He received a master's degree in mechanical engineering from Cornell University in 1912 and was offered a brief assistantship in the laboratory of Thomas Edison. In 1914, after 2 years as an instructor of geometry and drafting at the University of Minnesota, he enrolled as a graduate student in psychology at the University of Chicago. He later returned to the University of Chicago (1924–52) where he taught and conducted research (Fig. 11.3).

FIGURE 11.3   Louis Leon Thurstone (1887–1955) teaching at the University of Chicago. *With permission: Photo by George Skadding/The LIFE Picture Collection/Getty Images, 1947.*

Thurstone's full model of comparative judgment allows for the variances of the error terms to vary from object to object and for the variances of the different objects to be correlated. Having proposed the full model of comparative choice, Thurstone then offered a series of simplifying assumptions, the acceptance of which made computation of scale values from paired comparison data first of all feasible and then, with additional assumptions, easier. Accepting the full set of assumptions, Thurstone's Case V amounts to accepting what is now known as the independently and identically distributed errors condition, wherein the variances of the error terms are uncorrelated (in other words, independent) and identical across all the objects (Krabbe, 2013).

The discrimination mechanism underlying the LCJ is an extension of the "just noticeable difference" that played a major role in early psychophysical research, as initiated by Fechner and Weber in Germany. Later on, similar discrimination mechanisms were embedded in "signal detection theory," which was used to measure the way people make decisions under conditions of uncertainty. Much of the early work in this field was done by radar researchers (Marcum, 1947).

The LCJ makes possible a wide range of scaling operations, whether the data come from the method of paired comparisons or other methods from which comparative judgments may be inferred, such as rank order. The objects being compared may be of almost any kind. The advantages of paired comparisons as a method for eliciting human judgments include the method's simplicity and its use of comparative judgments.

Mathematical Psychology

Another way to analyze comparative data is with the Bradley—Terry—Luce (BTL) model, which was statistically formulated by Bradley and Terry in 1952 and extended by Luce in 1959. (It was later recognized that the German mathematician Ernst Zermelo had already published on a paired comparison model in 1929.) The BTL model extends the Thurstone model by allowing a person to choose among more than two objects. For mathematical reasons the BTL model is based on the simple logistic function instead of the normal distribution of the Thurstone model. It turns out that the Rasch model (Chapter 10) is very closely related to the BTL model with regard to measurement (Brogden, 1977).

Conjoint Measurement

Mathematical psychology was further advanced by fundamental measurement, a theory developed by Duncan Luce (Fig. 11.4) and John Tukey and published in 1964 under the name "conjoint measurement" (Luce, 1959; Luce and Tukey, 1964). This representational measurement theory postulates that measurement on a ratio scale level can be established if the data satisfy certain structural assumptions. Fundamental measurement theory is a mathematical framework based on logical (not normative) axioms. It concerns exclusively the qualitative conditions under which a particular representation holds. Given that the social sciences could not live up to the standards of objective measurement being applied in the physical sciences, conjoint measurement was developed to perform fundamental measurement with subjective phenomena. It can measure the joint effects of two or more independent variables on the ordering of a dependent variable (the property to be quantified). As Perline et al. (1979) put it: "The question is whether or not there exists a monotonic transformation of an ordinal measure of the dependent variable from which an additive representation can be constructed."

The axiomatization of conjoint measurement is complicated. Its full version includes technical axioms (e.g., consistency, transitivity, and double cancellation), which can often plausibly be assumed to hold approximately (Maas and Stalpers, 1992; Stalmeier et al., 1996). When the axioms hold, the result is that the observed but transformed dependent variable and the concomitantly constructed independent variables are simultaneously (hence the term "conjoint") represented on an interval scale with a common unit. Conjoint measurement, as a member of the class of fundamental measurement theories, is algebraic (designating an expression in which only numbers, letters, and arithmetic operations are contained or used) and therefore deterministic (as opposed to most models described in this book, which are probabilistic).

Although conjoint measurement is generally acknowledged as an important theoretical contribution, its practicality is in doubt because of its strict axiomatic assumptions. The Rasch model—independently developed from conjoint measurement—can be seen as a practical rendition of conjoint measurement with an underlying stochastic structure (Brogden, 1977). For this and other reasons, many scientists consider the Rasch model as the preeminent means to measure subjective phenomena "objectively."

BOX 11.2 Robert Duncan Luce

Duncan Luce was a pioneer in the field of mathematical behavioral sciences. His contributions, including his book *Individual Choice Behavior*, are recognized widely as groundbreaking in economics and psychology. His work in game and choice theory and fundamental measurement and applications in psychology helped shape the fields of psychology, the social sciences, and contemporary economics.

Luce entered the Massachusetts Institute of Technology (MIT) in 1942 studying aeronautical engineering. Later he decided to switch to applied mathematics.

At MIT he became fascinated by possible links between mathematics and psychology. Duncan Luce's work was cited in the 2000 Nobel Prize given to UC Berkeley economist Daniel McFadden, who said, "In a fully just world, there

Box 11.2 Robert Duncan Luce—cont'd

would be a Nobel Prize for psychology, and Duncan Luce would have long since received it." At the age of 80 years, he had a simple prescription for a long healthy life: "A balanced diet, a lively intellectual life, and by all means avoid strenuous exercise." (Fig. 11.4)

FIGURE 11.4   Robert Duncan Luce (1925–2012) in 2005 at the White House in Washington, as President George W. Bush bestows on him the National Medal of Science in behavioral and social science for his contributions to the field of mathematical psychology. *With permission: Photo by Chuck Kennedy/MCT/MCT via Getty Images.*

Conjoint Analysis

A professor in marketing, Paul Green (Green and Rao, 1971), recognized that Luce and Tukey's conjoint measurement article provided a new system to quantify rank order data. These types of data could be applied to marketing research (e.g., to forecast market response for new products). His more pragmatic approach (no formal checks and based on regression models) is what is now called conjoint analysis. Today this technique is used in many of the social and applied sciences.

The objective of conjoint analysis is to determine the separate contribution of a limited number of features (e.g., attributes) of an object on its overall value. Each object is composed of a unique combination of features. The respondents are generally shown a set of products, goods, services, scenarios, or pictures. Each example is similar enough to the others that respondents will see them as close substitutes but dissimilar enough that they can clearly determine a preference. The response task may consist of individual ratings, rank orders, or choices among alternative combinations. When choosing among alternative health conditions, conjoint analysis becomes roughly similar to the discrete choice model (see below). But besides having a range of possible response modes, the conjoint analysis shows variation in other ways. There are multiple models (full profile, partial/incomplete profile, hierarchical, Bayesian, and so on) and several designs (e.g., full factorial, fractional factorial, resolution III). In that regard, conjoint analysis can be taken as an umbrella term describing various methods to derive quantitative measures for (subjective) phenomena based on a combination of stimulus configuration, experimental design, response modes, and statistical analysis. The reader is referred to Louviere et al. (2010) for an excellent discussion of the differences between conjoint analysis, as described above, and discrete choice models, which will be discussed in the next section.

Discrete Choice

In Thurstone's LCJ, the perceived value of a health condition equals its objective level plus a random error. The probability that one health condition is judged better than another implies that this health condition has the higher perceived value. When the perceived values are interpreted as levels of satisfaction or utility (in economics, utility is a measure of preferences over some set of goods and services), this can be interpreted as a model for economic choice in which utility is modeled as a random variable. This assertion was made in 1960 by the economist Marschak, who thereby introduced Thurstone's work into economics. In fact, it was Thurstone who formulated the random utility model, though using different notation and terminology. Marschak called his own model the random utility maximization hypothesis or RUM (Marschak, 1960; Louviere and Woodworth, 1983). Like neoclassical economic theory, the RUM assumes that the decision-maker (e.g., policy maker, clinician, patient) has a perfect discrimination capability. But it also assumes that the decision maker has incomplete information, which implies that uncertainty (i.e., randomness) must be taken into account.

McFadden

So, the modern measurement theory inherent in choice models builds upon the early work and basic principles of Thurstone's LCJ. In fact, the class of choice- and rank-based models, with its lengthy history (1927—present), is one of the few areas in the social and behavioral sciences that has a strong underlying theory. Modern discrete choice models, which come from econometrics, build upon the work of McFadden, the 2000 Nobel Prize laureate in economics (McFadden, 1974).

In the mid-1960s, McFadden was working with a graduate student, Phoebe Cottingham, trying to analyze data on freeway routing decisions as a way to study economic decision-making behavior. He developed the first version of what he called the conditional multinomial logistic model (also known as the multinomial logistic model or conditional logistic model). Drawing upon the work of Thurstone, Marschak (1960) and Lancaster (1966), McFadden was able to show how his model was linked to the economic theory of choice behavior. McFadden proposed an econometric model in which the value of alternatives (objects) would depend on the weights assigned to their attributes, such as construction cost, route length, and areas of parkland, and open space taken up (McFadden, 2001). McFadden then investigated further the RUM foundations of the conditional multinomial logistic model. He developed a computer program that allowed him to estimate this model, which was based on an axiomatic theory of choice behavior developed by the mathematical psychologist Luce (1959). He also showed that the Luce model was consistent with the RUM model. The discrete choice strategy was conceived in transport economics and later disseminated into other research fields, especially marketing. There, discrete choice modeling was applied to analyze behavior that could be observed in real market contexts. The introduction of the discrete choice model as developed by McFadden is a successful example of multidisciplinary interaction and inspiration.

The Model

Discrete choice models belong to the class denoted in the statistical literature as (probabilistic) choice models. These techniques are applicable when individuals have the ability to choose between two or more distinct ("discrete") alternatives. What all these choice models have in common is that they are able to establish the relative merit of a phenomenon. More importantly, instead of one function, as in the classical Thurstone model (where only values for the objects can be estimated), the conditional multinomial logistic model comprises two functions. First, it contains a statistical model that describes the probability of ranking a particular object (e.g., health condition) as better than another, given the value associated with each object. Second, it contains a function that relates the value for a given object to a set of explanatory variables (i.e., health attributes).

Therefore, if a phenomenon is described according to characteristics (or attributes) with certain levels (e.g., no/some/severe pain), McFadden's choice model makes it possible to estimate the relative importance assigned to these attributes and their associated levels, and even to estimate overall value for different combinations of attribute levels. These types of choice models have been used widely to elicit values in a number of other research areas, notably in marketing, transportation, and environmental economics (Louviere et al., 2000).

Moreover, the conditional multinomial logistic statistical model developed by McFadden has two components. One takes an individual's characteristics as explanatory variables; the other component takes characteristics of the object of interest (i.e., a health condition) as explanatory variables:

$$v_{rs} = \underbrace{x_r \beta_s}_{\substack{A \\ measurement}} + \underbrace{z_{rs} \gamma}_{\substack{B \\ choice\ behavior}} . \qquad (11.5)$$

where v_{rs} is the value of the object s assigned by the respondent r. That value (v_{rs}) depends on both the object characteristics x and on the respondent's characteristics z. The probability that the respondent r chooses object s is:

$$P_{rs} = \frac{e^{(x_r \beta_s + z_{rs} \gamma)}}{\sum_{k=1}^{K} e^{(x_r \beta_s + z_{rs} \gamma)}} . \qquad (11.6)$$

Discrete Choice Experiments

Before the contribution of Louviere et al. (Louviere and Woodworth, 1983; Louviere et al., 2000), choice models had been used to analyze behavior that could be observed in real market contexts. Louviere et al. applied choice models to choices collected from respondents who were presented profiles with features of hypothetical products; this is what they called "simulated choice situations." So, instead of modeling the actual choices made by people, as McFadden did (in his revealed preferences approach), Louviere modeled the choices made by subjects in carefully constructed experimental studies (with discrete choice experiments). This new method (called the stated preferences approach) also made it possible to predict values for phenomena that could not be judged in the real world. Inspired by this experimental approach, choice models have been used as an alternative way to derive people's values for health conditions (Hakim and Pathak, 1999; Salomon, 2003; Stolk et al., 2010; Krabbe et al., 2014).

Similar to conjoint analysis, discrete choice models encompass a variety of experimental design techniques, data collection procedures, and statistical procedures that can be used to predict the choices that individuals will make between different objects (e.g., health aspects or conditions).

It should be noted that the purpose of discrete choice models differs from that of the other measurement methods presented in this book. Choice models are focused on explaining choice behavior, whereas measurement models serve to quantify one specific phenomenon such as length or temperature, or in the case of health to quantify severity, intensity, or quality. The discrete choice model was developed to explain choices (individual preferences), not to measure (place a value on) certain phenomena. Explaining choice behavior is exactly what McFadden and many others have been doing. However, with specific restrictions (mainly, the need to drop part B of Eq. (11.5)) the model turns into a measurement model (Arons and Krabbe, 2013).

Discrete choice experiments were introduced into health economics as a technique to explore and quantify the preferences of individuals for different treatment options and treatment regimes. Factors typically included in such discrete choice experiments are location of treatment, type of care (for example, surgical or medical), and staff providing care (consultant or specialist nurse). Often, process variables (e.g., waiting time) are added in such choice experiments, and individual characteristics are also collected. Choice models allow investigation of the trade-offs between such process and outcome attributes. For example, two or more hypothetical situations are described (choice set) and the respondent selects which situation is preferable (Ryan, 2004; Mol et al., 2015).

MEASUREMENT PROPERTIES

The conventional measures constructed under classical test theory consist of one or more health domains, each measured by multiple items (Chapter 9). The assumption is that high intercorrelations exist among the items within a domain. This assumption—actually, a requirement—can be evaluated by analyzing the internal consistency of a set of items belonging to a domain. The assumption of high intercorrelations is valid if items can be considered as reflective variables (Chapters 5 and 6).

In choice models the task is focused on the evaluation of compound health descriptions, which is associated with another type of assessment and another type of response task and therefore produces another type of data. What we actually want to measure, the object of interest, is a set of key elements that should reflect the phenomenon we want to measure. In the setting of health measurement, accordingly, the attributes of "pain," "mobility," "fatigue," and many others may be selected and used to designate various "health conditions" (e.g., health states). A crucial operational issue in the use of choice models (though the same holds for the economic valuation techniques and the clinimetric index approach) is that these models are usually based on a formative measurement model (Chapters 5 and 6). The principal difference between a reflective and a formative measurement model may be ascribed to their modes of measurement: measuring characteristics of subjects (patients) versus measuring characteristics of objects (health states; Krabbe et al., 1997). Each mode calls for its own way of managing the collected data to capture the aspects of health.

Furthermore, when using choice models, it is important to recognize that the tasks do not generate scores for a health state. Instead, they produce binary scores that express whether one health state was preferred to another. Only after analyzing the binary response data in a specific statistical model can estimates of the values (and the levels of the attributes) for the health states be made available.

Reliability

No simple, easy, and universally accepted criteria exist for assessing the reliability of choice models. If information about reliability is desired, it seems that the test–retest option is the only one left (Konerding, 2013).

Test–Retest

Test–retest reliability deals with the reproducibility of a measurement method. If a method is reliable, it should evoke the same outcomes on a second occasion if there is no alteration due to expected change. The most appropriate way of gauging reliability is by computing an intraclass correlation between the first measurement (test) and a subsequent measurement (retest). For preference-based choice models, test–retest reliabilities can be estimated for the computed values based on estimated coefficients for the levels of the various attributes. But in practice this procedure is rarely used for choice models, perhaps because rather complex (and expensive) studies would have to be reproduced.

Internal Consistency

To assess how well the composite of all the items measures certain health domains, Cronbach's α (internal consistency) is normally estimated. The insight that Cronbach's α can offer is how well the composite of all the items associated with the underlying domain or scale measures the construct. Substantial intercorrelation of these elements is interpreted to mean that the items do measure the same construct in a reliable manner. This type of internal consistency reliability, however, is not suitable for choice models, where the respondents are not scoring items. Instead, attributes are presented; these have to be read and mentally processed. Subsequently, the whole set of attributes of a description has to be compared to another description, followed by provoking a preference for one of the descriptions. The situation is more or less the same for the (economic) valuation techniques (Chapter 12) and the index approach as used in clinimetrics (Chapter 13). A theoretical assumption about interitem correlation between the attributes is not made for any of these measurement models.

Validity

Content validity is the extent to which an instrument assesses the relevant and important aspects of the concept it was designed to measure. It is generally established by gathering expert opinion or conducting qualitative research with patients. As there is no "gold standard" for the measurement of health-related quality of life, content validity seems to pose an unanswerable question. So what researchers fall back on instead are appeals to plausibility of the selection of the health attributes in the use of preference-based measurement methods.

These are all very subjective notions, and ultimately the answers rely heavily on intuition and professional judgment.

Content

It has been realized in the field of health instrument development that content validity cannot be established via quantitative psychometric analyses, as highlighted in PRO guidance of the Food and Drug Administration (2009). The solution might lie in establishing special procedures to retrieve the relevant attributes, perhaps by consulting expert opinion or reexamining the content of existing instruments. Increasingly, researchers recommend methods that allow attributes to come from the patients themselves (see also Chapter 14).

Construct

Tests of construct validity seem almost impossible in the context of health measurements based on choice models. The reason is that the health objects presented to the respondents for judgment may consist of unspecified (holistic) features. For example, the respondents may be asked to compare different outcomes after plastic surgery or photos of skin affected by psoriasis. If the object they have to judge consists of health descriptions that are built on a small and fixed set of health attributes with various levels, an analysis to compare these attributes with attributes of scales from other instruments is also impossible. These problems are related to the fact that we are dealing here with formative measurement. Other types of construct validity, such as known-groups validity, may be feasible for the estimated values derived with choice models.

INSTRUMENTS

There is considerable interest in using choice models, and many studies have applied this methodology. However, these have been mainly conducted to evaluate treatment regimens and services (Langenhoff et al., 2007; Poulos et al., 2016; Russo et al., 2016). Recently, studies have been done on the measurement of specific health outcomes, but most are still mainly concerned with exploring the possible benefits of choice models (Krabbe et al., 2014; Norman et al., 2014, 2015). One recent measurement model will be discussed in the next section and some novel work will be briefly presented in Chapter 14.

APPLICATION

Dementia has a major impact on a person's life, and its prevalence is expected to double or triple by 2050. Because of its rapid increase in the population and the poor prospects of a cure in the near future, some governments now seek to keep people in the community as long as possible. Currently, many interventions are directed to the antiamyloidal pathway (proteins that become folded into the wrong shape). Clinical trials investigating the effects of new treatments rarely use health status or health as an outcome measure, even though maintaining or improving health is a primary goal in dementia care.

Health instruments that enable the generic expression of the quality of patients' health in a single standardized value (index) are increasingly important in the evaluation of care interventions. Generic index instruments (Chapter 12) such as the EuroQol-5D (EQ-5D), the Short-Form 6-D (SF-6D), and the Health Utility Index are already available for this purpose. In the field of dementia, however, clinicians and researchers generally discredit the use of generic health index instruments because these do not specifically concern the predominant health aspects affected by the disease. The main objection is that their results are insufficiently valid. Instead, researchers suggest the use of disease-specific index instruments, which focus on the most important health domains affected by a certain disease.

The dementia quality-of-life instrument (DQI) describes dementia-specific health in six domains (Fig. 11.5). The DQI is intended for use in community-dwelling patients. Having six domains (attributes) with three levels each, a total of 3^6 (729) health states can be created. Each health state can be classified by a six-digit code consisting of one digit per domain.

Respondents

Conventionally, values for health states are derived from members of the general population. But respondents who evaluate hypothetical health states might not be familiar with dementia. It seems reasonable to assume that healthy people have insufficient information or imagination to make a valid judgment about the impact of dementia in its various stages. The best judges of a health state are presumably those who have actually experienced it. In the case of dementia, however, their judgments about their own health state are probably biased because of their declining intellectual capacity, semantic knowledge, and episodic memory, as well as varying deficits in judgment and insight. An alternative would be to question informal caregivers or professionals working with people with dementia; because of their regular contact, these other informants will be familiar with the impact of dementia on health. From that perspective, the developers of the DQI investigated values derived from professionals working with people with dementia but also from laypersons. In line with the literature, they hypothesized that it would be likely for the judgments of these two groups to differ substantially, in which case they would advocate using the values of the professionals.

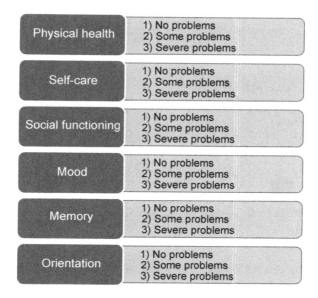

For example: Health state '121312' corresponds to

No problems with physical health
Some problems with self-care
No problems with social functioning
Severe mood problems
No memory problems
Some orientation problems

FIGURE 11.5   The dementia quality-of-life instrument (reprinted with permission Elsevier). *From Arons, A.M.M., Schölzel-Dorenbos, C.J.M., Olde Rikkert, M.G.M., Krabbe, P.F.M., 2016. A simple and practical index to measure dementia-related quality of life. Value in Health 19 (1), 60−65.*

Design

Respondents from the professional panel and the general population were repeatedly presented with two health-state descriptions (paired comparisons) and asked to indicate which one they preferred (Fig. 11.6). The DQI classification system allows for 729 health states, which makes it impractical to conduct a study in which all states are assessed (full factorial design). Instead, a near-orthogonal main effects discrete choice experiment design was generated (Sawtooth software, complete enumeration option) to meet certain methodological criteria (minimal overlap, level balance, and orthogonality). This orthogonal design allowed the estimation of main effects independent of one another. All presentations of health states were randomized while the order of the domains was kept constant.

Which health state is better, state A or state B?

Severe problems with physical health	Some problems with physical health
Severe problems with self-care	Some problems with self-care
No problems with social functioning	Some problems with social functioning
No mood problems	Some mood problems
No memory problems	Some memory problems
Some orientation problems	Severe orientation problems

FIGURE 11.6   Paired comparison task for the dementia quality-of-life instrument.

Statistical Model

This study was couched in the framework of modern discrete choice theory, which was deemed appropriate to ascertain the relative merit of health states. An established member of this class of choice models (conditional fixed-effects logit model) was applied to estimate the weights of the domain levels. As the aim was to apply the DQI health-state values to compute quality-adjusted life years, the values had to be calibrated in a separate task (Arons et al., 2016).

Sample

A convenience sample was drawn among professionals working in the field of dementia. They came from a wide range of professions: clinical geriatricians, elderly care physicians, nurses, social workers, researchers, psychologists, case managers, and (physical) therapists. No inclusion or exclusion criteria were used for this sample. In addition, a market research company recruited members of the general population, selecting them from its own respondent panel. Quota sampling ($n = 600$) was used to recruit respondents aged 18–75 years who were representative of the Dutch population with regard to age, sex, geographical area, and education.

Results

Professionals working with people with dementia attached the highest weight (i.e., burden) to severe mood problems (Table 11.1). In contrast, the general population attached the highest weight to severe problems with physical health and severe memory problems. Four domain levels were statistically significantly judged as more important by professionals working with people with dementia, namely, "severe problems with social functioning," "some mood

TABLE 11.1 Weights Given to the Domain Levels of the Dementia Quality-of-Life Instrument (Reprinted With Permission Elsevier)

Domain Levels[a]	Dementia Professionals[b,c] ($n = 207$)			General Population[b,d] ($n = 631$)		
	β	SE	p	β	SE	p
Some problems with physical health (2)	−0.011	0.012	0.419	−0.048	0.009	0.000
Severe problems with physical health (3)	−0.150	0.013	0.000	−0.223	0.010	0.000
Some problems with self-care (2)	−0.045	0.012	0.001	−0.059	0.009	0.000
Severe problems with self-care (3)	−0.121	0.013	0.000	−0.195	0.010	0.000
Some problems with social functioning (2)	−0.029	0.012	0.028	−0.046	0.009	0.000
Severe problems with social functioning (3)	−0.154	0.013	0.000	−0.162	0.010	0.000
Some mood problems (2)	−0.071	0.012	0.000	−0.035	0.009	0.000
Severe mood problems (3)	−0.343	0.015	0.000	−0.165	0.010	0.000
Some memory problems (2)	−0.060	0.012	0.000	−0.065	0.009	0.000
Severe memory problems (3)	−0.234	0.015	0.000	−0.225	0.011	0.000
Some orientation problems (2)	−0.021	0.014	0.178	−0.022	0.011	0.035
Severe orientation problems (3)	−0.101	0.015	0.000	−0.133	0.011	0.000

[a]Note that level 1 problems indicate "no problems" and therefore do not lead to a diminished health-related quality of life.
[b]This model is rescaled to the dead-full health scale.
[c]Pseudo $R^2 = 0.41$.
[d]Pseudo $R^2 = 0.19$.
From Arons, A.M.M., Schölzel-Dorenbos, C.J.M., Olde Rikkert, M.G.M., Krabbe, P.F.M., 2016. A simple and practical index to measure dementia-related quality of life. Value in Health 19 (1), 60–65.

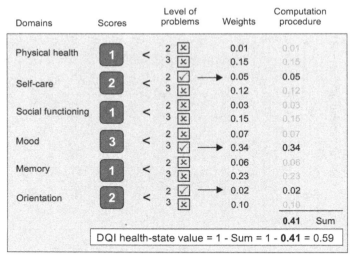

FIGURE 11.7   Dementia quality-of-life index (DQI) health-state value calculation example for the DQI state "121312" using the (rounded-up) values from professionals working with people with dementia (note that problems at level 1 indicate "no problems" and therefore do not lead to a diminished quality of life (reprinted with permission Elsevier)). *From Arons, A.M.M., Schölzel-Dorenbos, C.J.M., Olde Rikkert, M.G.M., Krabbe, P.F.M., 2016. A simple and practical index to measure dementia-related quality of life. Value in Health 19 (1), 60—65.*

problems," "severe mood problems," and "severe memory problems." The DQI health-state values can simply be calculated as displayed in Fig. 11.7. They range from 1 to −0.103 (0 = dead).

The current instrument was developed in the Netherlands; one might wonder whether the values of the DQI are equally valid elsewhere. Research on generic preference-based index instruments reveals slight systematic differences in health-state values between most countries. Such differences can arise because of variations in the valuation methodology, translational issues, and (cultural) specificities in each population's preferences for health states and health domains. The developers of the DQI believe that the instrument in its present version can be used in other countries as long as the researchers are aware that the values for the DQI states may be somewhat less precise. A better approach would be to properly translate the DQI for each country and then conduct a separate study to derive country-specific values.

DISCUSSION

Thurstone's articles have had a tremendous influence on the development of methods for collecting and analyzing judgments in the form of preferences, attitudes, or, more specifically, health outcomes. Because of the limited computational capacity available at the time, the original model (LCJ) as proposed by Thurstone could not be computed for large set of objects. Before the computer revolution, his method was thus unfeasible, and when Thurstone proposed his model more than 80 years ago only univariate integration could be done. Fortunately, these computational limitations have been overcome and Thurstone's model can be estimated in its full generality. As a result, the applicability of this approach has widened tremendously (Maydeu-Olivares and Böckenholt, 2008).

The emphasis that this chapter places on the importance of Thurstone and his scaling models may seem odd to readers who have read the explanation of Thurstone's methods in the recent edition of the book by Streiner et al. (2015). On page 55 they conclude: "As is obvious from this description, Thurstone scales are extremely labor-intensive to construct. Added to the fact that they are not noticeably better than Likert scales in terms of their interval properties and that most statistical techniques are fairly robust to violations of intervalness in any case, this technique exists only in museums of scale development (if such museums exist), although the odd example does occasionally pop up (e.g., Krabbe, 2008)." Readers of the present book may appreciate how surprised I was to read this in the 5th edition of the "flagship" handbook on health measurement scales. Careful reading makes it clear that Streiner et al. are mixing up another Thurstone scaling model (equal-appearing intervals) with his renowned paired comparison model. Luckily, my 2008 publication was about paired comparison and the law of comparative judgment. Nevertheless, their style of writing is cheerful as always.

All choice models estimate the relative contribution of attributes and attribute levels and are therefore able to produce relative health values. For many uses in the setting of health, it is inconsequential that values derived with the methods discussed in this chapter are relative and not absolute. The main objective of practical (clinical) research is to measure differences between and within patients, and choice methods will usually do the job. In some situations, however, health values need to be anchored in such a way that they become real index values and can be used in calculating quality-adjusted life years. Accordingly, on the health scale, 0.0 is death and 1.0 is full health. This condition of calibrating the values on a scale with fixed anchor points is important when performing cost-effectiveness analysis, which explains the special methods used by the field of health economics (Chapter 12). It might be desirable to rescale the values for health measured with one of the choice models in such a way that an absolute scale is obtained (for example, 0.0 = dead). To that end, research is ongoing to extend existing discrete choice models (see also Chapter 14).

Choice models present relatively simple and straightforward response tasks. While these are easy to perform from simulated (i.e., hypothetical but realistic) health descriptions or scenarios, they provide sufficient information to arrive at quantitative measures. One of the requirements of choice models is that the responses are somewhat different per individual but rather similar across the sample. Because all these choice models are statistical models, some variation in responses is required. However, to ensure that the measurement is robust and meaningful, all respondents have to come up with more or less the same responses. In a paired comparison between health conditions, it may be observed that 70% of the respondents are choosing condition A and 30% condition B. There is no right or wrong answer (judgment), so such a result is standard. If the two health conditions are less similar (one is quadriplegia, the other is rib fractures), it is likely that almost 100% of the responses will be that "rib fractures" is the "better" health condition. In case two health conditions are perceived as (almost) equally bad or good, then 50% of the respondents will probably choose A and 50% B. It is likely that most respondents will be indecisive and unable to decide between two conditions that are close substitutes. Nonetheless, they are forced to make a choice, and this choice process is probably arbitrary. All together this lead to spread (deviation) in the collected response data. As such, it is not "uncertainty" or making the wrong choice or arbitrary choice that pervades choice models but a degree of randomness in the responses. The random choices are generated by a diversity of response mechanisms, including attentional failures (slips) and failure in memory (lapses). All these response mechanisms are assumed to be based on a level of randomness (Roy, 1989; Shaban et al., 2004; Han et al., 2011).

REFERENCES

Arons, A.M.M., Krabbe, P.F.M., 2013. Probabilistic choice models in health-state valuation research: background, theories, assumptions and applications. Expert Review of Pharmacoeconomics and Outcomes Research 13 (1), 93−108.

Arons, A.M.M., Schölzel-Dorenbos, C.J.M., Olde Rikkert, M.G.M., Krabbe, P.F.M., 2016. A simple and practical index to measure dementia-related quality of life. Value in Health 19 (1), 60−65.

Attema, A.E., Edelaar-Peeters, Y., Versteegh, M.M., Stolk, E.A., 2013. Time trade-off: one methodology, different methods. European Journal of Health Economics 14 (Suppl. 1), S53−S54.

Bradley, R.A., Terry, M.E., 1952. Rank analysis of incomplete block designs: I. The method of paired comparisons. Biometrika 39 (3/4), 324−345.

Brogden, H.E., 1977. The Rasch model, the law of comparative judgment and additive conjoint measurement. Psychometrika 42 (4), 631−634.

Brown, T.C., Peterson, G.L., 2009. An enquiry into the method of paired comparison: reliability, scaling, and Thurstone's law of comparative judgment. General Technical Report RMRS-gtr-216WWW. U.S. Department of Agriculture, Forest Service, Rocky Mountain Research Station, Fort Collins, CO, p. 98.

Brown, T.C., Nannini, D., Gorter, R.B., Bell, P.A., Peterson, G.L., 2002. Judged seriousness of environmental losses: reliability and cause of loss. Ecological Economics 42 (3), 479−491.

Craig, B.M., Reeve, B.B., Brown, P.M., Cella, D., Hays, R.D., Lipscomb, J., Simon Pickard, A., Revicki, D.A., 2014. US valuation of health outcomes measured using the PROMIS-29. Value in Health 17 (8), 846−853.

Food and Drug Administration, 2009. Guidance for industry use in medical product development to support labeling claims guidance for industry. In: Guidance for Industry Patient-Reported Outcome Measures: Use in Medical Product Development to Support Labeling Claims. Silver Spring.

Froberg, D.G., Kane, R.L., 1989. Methodology for measuring health-state preferences. 2. Scaling methods. Journal of Clinical Epidemiology 42, 459e71.

Gafni, A., 1992. The Standard Gamble method — what is being measured and how it is interpreted. Health Services Research 29 (2), 207–224.

Green, P.E., Rao, V.R., 1971. Conjoint measurement for quantifying judgmental data. Journal of Marketing Research 8 (3), 355–363.

Hakim, Z., Pathak, D.S., 1999. Modelling the EuroQol data: a comparison of discrete choice conjoint and conditional preference modelling. Health Economics 8 (2), 103–116.

Han, P.K.J., Klein, W.M.P., Arora, N.K., 2011. Varieties of uncertainty in health care: a conceptual taxonomy. Medical Decision Making 31 (6), 828–838.

Keeney, R.L., Raiffa, H., 1976. Decision with Multiple Objectives: Preferences and Value Trade-offs. John Wiley, New York.

Konerding, U., 2013. What does Cronbach's alpha tell us about the EQ-5D? A methodological commentary to "psychometric properties of the EuroQol five-dimensional questionnaire (EQ-5D-3L) in caregivers of autistic children". Quality of Life Research 22 (10), 2939–2940.

Krabbe, P.F.M., 2008. Thurstone scaling as a measurement method to quantify subjective health outcomes. Medical Care 46 (4), 357–365.

Krabbe, P.F.M., 2013. A generalized measurement model to quantify health: the multi-attribute preference response model. PLoS One 8 (11), e79494.

Krabbe, P.F.M., Devlin, N.J., Stolk, E.A., Shah, K.K., Oppe, M., van Hout, B., Quik, E.H., Pickart, S.A., Xie, F., 2014. Multinational evidence on the feasibility and consistency of a discrete choice model in quantifying EQ-5D-5L health states. Medical Care 52 (11), 935–943.

Krabbe, P.F.M., Essink-Bot, M.L., Bonsel, G.J., 1997. The comparability and reliability of five health-state valuation methods. Social Science & Medicine 45 (11), 1641–1652.

Lancaster, K.J., 1966. A new approach to consumer theory. The Journal of Political Economy 74 (2), 132–157.

Langenhoff, B.S., Krabbe, P.F.M., Ruers, T.J.M., 2007. Computer-based decision making in medicine: a model for surgery of colorectal liver metastases. European Journal of Surgical Oncology 33.

Louviere, J.J., Flynn, T.N., Carson, R.T., 2010. Discrete choice experiments are not conjoint analysis. Journal of Choice Modeling 3 (3), 57–72.

Louviere, J.J., Hensher, D.A., Swait, J.D., 2000. Stated Choice Methods. Cambridge University press, Cambridge.

Louviere, J.J., Woodworth, G., 1983. Design and analysis of simulated consumer choice or allocation experiments — an approach based on aggregate data. Journal of Marketing Research 20 (4), 350–367.

Luce, R.D., 1959. Individual Choice Behavior: A Theoretical Analysis. Wiley, New York.

Luce, R.D., Tukey, J.W., 1964. Simultaneous conjoint-measurement — a new type of fundamental measurement. Journal of Mathematical Psychology 1 (1), 1–27.

Maas, A., Stalpers, L., 1992. Assessing utilities by means of conjoint measurement: an application in medical decision analysis. Medical Decision Making 12 (4), 288–297.

Marcum, J.L., 1947. A Statistical Theory of Target Detection by Pulsed Radar. Research Memorandum RM-754. The Rand Corporation, Santa Monica, CA.

Marschak, J., 1960. Binary-choice constraints and random utility indicators. In: Arrow, K.J., Karlin, S., Suppes, P. (Eds.), Mathematical Methods in the Social Sciences. Stanford University Press, Stanford, pp. 312–329.

Maydeu-Olivares, A., Böckenholt, U., 2008. Modeling subjective health outcomes: top 10 reasons to use Thurstone's method. Medical Care 46 (4), 346–348.

McFadden, D., 1974. Conditional logit analysis of qualitative choice behavior. In: Zarembka, P. (Ed.), Frontiers in Econometrics. Academic Press, New York, pp. 105–142.

McFadden, D., 2001. Economic choices. American Economic Review 91 (3), 351–378. Nobel Lecture, December 2000.

Miller, G.A., 1956. The magical number seven, plus or minus two: some limits on our capacity for processing information. The Psychological Review 63 (2), 91–97.

Mol, P.G.M., Arnardottir, A.H., Straus, S.M.J., de Graeff, P.A., Haaijer-Ruskamp, F.M., Quik, E.H., Krabbe, P.F.M., Denig, P., 2015. Understanding drug preferences, different perspectives. British Journal of Clinical Pharmacology 79 (6), 978–987.

Norman, R., Viney, R., Aaronson, N.K., Brazier, J.E., Cella, D., Costa, D.S.J., Fayers, P.M., Kemmler, G., Peacock, S., Pickard, A.S., Rowen, D., Street, D.J., Velikova, G., Young, T.A., King, M.T., 2015. Using a discrete choice experiment to value the QLU-C10D: feasibility and sensitivity to presentation format. Quality of Life Research 25 (3), 637–649.

Norman, R., Viney, R., Brazier, J., Burgess, L., Cronin, P., King, M., Ratcliffe, J., Street, D., 2014. Valuing SF-6D health states using a discrete choice experiment. Medical Decision Making 34 (6), 773–786.

Perline, R., Wright, B.D., Wainer, H., 1979. The Rasch model as additive conjoint measurement. Applied Psychological Measurement 3 (2), 237–255.

Peterson, R., Thurstone, L.L., 1932. The effect of a motion picture film on children's attitudes toward Germans. Journal of Educational Psychology 23 (4), 241–246.

Poulos, C., Kinter, E., Yang, J.-C., Bridges, J.F.P., Posner, J., Gleissner, E., Mühlbacher, A., Kieseier, B., 2016. A discrete-choice experiment to determine patient preferences for injectable multiple sclerosis treatments in Germany. Therapeutic Advances in Neurological Disorders 9 (2), 95–104.

Roy, B., 1989. Main sources of inaccurate determination, uncertainty and imprecision in decision models. Mathematical and Computer Modelling 12 (10/11), 1245–1254.

Roy, B., 1996. Multicriteria Methodology for Decision Aiding: Nonconvex Optimization and Its Application. Kluwer Academic Publishers, Dordrecht.

Russo, P.L., Chen, G., Cheng, A.C., Richards, M., Graves, N., Ratcliffe, J., Hall, L., 2016. Novel application of a discrete choice experiment to identify preferences for a national healthcare-associated infection surveillance programme: a cross-sectional study. BMJ Open 6, e011397.

Ryan, M., 2004. Discrete choice experiments in health care. BMJ 328 (7436), 360–361.

Salomon, J.A., 2003. Reconsidering the use of rankings in the valuation of health states: a model for estimating cardinal values from ordinal data. Population Health Metrics 1, 1–12.

Shaban, R., Wyatt-Smith, C., Cumming, J., 2004. Uncertainty, error and risk in human clinical judgment: introductory theoretical frameworks in paramedic practice. Australasian Journal of Paramedicine 2 (1), 1–11.

Stalmeier, P.F.M., Bezembinder, T.G., Unic, I.J., 1996. Proportional heuristics in time tradeoff and conjoint measurement. Medical Decision Making 16 (1), 36–44.

Stolk, E., Oppe, M., Scalone, L., Krabbe, P.F.M., 2010. Discrete choice modeling for the quantification of health states: the case of the EQ-5D. Value in Health 13 (8), 1005–1013.

Streiner, D.L., Norman, G.R., Cairney, J., 2015. Health Measurement Scales: A Practical Guide to Their Development and Use, fifth ed. Oxford University Press, Oxford.

Thurstone, L.L., 1927. The method of paired comparisons for social values. Journal of Abnormal and Social Psychology 21 (4), 384—400.

Thurstone, L.L., 1994. A law of comparative judgment. Psychological Review 101 (2), 266—270.

Versteegh, M.M., Attema, A.E., Oppe, M., Devlin, N.J., Stolk, E.A., 2013. Time to tweak the TTO: results from a comparison of alternative specifications of the TTO. European Journal of Health Economics 14 (Suppl. 1), S43—S51.

Zermelo, E., 1929. Die Berechnung der Turnier-Ergebnisse als ein Maximumproblem der Wahrscheinlichkeitsrechnung. Mathematische Zeitschrift 29, 436—460.

Chapter 12

Valuation Techniques

Chapter Outline

INTRODUCTION

A key area of health economics deals with the value of medical treatments and other interventions (e.g., diagnosis, devices). Various institutions across Europe perform economic evaluations to appraise new pharmaceuticals and devices. In Germany, Europe's largest health market, this work is done by the Institute for Quality and Economy in Health Services (Institut für Qualität und

Wirtschaftlichkeit im Gesundheitswesen—IQWiG). In the United Kingdom, it is the responsibility of the National Institute for Health and Clinical Excellence.

Economic evaluation is the comparison of two or more alternative courses of action in terms of costs and consequences (Drummond et al., 2005). One frequently applied type is cost-effectiveness analysis. In its most generic form (cost-utility analysis), the outcomes of medical interventions are expressed in a composite metric of both length and quality of life, namely the quality-adjusted life year (QALY) (Chapter 3).

None of the measurement methods discussed in this chapter belongs to the class of probabilistic choice models, as discussed in Chapter 11. Some of the methods may seem to resemble choice models in the way the judgmental tasks are performed, and they may even appear to be a typical choice task. However, there is a crucial difference between the methods introduced below and the methods presented in the preceding chapter. The valuation task in the methods discussed here consists of a health description and another element. The underlying mechanism is to make a trade-off (give up one thing to gain something else) between two different substances or commodities (health status versus risk to die or life years). A further distinction between valuation techniques and choice models is that the former ask respondents for a direct and absolute numerical expression. Nonetheless, the valuation techniques discussed in this chapter and the choice models discussed in Chapter 11 have something in common: both are preference based. For a deeper understanding of the content of this chapter, readers are advised to first look at Chapter 11, in particular at the "preference-based" section.

TERMINOLOGY

Before we proceed, we should explain several terms that have a special meaning in the measurement approaches discussed in this chapter.

Measurement, Valuation, and Technique

In health economics, the term measurement denotes the way respondents classify their own health status, whereas the term valuation denotes an exercise to produce values for (hypothetical) health-state descriptions. Scientists from other fields would probably refer to the latter as measurement instead of valuation and would call the former classification (i.e., nonmetric responses).

The methods introduced by health economists to elicit values for health states are generally called valuation techniques. One might ask, what is the difference between a method and a technique? Probably not much; roughly, a method has been formalized, whereas a technique has features and conditions that can be modified.

Utilities, Values, and Indices

Utility refers to a measure of the level of satisfaction gained from the consumption of a good or service (such as healthcare) (Drummond et al., 2005). When someone is faced with a number of alternatives, the choice falls on the one he or she prefers the most. In economic theory, this amounts to choosing the option that gives the individual the highest utility, i.e., what gives an individual the highest benefit, what makes him or her feel better. These preferences need to obey some distinct axioms to fit into economic theory.

Utilities are always represented as lying between 0.00 and 1.00. Regarding health, 0.00 is generally associated with death and 1.00 with full health. Only the standard gamble (SG) valuation technique (see below) produces utilities. The values deduced by another valuation technique, the time trade-off (TTO), are often referred to as utilities. However, the TTO (see below) does not conform to the expected utility theory in the sense that there is no element of risk in the assessment. Apart from the SG technique, the TTO and all other valuation techniques produce values, not utilities. Even so, these two terms are used interchangeably.

Just like utilities, an index is an absolute number that can be expressed in proportions (ranging from 0% to 100%). But unlike utilities, an index can aggregate various aspects (e.g., indicators, variables, attributes) into a single score (or index measure). As some of these are not preference based (many come from the field of clinimetrics; see Chapter 13), this book will use the term "index value" if a preference-based index measure is similar to a utility. In our case, this means that an index value is an absolute measure expressed in proportions; in most situations, it will also be based on aggregating several distinct health attributes (composite measure).

PREFERENCE-BASED MEASUREMENT SYSTEMS

From the mid-1970s, considerable progress has been made, mainly by health economists and mathematically trained researchers (e.g., process engineers), in developing empirically derived health functions. Clinical researchers no longer have to conduct their own exercises with valuation techniques that are difficult to perform. Instead, this part is done once by other researchers for a set of health states based on a classification system. The result is a connected set of systems: the health classification (or descriptive) system and the valuation system are referred to collectively as preference-based measurement systems (Torrance et al., 1996; Dolan, 1997). A distinguishing feature of these systems (which are also known as multiattribute utility instruments, multiattribute preference-based health classification instruments, or preference-based measures) is their potential for assigning a single numerical value to any health state that is defined by the classification system. Such preference-based measurement systems have also been developed for choice models (Chapter 11).

Each of these systems has two components. One is the classification (or descriptive system). It consists of a set of questions (attributes) with response categories (for example, "mobility": excellent physical functioning, in a wheelchair, quadriplegia) or levels (for example, "mobility": no problems walking, some problems walking, severe problems walking) to describe a person's health (Fig. 12.1). The other is the formula (or algorithm), which converts the attribute responses into an index value (utility). Each instrument [e.g., EQ-5D, HUI-3, Quality of Well-Being scale (QWB)] has its own health classification system and a unique preference-based scoring formula. These formulas result from dedicated and complex valuation studies in which numerous health-state descriptions based on the classification system are subjected to specific valuation techniques [e.g., SG, TTO, visual analogue scale (VAS)]. The advantage of health descriptions based on the attributes of the classification is that, if a suitable subset of health states is empirically valued, the index value of the remaining states can be estimated.

The EQ-5D, SF-6D, HUI-3, AQoL, and 15D produce index values that are widely employed to estimate QALYs (Chapter 3) for economic evaluations (Richardson et al., 2015a,b). They are applicable across all health conditions but differ in the size and content of the classification systems (Table 12.1). Numerous studies have investigated their empirical validity (as far as possible),

Mobility
No problems in walking about ○
Some problems in walking about ○
Confined to bed ○

Self-Care
No problems with washing or dressing self ○
Some problems with washing or dressing self ○
Unable to wash or dress self ○

Usual Activities
No problems with performing usual activities ○
 (eg. work. study, housework, family or leisure activities)
Some problems with performing usual activities ○
 (eg. work. study, housework, family or leisure activities)
Unable to perform usual activities ○
 (eg. work. study, housework, family or leisure activities)

Pain/discomfort
No pain or discomfort ○
Moderate pain or discomfort ○
Extreme pain or discomfort ○

Anxiety/depression
Not anxious or depressed ○
Moderately anxious or depressed ○
Extremely anxious or depressed ○

FIGURE 12.1 Classification system of the EQ-5D-3L (five attributes, each with three levels).

TABLE 12.1 Properties of the Major Preference-Based Measurement Systems

	QWB	15D	EQ-5D	HUI 3	SF-6D	AQoL-8D
Theory	MAUT	MAUT	Statistical	MAUT	Statistical	MAUT/statistical
Scaling	RS	RS	TTO and RS	SG/RS	SG	TTO
Best health	1.00	1.00	1.00	1.00	1.00	1.00
Worst health	0.32	0.11	−0.59	−0.36	0.20	−0.04
Valuing method	VAS	VAS	TTO	VAS/SG	SG	TTO
Combination model	Additive	Additive	Additive	Multiplicative	Additive	Multiplicative

MAUT, multiattribute utility theory; *RS*, rating scale; *SG*, standard gamble; *TTO*, time trade-off; *VAS*, visual analogue scale.
Adapted with permission from Richardson, J., Khan, M.A., Iezzi, A., Maxwell, A., 2014a. Comparing and explaining differences in the content, sensitivity and magnitude of incremental utilities predicted by the EQ-5D, SF-6D, HUI 3, 15D, QWB and AQoL-8D multi attribute utility instruments. Medical Decision Making 35 (3), 276–291; Richardson, J., McKie, J., Bariola, E., 2014b. Multi attribute utility instruments and their use. In: Culyer, A.J. (Ed.), Encyclopedia of Health Economics. Elsevier, San Diego, pp. 341–357; and Richardson, J., Sinha, K., Iezzi, A., Khan, M.A., 2014c. Modelling utility weights for the Assessment of Quality of Life (AQoL) 8D. Quality of Life Research 23, 2395–2404.

and the outcomes have been summarized in a number of systematic reviews (Richardson et al., 2014a,b,c).

The literature is dominated by publications on a small number of preference-based measurement systems. A review of articles listed on the Web of Science between 2005 and 2010 found 1663 studies that had employed a preference-based measurement system. Of these, 61% used the EQ-5D; 15% the HUI-2 or HUI-3; 9% the SF-6D; and the remaining 15% used the 15D, QWB, or one of the new Assessment of Quality of Life (AQoL) instruments (Richardson et al., 2014a,b,c).

Inspection of the preference-based measurement systems reveals a fair amount of heterogeneity. The systems vary in terms of the number and type of attributes included, the number of response categories a respondent has to review, the abstractness and concreteness of the phrasing representing the attributes, the response task, and so on.

Content of Classification Systems

Any classification system is a compromise between, on one hand, the comprehensiveness (attributes) and refinement (number of response categories or levels) of the classification, and, on the other hand, the feasibility and accuracy of the evoked assessments. Methodologically, this compromise is about the validity (and to some extent indirectly the reliability) of classification systems. A recurrent question is whether the number of attributes of a classification system should be extended or reduced. Similarly, the number of levels within an attribute is often questioned.

Any preference-based measurement system should be based on a limited set of key attributes because respondents can process only a small amount of information simultaneously. As empirical studies show (and theories underpin this), a limited set of key attributes may be sufficient to describe overall health (Brazier et al., 2007; Arons et al., 2016). Crucially, only the most important and relevant health attributes should be included; while nonkey attributes might increase the content validity marginally, they would make the assessment tasks substantially more difficult.

Like the preference-based choice methods described in Chapter 11, valuation techniques also require respondents to assess compound health-state descriptions (containing all the attributes of the classification system). There is an understandable tendency to include everything that might be of any interest to anyone and to work with fine enough gradations of "severity" within each attribute of health to pick up any effects of healthcare treatment that might be of interest to a practitioner. Ideally, each respondent should assess as many states as possible, if not all potential health states. However, the number of potential health states rises rapidly by adding attributes or categories. An instrument with two categories in each of three attributes generates eight (2^3) health states, whereas one with six attributes, each with four categories, generates 4096 (4^6) states. It is important to recall that expanding the classification system will explode the number of combinations and thus the number of health states. A classification based on ten attributes, with ten categories or levels for each attribute, would produce 10,000 million (10^{10}) different health states. The classification system therefore needs to be simple, using as few attributes as possible and limiting the categories for each attribute. As a further consideration, the description of a health state needs to be fairly short. Moreover, it should be sufficiently clear so that the respondent can identify differences between the states, particularly those with only one level of difference in categories. It is therefore advisable to present the categories or levels of the attributes for each health state as bullet points rather than in a more narrative style (Gudex, 2005).

To create a generic instrument, the attribute should be relevant to patients across the healthcare spectrum as well as to members of the general population. Thus, there should be no mention of specific diagnoses, diseases, or treatments, while disease-specific items, such as symptoms, would not be included. These are not hard and fast rules, but it is important to keep these considerations in mind.

Three of the instruments, namely EQ-5D, HUI-3, and 15D, are predominantly constructed of attributes that relate to physical health (Table 12.2). The SF-6D has an equal number of items in the two broad domains of physical and psychosocial health, and the AQoL-8D has most of its attributes in the psychosocial domain. Conceptually, HUI-3 has a "within the skin" classification system: it focuses upon an individual's bodily functions. The other instruments are conceptualized primarily, but not exclusively, in terms of handicap (more recently described by the WHO as activity and participation), i.e., the effect of a health state on a person's ability to function in a social environment. Yet, in three large-scale surveys containing five preference-based measurement systems, it was found that, on average, only 56, 42, and 57%, respectively, of the variance of one instrument could be explained by another instrument (Fryback et al., 2010; Hawthorne et al., 2001; Richardson et al., 2014a,b,c). Richardson et al. (2015a,b) therefore attribute these differences primarily to differences in the classification systems.

Analytical Steps

Five steps are necessary to arrive at index values for health states constructed for a preference-based measurement system. Each step represents a distinct analytical element in the development and application of such systems. For the EQ-5D-3L, these steps will be briefly explained below (Fig. 12.2).

Step I: Defining Health

The EQ-5D describes health according to five attributes: mobility, self-care, usual activities, pain/discomfort, and anxiety/depression. Each attribute has three levels, i.e., "no problems" ("1"), "some problems" ("2"), "severe problems" ("3"). Health states are constructed by taking one level for each attribute

TABLE 12.2 Comparison of the Attributes and Content of Six Preference-Based Measurement Systems

	QWB[a]	15D	EQ-5D-5L	HUI 3	SF-6D	AQoL-8D
Physical						
Physical ability/mobility/vitality/(coping/control)	••	••	•	••	•	••
Bodily function/self-care		•••	•			•
Pain/discomfort		•	•	•	•	••
Senses		••		••		•••
Usual activities/work		•	•		•	••
Communication		•		•		•
Psychosocial						
Sleeping			•			••
Depression/anxiety/anger		•	•	•	•	•••••
General satisfaction		•••				••••
Self-esteem						•••••
Cognition/memory ability				•		
Social function/relationships	•				•	••••
(Family) Role					•	•
Intimacy/sexual relationships		•				•
Total items/symptoms		15	5	8	6	35
Health states described		3.1×10^{10}	3125	972,000	18,000	2.4×10^{23}

[a]QWB has three items relating to mobility and physical and social health, plus 27 symptom groups, which include, inter alia, "spells of feeling upset, being depressed or crying," "trouble sleeping," and "excessive worry or anxiety."

With permission from Richardson, J., Khan, M.A., Iezzi, A., Maxwell, A., 2014a. Comparing and explaining differences in the content, sensitivity and magnitude of incremental utilities predicted by the EQ-5D, SF-6D, HUI3, 15D, QWB and AQoL-8D multi attribute utility instruments. Medical Decision Making 35 (3), 276–291; Richardson, J., McKie, J., Bariola, E., 2014b. Multi attribute utility instruments and their use. In: Culyer, A.J. (Ed.), Encyclopedia of Health Economics. Elsevier, San Diego, pp. 341–357; and Richardson, J., Sinha, K., Iezzi, A., Khan, M.A., 2014c. Modelling utility weights for the Assessment of Quality of Life (AQoL) 8D. Quality of Life Research 23 (8), 2395–2404.

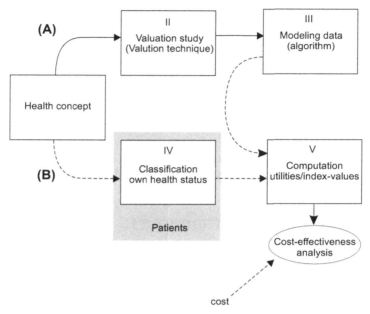

FIGURE 12.2 Steps in the construction of a preference-based measurement system (A, research study; B, application, shaded area is stage in which patients are involved).

(e.g., the best health state is represented by 11111). Theoretically, this set of attributes and levels of the EQ-5D allows for 243 (3^5) different health states.

Step II: Eliciting Values

Assigning numerical values to EQ-5D health states is the second step in the process of arriving at index values for health states. Several valuation techniques are available to perform this task. This process also involves deciding which subjects should perform the valuation task (e.g., physicians, patients, or the general population). A large study conducted in the United Kingdom entailed the valuation of a subset of 45 EQ-5D health states by 2997 members of the general public (Dolan, 1997). In this interviewer-based study, values were elicited by the TTO technique using visual props.

Step III: Estimating Value Function

In the next step, a mathematical function is estimated to predict the index values for the complete array of possible health states (all the combinations allowed by the classification system), as illustrated by the empirically valued sample in step II. In 1997, Dolan published the now widely applied formal mathematical model to convert EQ-5D scores to a single index value. The data were analyzed by a generalized least-squares regression technique in which the functional form is additive with the addition of a restrictive interaction term (the so-called N3 term).

Step IV: Classifying Health

The EQ-5D classification (Fig. 12.1) is the standard layout for the five attributes when recording an individual's current EQ-5D health state. Often the classification is referred to as a "questionnaire."

Step V: Computing Values

The responses derived with the EQ-5D classification in the previous step are merged with the algorithm estimated in step III, which produces the EQ-5D index values.

ECONOMIC VALUATION TECHNIQUES

A basic tenet of economics is that there are unlimited human wants that are to be met by limited resources. This amounts to what economists call scarcity, a central concept in economic reasoning. Scarcity is closely related to choices between goods and services, and it includes the need to sacrifice one alternative to receive another (i.e., opportunity costs). Trade-off methods elicit utilities or index values expressing an individual's preference for a particular health state under a condition where something has to be sacrificed (e.g., health, life years, budget). This mechanism has been used in valuation techniques developed by economists.

In this overview, we only deal with valuations of health that are derived from a sample of respondents. In the area of clinical decision making, individual patients are often involved in eliciting values for health states that concern possible outcomes related to their own disease and optional treatment modalities. The measurement properties of patient-derived values of this type will not be discussed or presented in this book. The main reason to refrain from incorporating such values is that clinical decision making is an entirely different area of research, with different goals and different premises (McNeil et al., 1981, 1982; Hunink et al., 2014).

The two most widely accepted ways to elicit index values are the SG and the TTO techniques. The former emerged from the field of economics (von Neumann and Morgenstern, 2004), the latter from the area of operations research (Torrance, 1976). Health index values have also been derived by another technique, the VAS (Torrance et al., 2001; Krabbe et al., 2006; Parkin and Devlin, 2006).

Standard Gamble

The standard decision model utilized by scientists in analyzing decision problems is derived from expected utility theory. The theory and its measurement approaches were developed in the fields of microeconomics and operations research as a model for individual decision making under uncertainty. Utilities, in this model, are (cardinal) numbers representing the strength of preferences for particular outcomes. This axiomatic theory of rational decision making under uncertainty was presented in 1944 by the mathematician John von Neumann and the economist Oskar Morgenstern. Their theory is normative or prescriptive; that is, it prescribes how a rational individual should make decisions when faced with uncertain outcomes to increase his/her welfare in the most efficient way. Known as expected utility theory, it has been widely applied in operations research for business, government, policy, and many other fields for more than 3 decades. Medical decision science has adopted expected utility theory and its accompanying methodology to derive preference measures (utilities) of patients for medical treatment decisions (Gafni, 1994).

The SG valuation technique is based on choices between lotteries with uncertain prospects. The values derived from the SG are usually referred to as "utilities" and are regarded as measures of the strength of preferences if the respondents adhere to the von Neumann-Morgenstern axioms of expected utility (von Neumann and Morgenstern, 2004). When measuring the preferences for a chronic health state, the risk or probability of dying of a treatment, which if successful would result in perfect health, is traded until the respondent is indifferent between the choice between accepting the treatment or not (Fig. 12.3).

For years, the SG was considered the gold standard because it had been developed under expected utility theory. One of its earliest published applications in the context of measuring health status involved an experiment where two physicians would draw imaginary pills that would cure their patient but with a risk of drawing a pill that would kill the patient (Ginsberg and Offensend, 1968). Other applications involve choosing to remain in less than full health for a period of time or to undergo a medical intervention with a chance of either being restored to perfect health or being killed by it. The probability indicating the chance that the respondent is returned to full health (or 1 minus the risk of death the respondent would take in order to regain full health) is interpreted as the value of the lesser health state. This method has been used for clinical or patient decision making but also for eliciting index values for hypothetical health-states descriptions (McNeil et al., 1981, 1982).

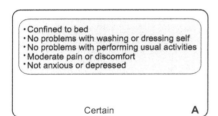
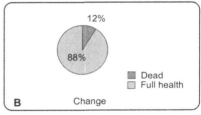

FIGURE 12.3 Example of the standard gamble valuation technique.

However, as empirical research has shown, people's behavior systematically violates the underlying assumptions of the von Neumann and Morgenstern utilities. People have difficulty working with probabilities, and they may be averse to taking risks (Bleichrodt, 2002). Richardson (1994) has mentioned the mismatch between the experimental (p) of the SG technique and the empirical probability (p^*) of the risk of an intervention.

Time Trade-Off

The TTO was developed by George Torrance (Torrance et al., 1972, 1973; Torrance, 1976) as a less complicated but conceptually different alternative to SG that would be easier to administer (Box 12.1). While the principle of trade-offs is similar to the SG, the concept of probability is replaced by time. The TTO valuation task requires respondents to trade longevity of life for improved health when making choices between certain prospects. The duration of 10 years is conventionally used in TTO tasks as a compromise between avoiding proximal mortality (i.e., not dying too soon) and projecting a realistic horizon for older respondents, whose life expectancy might not exceed 10 years. For the TTO as a basis for constructing QALYs, researchers have developed an axiomatic theory analogous to the von Neumann–Morgenstern (Pliskin et al., 1980; Miyamoto and Eraker, 1985).

The values deduced by the TTO are often called utilities. However, as already mentioned, the TTO does not conform to the expected utility theory in the sense that no element of risk is involved in the assessment. As such, TTO tasks produce (index) values and not utilities. In the TTO task, the respondents are required to trade-off life years against health. The first alternative specifies a (suboptimal) health state with a given duration, often 10 years. The competing alternative offers a better health status (conventionally, optimal or full health) of shorter duration (Fig. 12.4). The point of indifference is reached by varying the number of life years spent in full health. At this point the respondents cannot choose between 10 years in the suboptimal state and years in full health. In the standard case, where the better health state is "full health," the number of life years in full health divided by 10 is the "utility" of the suboptimal state.

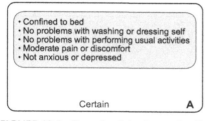
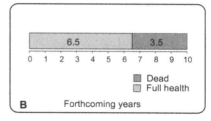

FIGURE 12.4   Example of the time trade-off valuation technique.

The TTO technique may not be effective in the upper/healthy range of the scale for valuing and differentiating among health states. When respondents (either persons from the general population or patients) perceive medical conditions or health states as relatively mild, they may be unlikely to accept any reduction in lifespan. If a respondent is not willing to trade time, the TTO task suffers from a ceiling effect. The ability to provide valid TTO responses implies that the respondent comprehends the notions of extended time horizons and mortality. Several studies have reported that a small number of participants cited religious beliefs as the reason why they would not trade time or complete the TTO task. TTO index values have occasionally been found to differ by characteristics such as education level, quantitative skills, gender, marital status, and being a parent or caregiver (Boye et al., 2014). While some problems with the TTO method exist regardless of the time horizon, other limitations may be linked to the commonly used 10-year horizon. Researchers have noted that this length of time presents most respondents with unrealistic or implausible choices. For respondents in their 60s or 70s, the 10-year time horizon may seem reasonable. However, it may seem unrealistically short to younger respondents and unexpectedly long to respondents in their 80s or 90s. It has been suggested that this problem can be avoided by adjusting the time horizon of the TTO task to align with each individual's additional life expectancy (Heintz et al., 2013; Matza et al., 2015). However, such an alteration is likely to induce bias again. Another complication of the TTO is that the relationship between a health state, its duration, and its value. The problem is that this requires the values for health states to be independent of the duration of these states (see also Chapter 14).

Despite these impediments, the prevailing approach to quantify health states, certainly in the field of health economics, is the TTO. This technique may be intuitively appealing for three reasons. First, it seems to reflect the actual medical situation. Second, it shows some correspondence to the general health-outcome framework (since TTO is essentially a QALY equivalence statement). And third, it is grounded in economic thinking (the trade-off principle).

BOX 12.1 George Torrance

George Torrance is one of the founding fathers of economic evaluation in health and was the first researcher at McMaster University (Hamilton, Canada) in this field. He joined the University in 1967 and is now professor emeritus in the Department of Clinical Epidemiology and Biostatistics. Torrance was a pioneer in the development and application of methods to improve healthcare decision making, and he reformed the science and practice of assessing and determining health status (Fig. 12.5).

Continued

Box 12.1 George Torrance—cont'd

FIGURE 12.5   George Torrance

OTHER VALUATION TECHNIQUES

There are other ways to arrive at values that express the quality of health. Two of the three introduced below are based on a trade-off mechanism. The third is the VAS, which has been discussed earlier in a measurement context (Chapter 5). A brief overview of the main measurement characteristics of the most prominent preference-based methods is presented in Table 12.3.

Magnitude Estimation

Magnitude estimation was initially adopted in the field of psychophysics (Wegener, 1982), where it was used to reveal the relation between physical sensory attributes (e.g., light intensity, loudness) and their subjective sensation (Stevens, 1951). Basically, a subject is asked to provide a subjective ratio by assigning numbers to a certain object in proportion to the number that has been assigned to a reference object (Stevens, 1957). The objective of magnitude estimation is to make a comparable subjective quantification for identical objects. An adaptation of magnitude estimation in the field of health research is the equivalence technique, first introduced by Patrick et al. (1973). While it has been applied in various forms, there is a common underlying task: for example, to decide how many people in health condition A are equivalent to a specified number of people in condition B. One of the earliest applications of an equivalence technique ("How many times more ill is a person described as being in state 2 as compared with state 1") was used by Rosser and Kind (1978).

TABLE 12.3 Characteristics of Various (Preference-Based) Measurement Methods

Measurement Method	Type of Judgment	Outcomes [on Group Level]	Response
Standard gamble	Absolute judgment	Utilities	Preference
Time trade-off	Absolute judgment	Index values	Preference
Person trade-off	Absolute judgment	Index values	Preference
Visual analogue scale	Absolute judgment	(Index) values	Preference/ Scaling
Magnitude estimation	Absolute judgment	Values	Scaling
Ranking	Ordinal judgment	Proportions [values]	Preference
Paired comparisons	Ordinal judgment	Proportions [values]	Preference

Person Trade-Off

More recently, extended versions of equivalence techniques have been advanced. One, known as "person trade-off" (PTO), has been used mainly in the area of policy-making (Murray and Lopez, 1996a,b). It was named by Nord (1992, 1995), but the technique itself had been applied earlier by Patrick et al. (1973). The PTO asks respondents to answer from the perspective of a social decision maker considering alternative policy choices. It has been adopted for the assessment of disease severity by the Global Burden of Disease researchers, asking health workers to make judgments about the trade-off between quality and quantity of life. They asked, for instance: Would you prefer to save 1 life year for 1000 healthy people or 1 life year for a larger number of people in less perfect health?

Severity weights are determined by the number of people with a specified disorder whose total health is thought to be equal to that of 1000 healthy people. For example, if participants estimate that 1000 healthy people should have the same claim on resources as 8000 people with a disorder causing a severe disability, then the severity weight assigned to that disorder is equal to 1 minus (1000/8000), or 0.875 (the utility or index value would be: 1 minus 0.875). Likewise, if 1000 healthy people were thought to have a claim on resources equal to that of 2000 people with a disorder causing a less severe disability, then the severity weight assigned to this disorder would be 1 minus (1000/2000), or 0.5.

It is important to recognize that equivalence techniques do not necessarily yield genuine preferences for health because distributive justice considerations are introduced into the valuation task. Therefore, it is debatable whether the equivalence technique and the person trade-off technique can be regarded as measurement methods or instead should be classified as value judgment methods. To some extent, the TTO technique can also be considered a value judgment method, as it involves assigning a value to health conditions (health states) and to (remaining) life years. Only if all the assumptions about TTO hold, does it produce real values for health states.

Visual Analogue Scale

Alternative techniques to assess the value of health have been adopted from the social sciences. Probably the most renowned and popular one is the VAS. Conventionally, an individual uses it to indicate (i.e., to score or value) his or her own actual health. Alternatively, it is used to derive values for a set of hypothetical health states that are simultaneously assessed on one single VAS (multiitem VAS). When applied to health states, the respondents are presented with a scale in the form of a line, often resembling a thermometer, which has clearly defined endpoints, say 0.00 and 1.00 or 0 and 100. These endpoints are "fixed" in the sense that one end of the scale should represent the least preferred or "worst" health state, and the other end should represent the most preferred or "best" health state. Typical endpoints are "full health" and either "dead" or "worst imaginable health status." With these endpoints in place, health-state descriptions are then inserted on the scale such that a respondent evaluating health states A, B, and C must consider whether A is preferable to B, B to C, and A to C. The respondent also has to decide on the strength of these preferences. When respondents position numerous health-state descriptions on a single VAS, the assumption is that they are implicitly making comparisons and subsequently making decisions about which states they prefer and, in the next step, are adjusting the distances between the states accordingly. This would mean that the respondent views a change in health from 40 to 50 on the VAS as equivalent to a change from 70 to 80, thus meeting at least the interval scale requirement so that intervals or differences can be equated (Brooks, 1995). In contrast to economic trade-off techniques, the VAS does not trade off any secondary object or commodity (e.g., risk or life years) but instead scales health directly.

There is a longstanding theoretical debate about whether or not interval properties can be ascribed to the VAS. Economists claim that responses to the SG and TTO have interval scale properties, whereas responses to rating scales, including the VAS, do not. Parkin and Devlin (2006) argued that there is no more evidence for interval scale properties in TTO and SG responses than in VAS responses. In search of empirical evidence that mean health-state values collected with a multiitem VAS can be characterized roughly as interval data, a study using a rank-based scaling method (unfolding) observed a very strong relationship that supports the interval scale property of the VAS data (van Agt et al., 1994). Confirmation was found in a study that applied nonmetric multidimensional scaling on data (metric and ranks) that were derived from VAS values (Krabbe et al., 2007).

Empirical studies have exposed a number of potential defects in the VAS. It is sensitive to certain biases (e.g., anchor-point, context, end-aversion). To some extent, these biases can be controlled in carefully designed studies (Anderson, 1979). Context bias (also called "central tendency bias," "response spreading," or "spacing-out bias") means that the VAS values for health states depend on the range spanned by the health states and the relative frequencies of health states within this range (Parducci, 1968). The presumed independence of the set of health states to be positioned on the VAS has been rejected in two Dutch studies and one Australian study (Bleichrodt and Johannesson, 1997a,b; Krabbe et al., 2006; McPhail et al., 2010). They clearly show that different values will be collected with a multiitem VAS for a fixed set of health states if these are presented along with varying other states. Another issue, by no means confined to this technique, is the "framing" problem, where scale values may depend on the precise way in which questions are asked. In addition, the choice and phrasing of the anchors will lead to different results (Ubel et al., 2005). Anchor-point bias occurs when people interpret anchor points (e.g., dead, best imaginable health state, full health) differently and thus assign them different values. Torrance et al. (2001) showed the systematic difference between values derived with the SG and with the VAS (Fig. 12.6). Systematic differences can be observed between other valuation techniques too, so choosing one over another has consequences for the outcomes.

WHOSE VALUES?

Conventionally, index values for the health states used in economic evaluations are derived from a representative community sample (Gold et al., 1996). In the original DALY approach, index values for disease states were obtained from medical experts (Murray, 1996). There is broad consensus that, when possible, a patient perspective should be adopted for measuring health outcomes (Coon and McLeod, 2013; Hodgkin and Taylor, 2013). Nonetheless, health economists usually derive their measures from a cross-section of the population, arguing that it is the community, as taxpayers, who pay for health programs. However, this argument does not seem compelling (Richardson, 2002). Taxpayers do not specify how their money should be spent in other contexts. For instance, we do not vote on the composition of the armed forces or the location of roads. Another argument is that the most frequently used

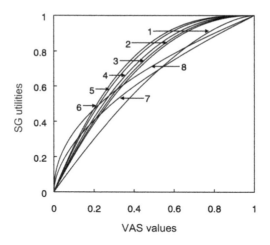

FIGURE 12.6   Relationship between mean standard gamble values and mean visual analogue scale values for health states from eight different studies. *Reproduced with permission from Torrance, G.W., Feeny, D., Furlong, W., 2001. Visual analog scales: do they have a role in the measurement of preferences for health states? Medical Decision Making 21, 329–334.*

valuation technique, the TTO, is susceptible to adaptation effects, which may be a reason not to use input from patients. For example, it has been noted that patients may adapt to their health state over a period of time. As a result, they may assign higher values to their own poor health state (e.g., chronically ill patients may consider themselves healthy). Patients may also strategically underrate the quality of their health state, knowing they will directly benefit from doing so. As a result certain patient groups may be considered more relevant by policy makers, or cost-effectiveness studies may show more favorable results.

The proposition upheld here is that while adaptation is a real phenomenon among patients, it can largely be reduced and eventually eliminated if health values are derived in a fitting measurement framework. For this reason, it is pertinent to develop alternative measurement approaches based on patient responses that are not sensitive to an adaptation mechanism.

One very important (though implicit) assumption regarding the valuation of health is that people not only understand the task but are also capable of imagining a broad range of different health states and could thereby arrive at comparable appraisals. It presumably holds even if people do not have personal experience with the health states to be valued. If this assumption does not hold, the validity of the elicited values will be uncertain at the valuation stage and subsequently in all of the computational steps.

It is reasonable to assume that healthy people are inadequately informed or lack the imagination to make appropriate judgments on the impact of (severe) health conditions. This is one reason why health researchers are debating which values are more valid (Brazier et al., 2009; Krabbe et al., 2011). Nevertheless, many researchers assert that individuals are the best judges of their own health (Gandjour, 2010; Krabbe et al., 2011). Therefore, in a health-care context, it is sensible to defend the position that it is the patient's judgment that should be elicited to arrive at health-state values, not that of a sample of unaffected members of the general population. Voices from another area have also stressed that assessments by patients (experienced utility) should get more attention (Kahneman and Sugden, 2005; Dolan and Kahneman, 2008).

Of course, it may turn out that well-informed healthy people come up with the same value for health states as patients who are really in such a health state. Many studies have investigated the potential differences in valuing health states between the general population and people who actually experience that particular illness (see De Wit et al., 2000 for an overview). Some indicate that this is not the case, but so far, no carefully planned, systematic studies on this issue have been conducted (de Wit et al., 2000; Ubel et al., 2003; Krabbe et al., 2011). Several authors attribute these differences largely to the measurement methods (Hakim and Pathak, 1999; Krabbe et al., 2011). Guidelines on cost-effectiveness research in health recommend using values that are representative of the general population, although the recent reprint of a standard textbook reveals that the arguments in favor of this choice are under attack (Gold et al., 1996).

ASSUMPTIONS AND ARGUMENTS

Most studies on the elicitation of utilities or index values for health states use a response modality that mirrors the clinical situation. Response biases can be introduced by configuring valuation tasks in a manner that resembles/mimics the clinical problem. For instance, in regard to hypertension treatment (Johannesson et al., 1991), patients were asked to indicate the highest user fees they were prepared to pay (see also Chestnut et al., 1996). Willingness-to-pay operationalizations, represented as an insurance problem, are based on the same disputable assumption that underlies the combination of SG with surgery (risk situations) and TTO with chronic states. For example, an SG response mode resembles the direct, short-term risk of a surgical procedure. TTO resembles the chronic health condition that can be improved, though with the consequence of fewer life years. The implicit assumption is that this resemblance to clinical situations adds to the validity of the assessments.

One of the theoretical arguments often raised to justify using SG in cost-effectiveness (i.e., cost-utility) studies is somewhat puzzling. The use of SG has been defended on the premise that many medical decisions involve choices between different outcomes under conditions of uncertainty. It is clear, however, that the context and perspective of patient decisions and allocation decisions (i.e., cost-effectiveness analyses) are different and largely unrelated.

So, for individual decision making, the SG (or TTO) may be preferable and even theoretically superior to other valuation techniques. But for allocation matters and other purposes, this is not self-evident. Nor is it clear that SG is relevant to individual clinical decisions. The fact that the SG involves uncertainty as well as many clinical decisions, however, does not mean they correspond perfectly (Richardson, 1994). The types of risks encountered in the clinical setting may be quite different from those presented in the valuation technique of the SG.

Assessment by means of TTO and SG requires the presence of a trained interviewer or specialized computer programs. Because of these resource needs, these techniques are generally time and resource intensive. Therefore, such valuation tasks are not applicable in clinical studies. A more serious concern is the observation that major differences in TTO values result from such interviewers (Shah et al., 2015; Yang, 2015).

MEASUREMENT PROPERTIES

It is not surprising that the results that can be found about the (psychometric) measurement properties of preference-based measurement systems and the underlying valuation techniques are heterogeneous. Most comparisons of valuation techniques were conducted years ago. Certainly in the beginning, the studies were relatively small, often clinically oriented, and there was less agreement on how valuation techniques should be performed. Regarding the use of valuation techniques in clinical decision making (not discussed in this book), many measurement properties cannot be (directly) estimated on an individual basis or are difficult to perform (Roest et al., 1997). Moreover, each valuation technique confronts a person with a cognitive task that differs from that used in other techniques. In addition, several of the techniques have multiple versions that frame the tasks differently (Krabbe, 2014). In general, studies comparing valuation techniques can be differentiated in terms of the type of descriptions of the health states, selection of study population, number of health states, and types of health states. And the health states themselves can be divided into hypothetical states and actual or optional health states pertaining to treatment outcomes or particular stages of disease.

Reliability

Health profile instruments (e.g., SF-36) are usually used to determine the position of some characteristic on specific domains, which are measured by multiple items (Chapter 9). Preference-based measurement systems, in contrast, are used to place a value on health conditions (i.e., health outcomes), which is an entirely different task and therefore produces another type of data. As mentioned in Chapter 11, the principal difference between these two approaches is that the former (i.e., profile) measures characteristics of subjects (patients), whereas the latter (i.e., preference) measures characteristics of objects (health states) (Krabbe et al., 1997). Each is associated with a particular way of managing the collected data to investigate psychometric aspects.

A complicating factor is that preference-based measurement systems have two components: (1) the classification system in which patients have to check the categories or levels of each of the attributes; and (2) the values for a set of health states that formed part of the valuation study. For each of these two components, only a few reliability and validity statistics can be estimated.

Test—Retest

Test—retest reliability concerns the reproducibility of a measurement method. If a method is reliable, it should evoke the same outcomes on a second occasion if there is no change expected. The most appropriate way to ascertain reliability is by computing an intraclass correlation between the first measurement (test) and a subsequent measurement (retest). For preference-based measurement systems, test—retest reliabilities can be estimated for the scores on the separate attributes of the classification system and for the computed index values based on these scores.

Internal Consistency

In a preference-based measurement system, the set of attributes must cover all relevant aspects of the object under investigation. Ideally, the set will reflect independent health aspects because the information provided by every single attribute has to be maximal. Selecting attributes in keeping with this ideal will produce a heterogeneous set. Consequently, neither Cronbach's α nor item—total correlations for such an instrument can be particularly high. It would even be unwise to strive for a high Cronbach's α and high item—total correlations. Selecting attributes with an eye to maximizing these statistics would in fact impede the construction of a measurement instrument that covers all relevant aspects of health (Konerding, 2013). Therefore, estimation of internal consistency reliability is not suitable for the scores of the respondents on the attributes (classification). On the valuation part of the instrument (e.g., TTO), the respondents are not scoring distinct attributes. Instead, the attributes are presented to them as narratives in the form of a text, which should be read and mentally processed as a whole.

For example, the attributes of the EQ-5D instrument (mobility, self-care, usual activities, pain/discomfort, and anxiety/depression) were selected to cover all the different aspects of health that were at that time, somewhere around 1994 (van Dalen et al.), considered relevant to measure health-related quality of life. It is not assumed, nor is it very likely, that someone who is experiencing problems with self-care will also experience pain. Some attributes may be largely independent of each other. Consequently, a high Cronbach's α is neither expected nor desired for the scores on this set of attributes.

Interrater

With multiitem profile health instruments, the goal is usually to measure a specific position of a person on a domain (Chapter 9). Assessment of health with one of the valuation techniques is based on another response task and yields another type of data. Instead of persons, objects (health conditions or health states) are positioned on a scale.

However, from a more fundamental measurement perspective, interrater reliability can indicate whether basic (measurement) requirements are fulfilled. It may be important to determine how similar people's judgments actually are. Heterogeneous responses (or even distinct response structures) suggest that a measurement method is less appropriate, as it may not yield largely comparable or unidimensional responses. This leads to an important question: how well do the responses on health conditions represent the value of these conditions?

That question can be addressed with intraclass correlation statistics or mathematical routines closely related to factor analysis (Krabbe, 2006). We may want to assess how consistently individuals perform their tasks when valuing health states, raising the issue of interrater reliability. To compute the reliability coefficient for all health states together, a global interrater coefficient can be estimated with a particular analysis of variance. Formally, this coefficient is a simple adaptation of the conventional Cronbach's α (internal consistency measure); instead of multiple items, multiple raters are now being investigated with respect to their values for a set of health states (Krabbe et al., 1996). Therefore, Eq. (12.1) shows index p (persons) instead of i (item).

$$\alpha = \frac{n}{n-1}\left(1 - \frac{\sum_i \sigma_p^2}{\sigma_t^2}\right) \tag{12.1}$$

Although interrater reliability is formally a statistic that expresses the homogeneity of the responses among raters, it may also be taken as evidence of content validity. A high interrater coefficient may only be expected if most of the raters have a similar understanding of the valuation task as well as similar judgments for health states. As a consequence, they will come up with comparable health values. The analysis of interrater reliability is only appropriate for health states that are actually assessed.

Interrater analysis is not feasible for the separate attributes (classification). If respondents are classifying their own health condition by checking a level for each attribute, they are not rating a presented health state but rating the levels of the separate attributes for their own health condition. On the valuation part of the instrument (e.g., TTO), respondents are not scoring distinct attributes. Instead, health-state descriptions consisting of a combination of levels for the set of attributes are presented to them as narratives in the form of a text, which should be read and mentally processed. Then the full health-state descriptions are valued.

Validity

Years ago, Alan Williams proposed a radical approach by calling for abandonment of the term "validity" in the context of preference-based health measurement. His stance was that there is no "correct" concept of health against which to measure what is happening, so the question how to proof validity for preference-based health instruments is unanswerable. At a conference of the International Society for Quality of Life Research in 1999, he stated "the best we can do is to judge whether any particular approach is intuitively plausible and reasonably appropriate for its purpose."

Content

Although the content validity of preference-based measurement systems is seldom examined, it is a critical measurement property and should be documented as part of the development and validation of any outcome measure. Content validity is the extent to which an instrument assesses the relevant and important aspects of the concept it was designed to measure. Previously, various approaches were used to develop content for preference-based measurement systems. Most developers turned to the literature for attributes; for two systems (QWB and EQ-5D), they reviewed other health status measures and for one (15D), they reviewed policy documents. For most of these instruments, it is not clear how the final attributes emerged from the literature review (Grewal et al., 2006). Researchers in the field of patient-reported outcome instrument development have realized that content validity cannot be established via quantitative psychometric analyses, as highlighted in the FDA guidance (2009).

Proving the content validity of health instruments in general, and of preference-based measurement systems in particular, seems almost impossible. Nonetheless, a unique type of "evidence" to support the validity of a preference-based measurement system was observed in a small (unpublished) pain study. Patients with posttraumatic dystrophy (also known as complex regional pain syndrome or reflex sympathetic dystrophy syndrome) filled in the EQ-5D instrument. Posttraumatic dystrophy can produce unbearable pain, and medicine does not have much to offer to patients with this chronic systemic disease. The principal investigator in this study announced that two of his patients had committed suicide. Indeed, for several patients, we observed index values for their health conditions on the EQ-5D that were below zero (worse than dead).

Construct

It seems almost impossible to ascertain construct validity for health measurements that have been obtained with preference-based measurement systems. The reason is that the health objects presented to the respondents may consist of unspecified features (holistic). For example, the respondents may be asked to compare different outcomes after plastic surgery or to assess photos of skin affected by psoriasis. When the object they have to judge consists of health descriptions built on a small and fixed set of health attributes with levels, it is also impossible to compare these attributes with attributes of scales from other instruments. All of these obstacles are related to formative measurement, whereby instead of scoring a set of separate attributes, a whole set of attributes is valued. Other tests, such as for known-groups validity, may be feasible for the estimated values derived with preference-based measurement systems. From the field of economic evaluation, Brazier and Deverill (1999) adopted a modified concept (descriptive validity) to deal with formative measurement. These authors conclude that classical validity methods stemming from psychometrics, such as construct validation as applied to profile measures, is not appropriate for preference-based measures.

INSTRUMENTS

Operations research (e.g., systems analysis) provides mathematical models or algorithms that simulate the conditions needed to optimize a desired outcome. The quality of an object, event, or person's life can be one of these desired outcomes. Interestingly, operations research was a critical element in the early development of the field of health and quality-of-life assessment. Fanshel, an operations researcher, and Bush (1970), a medical doctor, developed a model of health, which subsequently evolved into the QWB (Kaplan and Anderson, 1990). Before he became interested in health and QALYs, Torrance (1976, 2006) was an operations research engineer working for a commercial firm. He used his training to develop the Health Utility Index with his colleagues. Also influenced by operations research is the work of Rachel Rosser, who developed the Rosser and Kind index (1978) that later inspired the development of the EuroQol-5D (EQ-5D).

There are only a few generic preference-based measurement systems for measuring health status or health-related quality of life. Here, we briefly describe the development and properties of the most well-known preference-based measurement systems.

QWB

The QWB combines values for symptoms and functioning. This instrument was mainly envisioned by James Bush, the second author of the initial publication (Kaplan et al., 1976). A physician with a passion for public policy, Bush, recognized the need to evaluate public health programs, and he proposed to do so using health outcomes. This was in the 1960s, and many people at that time thought he was crazy (Whiteside, 1984; Kaplan, 2005).

Values for the QWB were obtained from ratings by 867 people from the general population. They were asked to rate the desirability of health conditions, placing each on a continuum between dead (0.00) and optimum health

(1.00). Symptoms were assessed by questions about the presence or absence of different symptom complexes (e.g., trouble sleeping; burning or itching rash in large areas of face, body, arms, or legs; death). Functional limitations were recorded over the previous 3 days within three separate domains (mobility, physical activity, and social activity). Combining the four domain scores into a total index value yields a numerical point-in-time expression of well-being (health) that ranges from zero for dead to one for asymptomatic optimal functioning.

The QWB has been used in numerous clinical trials and studies to evaluate medical and surgical therapies (Kaplan and Anderson, 1988). Further, the method has been used for health resource allocation modeling. It has served as the basis for an innovative experiment on rationing of healthcare by the State of Oregon in the United States (Kaplan, 1994).

15D

Development of what is now known as the 15D (www.15D-instrument.net) started in the late 1970s with the publication of a piece of scientific work that is still inspiring (Sintonen, 1981). The basic goal was to develop a generic, multidimensional (or multiattribute), standardized, self-administered measure of health. It was mainly intended to derive health index values but was also supposed to adequately describe the respondent's present health status as a profile.

Conceptually, the 15D subscribes to the definition of health formulated by the World Health Organization (WHO), which states that health is composed of physical, mental, and social well-being. This concept was first operationalized in 1981 in the form of a 12-domain (attribute) instrument. After feedback from the medical profession, it was revised in 1986, when the first 15-domain instrument was established (Sintonen and Pekurinen, 1993). The coverage corresponds to the domains suggested by Fallowfield (1990) after conducting a broadly based analysis of literary, philosophical, and scientific sources. The developers of the 15D claim that the instrument corresponds almost one-to-one to the most important domains of health defined by the WHO (Chatterji et al., 2002) and over 80% of the domains in the new International Classification of Functioning, Disability and Health (ICF) published by the WHO (2001). On the basis of two extensive patient surveys, feedback from many instrument users, and factor analyses of empirical data from various patient groups, the classification system was revised in 1992 to its present form (Sintonen, 1994).

The present instrument includes 15 attributes, each covering a specific health domain: mobility, vision, hearing, breathing, sleeping, eating, speech (communication), excretion, usual activities, mental function, discomfort and symptoms, depression, distress, vitality, and sexual activity. Each domain is divided into five levels.

It usually takes around 5 min to fill the instrument. The 15D shows high response and completion rates. For example, the Finnish National Health

Survey 1995/96 had a response rate of 94%. In different studies, the completion rates have varied over the health domains from 96% to 100%, except for sexual activity (where it was 90−92%), which is usually lower in the elderly. The test−retest repeatability coefficients have varied by attribute from 92% to 100%. It has been shown that the agreement between the 15D values and TTO values for one's own health is quite good in a large group of heterogeneous patients. The minimum (clinically) important change in the 15D measure is estimated to be 0.015.

Apart from AQoL-8D (35 attributes condensed to eight domains), the classification system of the 15D is richer than that of other well-known measures of a similar type: EQ-5D has five, SF-6D has six, and HUI Mark III has eight attributes. Theoretically, this should lead to a more sensitive instrument in terms of both discriminatory power and responsiveness. In several empirical head-to-head comparisons with the instruments mentioned above, this improvement has been confirmed in most cases (Hawthorne et al., 2001; Moock and Kohlmann, 2008).

As the classification system defines an enormous number of health states, the direct valuation of all health states is impossible. Therefore, the valuation procedure is based on an application of the multiattribute utility theory. The index values (0.00 = being dead; 1.00 = full health) are calculated from the health-state classification system by using a set of population-based values in an additive aggregation formula (see Box 12.2). The importance weights and level values have been elicited with rating scale/magnitude estimation from representative population samples in several countries (Sintonen, 1995; Wittrup-Jensen and Pedersen, 2008; Aletras et al., 2009).

At the patient level, the 15D can be used as a diagnostic tool. The 15D profile scores of the patient are then compared to those of the age- and gender-standardized general population. By comparing the profile scores after treatment, one can ascertain how patients have been benefited from treatment, both on particular health domains and overall (Fig. 12.7).

HUI

The Health Utilities Index (HUI) is a family of preference-based health measures suitable for use in clinical and population studies (Torrance et al., 1995). Initially called the Health State Classification System, it was later renamed the Health Utility Index Mark I (Torrance et al., 1982) and had two successors (HUI-Mark II and III), all of which were developed at McMaster University (Torrance et al., 1995, 1996). Each member of the family includes a health classification and algorithms (preference-based multiattribute utility functions) for deriving HUI values from classification responses. The Health Utilities Index Mark II (HUI-2) consists of seven domains/attributes of health status: sensation (vision, hearing, speech), mobility, emotion, cognition, self-care, pain, and fertility. There are five or six levels per attribute. HUI-2

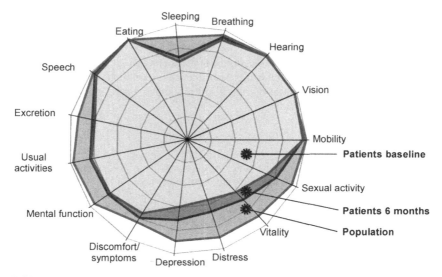

FIGURE 12.7   The mean 15D profile of bulimia nervosa patients at baseline before treatment and after 6 months of treatment, and the profile of a sample of the age-standardized general female population. *With permission based on data reported by Pohjolainen, V., Räsänen, P., Roine, R.P., Sintonen, H., Wahlbeck, K., Karlsson, H., 2010. Cost-utility of treatment of bulimia nervosa. The International Journal of Eating Disorders 43 (7), 596–602.*

focuses on capacity rather than performance. Multiplicative multiattribute utility functions based on community preferences have been estimated for HUI-2 (Torrance et al., 1995) and HUI-3 (Feeny et al., 2002). This form of function can represent a simple type of interaction among the attributes. The HUI-1 was based on the TTO technique, but the HUI-2 and the HUI-3 are both based on health-state assessments performed with the SG in combination with VAS assessments. Thus, although Torrance introduced the TTO technique, his technique was not used for the later HUI instruments. Instead, later versions of the HUI were worked out under (expected) utility theory.

Multiattribute Utility Functions

Multiattribute utility theory extends the von Neumann–Morgenstern theory to consider utility functions with more than one argument (attribute). One of the first applications of multiattribute utility models was a study of alternative locations for a new airport in Mexico City in the early 1970s. The factors that were considered included cost, capacity, access time to the airport, safety, social disruption, and noise pollution. The military is also a leading user of this technique. The design of major new weapons systems always involves trade-offs of cost, weight, durability, lethality, and survivability.

BOX 12.2 Multi-Attribute Utility Models

Additive Function (15D)
The utility or value (*u*), representing overall health on a 0.00–1.00 scale (0.00 = being dead; 1.00 = full health), is calculated from the health-state classification system by using a set of values in an additive aggregation formula as follows:

$$u(x) = \sum_{n}^{j=1} k_j(u_j(x_j)),$$

where *u*(x) is the utility for health state *x* represented by a 15-element vector (15 attributes of the 15D); $u_j(x)$ is the single-attribute utility function for attribute *j*; k_j is a model parameter.

The $u_j(x_j)$ is the single-attribute utility function for attribute *j*. For each attribute in the system, the single-attribute utility for level one (normal) is assigned a score of 1.00, and the lowest level (level five for the 15D) for that attribute is assigned a score of 0.00. The single-attribute utility function reflects the utility attached to each of the intermediate levels on the 0.00 to 1.00 scale. Crudely, k_j represents the average relative importance respondents attach to the various attributes *j* (*j* = 1, 2,...,15). The relative importance weights k_j sum up to 1.00.

Multiplicative Function (HUI-3)
For the multiplicative function, the same parameters as for the additive function have to be estimated in a different estimation procedure added with an additional parameter *k*. The parameter *k* (extra restrictions are imposed on this parameter at the estimation stage) captures the interaction in preferences among attributes. If *k* is positive, the attributes are all preference complements; if *k* is negative, the attributes are all preference substitutes; if *k* equals 0, the utility function is linear additive.

$$u(x) = (1/k)\left[\prod_{n}^{j=1}(1 + kk_j u_j(x_j)) - 1\right]$$

There are three fundamental forms of multiattribute utility functions: linear-additive, multiplicative, and multilinear. The linear-additive form does not allow for interactions among attributes. Thus, the change in utility caused by a problem in one attribute does not depend on whether there are any problems in other attributes. If there are no interactions between attributes, the effect on health of being both blind and deaf would be the sum of the effect of being blind and the effect of being deaf.

If vision and hearing are instead preference complements (which is the case for all three versions of the HUI), the effect on health of being both blind and deaf would be greater than the effect of being blind or deaf but less than the sum of the two individual effects. Finally, if vision and hearing are preference substitutes, the effect of being both blind and deaf would exceed the sum of the two individual effects.

The multiplicative form allows for one type of preference interaction among attributes. Attributes are either preference substitutes or preference complements (or there are no interactions, in which case the multiplicative function simplifies to a linear-additive function). The more complex multilinear function allows for various types of preference interactions among attributes (Feeny et al., 2002).

The 15D is based on the additive functional form and has no interactions among the attributes. This function form makes the strongest assumption (no interactions: most difficult to fulfill) but leads to the simplest function. The HUI instruments, as well as the AQoL (see below), are based on the multiplicative functional form, as estimations based on the additive case provided inconsistent results and strongly favored the multiplicative functional form. Interactions under the multiplicative functional form are highly constrained. The model forces the interactions among attributes to be the same among all attributes (Furlong et al., 1998).

Multiattribute utility theory is one methodology in the broader field of multicriteria decision analysis. The way these models are used when deriving a function for health states is largely a decomposed approach that reflects how multicriteria decision analysis operates (Froberg and Kane, 1989). Briefly, a small group of experts perform several assessment tasks, and the decision methods are worked out on the basis of these few responses. That is quite different from having a large pool of respondents (general population, patients) available, as is the case in health-state valuation research. In a decomposed approach, specific elements are separately measured or rated and then combined to arrive at a final measure or function. In most other approaches, such as the choice models (Chapter 11), this is done integrally instead of stepwise. There are subtle differences between the multiattribute utility approach and other preference-based methods. Multicriteria decision analysis is focused on optimization, whereas choice models are focused on measurement (i.e., quantification).

The underlying theory of the multiplicative, multi-attribute utility function is described in Keeney and Raiffa (1976) and in von Winterfeldt and Edwards (1986). An early example of an application in the area of health can be found in Gustafson and Holloway (1975).

EQ-5D

The EuroQol Group, which was formed in 1987, comprises a network of international, multicenter, multidisciplinary researchers, originally from seven centers in England, Finland, the Netherlands, Norway, and Sweden. Later, researchers from Spain as well as from Germany, Greece, Canada, the United

States, and Japan joined the group. The intention was to develop a generic value function for health that could be used in common across Europe. The initial version of the EuroQol had six different domains (attributes). A more recent version, known as the EQ-5D, consists of five attributes, each with three levels (EQ-5D-3L). The EQ-5D instruments are now used in a substantial number of clinical and population studies (Hurst et al., 1997; Xu et al., 2011).

The EQ-5D (www.euroqol.org) was designed to be an "abstracting" tool for medical records but was later transformed into a patient-reported outcome instrument (Brooks, 2012). Today, the EQ-5D is a prominent instrument within the class of preference-based measures. From the outset, the key advantage of the EQ-5D has been its simplicity. Although the developers realized that this might compromise its comprehensiveness, one of the objectives of the EuroQol Group (recently renamed EuroQol Research Foundation) was to construct a simple instrument with sufficient scope for full coverage across the health spectrum (operationalized in the context of health). In addition, the EQ-5D classification was deliberately constructed in straightforward wording to facilitate detailed and systematic methodological research.

Health-state descriptions based on the EQ-5D system consist of five attributes, each of which can be varied on three levels (Brooks, 1996). The five attributes of this preference-based measurement system are mobility, self-care, usual activities, pain/discomfort, and anxiety/depression. Their levels are expressed as: (1) no problems, (2) some moderate problems, (3) and extreme problems. Assuming independency, 3^5 (243) different health states are defined by the EQ-5D classification, and every health state is assumed to be approximately covered by one of the 243 generic health descriptions. Of course, the decision to limit the number of attributes and levels predetermined the scope of the EQ-5D. Because of the Group's choice for the health concept with these five attributes, precise distinction among rather comparable health states was precluded. Therefore, the ontology of the EQ-5D can probably best be described as scientific and pragmatic.

Each attribute level for the EQ-5D has an associated weight that was obtained from a random sample of the adult population in the United Kingdom (Dolan, 1997). These weights were assembled by performing basic regression analysis (see Application). Based on these weights, all EQ-5D health states have specific index values (e.g., state 11223 leads to a value of 0.26). The scores range from -0.54 (33333) to 1.00 (perfect health) with "as bad as being dead" at zero. The EQ-5D also contains a EuroQol VAS in which patients are asked to rate their present general health status. This VAS consists of a 20-cm vertical line marked from 0 to 100, where 0 represents the "worst imaginable health state" and 100 the "best imaginable health state."

Responsiveness is an important issue for the EQ-5D. There is some evidence that the instrument registers a "ceiling effect," especially in general population health studies. However, the purpose for which the EQ-5D was developed should be taken into account; it was never intended to determine small health effects in a relatively healthy population. Nevertheless, it is important to know whether the EQ-5D is not only capable of distinguishing substantial health improvements but can also detect smaller changes. Not much has been published on responsiveness or minimal important (clinical) difference (concepts that will be explained in the next chapter). A literature review provides insight into the remarkable heterogeneity among methods to estimate the minimal important differences for the EQ-5D. The estimates range from 0.03 to 0.52 (Coretti et al., 2014).

One of the recent developments is a five-level version, the EQ-5D-5L. For this extended classification system, new value functions had to be estimated. As it was recognized that the conventional TTO approach had some problems (for example, its approaches to the valuation of states better than dead and worse than dead are conceptually different), the EuroQol Research Foundation initiated research to develop new methods, resulting in the lead-time TTO (Devlin and Krabbe, 2013). Further study demonstrated that the lead-time approach produced results with low face validity for states better than death. Therefore, it was decided to combine the conventional TTO (for states considered better than death) with the lead-TTO (for states worse than death). This new approach was named the "composite TTO" (Oppe et al., 2014). Apart from deriving values with the composite TTO, paired comparison tasks (discrete choice model) are also performed in a computer-assisted personal interview mode: the EuroQol Valuation Technology (EQ-VT). The data from these two preference-based methods are processed in a hybrid regression analysis to produce an algorithm by which values for all EQ-5D-5L health states can be computed. It turned out that this analysis is so complex (e.g., combining responses collected by different techniques, censoring of data), restricted (imposing constraints on the estimation procedure to remove inconsistencies, having no interaction parameters), and idiosyncratic that its transparency has been lost. These latest ad hoc adjustments and tinkering of the data seem to be largely a consequence of using rather complicated valuation techniques.

SF-6D

The SF-6D was developed by a team at the University of Sheffield (Brazier et al., 1998, 2002) to take advantage of the world's most widely used health-status profile instrument at the time, the SF-36 (Chapter 9). The original SF-36 yields measures across eight domains and two summary measures. The British team developed the SF-6D health-state classification from the SF-36 to make it amenable to valuation studies. There is an SF-6D based on the SF-36 (Brazier et al., 2002) and another based on the 12-item (SF-12) version (Brazier and Roberts, 2004).

The SF-6D has six attributes: physical functioning, role limitation, social functioning, pain, mental health, and vitality (Table 12.4). The number of levels per attribute is between four and six, depending on the response choice categories of the original items from the SF-36. The SF-6D based on the SF-36 defines 18,000 health states, and the instrument based on the SF-12 defines 7500 states. These can be derived from 11 items of the SF-36 and 7 items of the SF-12, respectively. The SF-36 and SF-12 are copyrighted and can be obtained from the Medical Outcomes Trust and Quality Metric. The SF-6D algorithms are readily available and free for noncommercial applications (www.shef.ac.uk/scharr/sections/heds/mvh/sf-6d.).

A representative sample consisting of 836 members of the United Kingdom's general population was drawn. These persons were interviewed and asked to value a total of 249 states defined by the SF-6D using the SG (each respondent valued six states). An algorithm was estimated for the SF-6D by random effects regression methods for the 36-item and 12-item versions. Valuation studies have also been undertaken in Japan, Hong Kong, Portugal, and Brazil using the same version of the SG. There were significant differences found in the results between these countries.

SF-6D Version 2 was developed in response to concerns that the index values would not cover the same range as other widely used measures like EQ-5D and HUI-3. This seems to be a result of the classification system and the use of the two-stage SG valuation methodology (Abellán Perpiñán et al., 2012). It was decided to revisit the attribute selection from the SF-36 and also address the ambiguity in the levels of physical functioning (distinguishing between "moderate activities" and "bathing and dressing"), the confusion of positive wording on one attribute (i.e., vitality) and the crudeness of using two-level attributes for role limitation.

The researchers also adopted a variant of discrete choice modeling wherein duration statements (set at 1, 4, 7, and 10 years) had been added to the attributes (see also Chapter 14) to estimate a scoring algorithm (Norman et al., 2013). This indirect measurement replaces the earlier SG method. Such discrete choice methods had already been successfully applied to value the EQ-5D (Bansback et al., 2012; Mulhern et al., 2014). A preliminary survey in the United Kingdom used 300 health-state pairs in combination with duration. The survey was completed by 3000 persons who were representative of the general population in terms of age and gender. The algorithm is still being finalized for the United Kingdom, but initial results suggest that the co-efficients are generally larger, resulting in a wider range of index values (from 1.00 down to −0.72).

AQoL

There are now four AQoL instruments: the AQoL-4D, AQoL-6D, AQoL-7D, and AQoL-8D. These consist respectively of four, six, seven, and eight domains with multiple attributes. The four classification systems were each

TABLE 12.4 The Domains and Their Levels in the SF-6D (Version 2)

Level		Level	
	Physical Functioning		**Pain**
1	Limited in vigorous activities (such as running, lifting heavy objects, participating in strenuous sports) not at all	1	No pain
2	Limited in vigorous activities a little	2	Very mild pain
3	Limited in moderate activities (such as moving a table, pushing a vacuum cleaner, bowling or playing golf) a little	3	Mild pain
4	Limited in moderate activities a lot	4	Moderate pain
5	Limited in bathing a dressing a lot	5	Severe pain
		6	Very severe pain
	Role Limitation		**Mental Health**
1	Accomplish less than you would like (at work or during other regular daily activities as a result of your physical health or emotional problems) none of the time	1	Depressed or nervous none of the time
	Accomplish less than you would like a little of the time	2	Depressed or nervous a little of the time
3	Accomplish less than you would like some of the time	3	Depressed or nervous some of the time
4	Accomplish less than you would like most of the time	4	Depressed or nervous most of the time
5	Accomplish less than you would like all of the time	5	Depressed or nervous all of the time
	Social Functioning		**Vitality**
1	Social activities are limited none of the time	1	Tired none of the time
2	Social activities are limited a little of the time	2	Tired a little of the time
3	Social activities are limited some of the time	3	Tired some of the time
4	Social activities are limited most of the time	4	Tired most of the time
5	Social activities are limited all of the time	5	Tired all of the time

(uniquely) created using psychometric methods for instrument construction. Commencing with a hypothesized model that defined the instrument's scope and focus, attributes were derived from the literature and from population, patient, and provider focus groups. The resulting attributes were subjected to exploratory and confirmatory factor analysis (Hawthorne et al., 1999).

AQoL-4D, first published in 1999, is a relatively parsimonious instrument to capture independent living, mental health, relationships, and physical senses. It is conceptualized in terms of "handicap": the effect of health upon well-being in a social context. Index values were derived, as with the HUI instruments, using multiplicative modeling. AQoL-6D expanded upon AQoL-4D by adding separate domains for pain and coping and by increasing the response levels in the vicinity of full health, an area where AQoL-4D is relatively insensitive. AQoL-7D added a "bolt-on" domain for vision (Misajon et al., 2005), while AQoL-8D increased the content of the mental health and relationship domains and added new domains for self-worth and life satisfaction. Each of the eight, except senses, represents a subscale and, like the SF-36, these may be combined into two larger scales: the physical and mental "super domains."

Following psychometric practice, the need for sensitivity and validity resulted in three to five attributes per domain in each instrument. Consequently, relative to other preference-based instruments, the AQoL-8D with its 36 attributes is comparatively large. It takes an average of 5.5 min to complete. Its size also required the adoption of a hybrid decision analytic methodology for deriving values (Richardson et al., 2014a,b,c). In two surveys (one for the classification system and one valuing the health-state descriptions), half of the respondents were a representative sample of the general public, and half were mental-health patients selected for their greater capacity to appreciate the psychosocial health states.

A major validation study was undertaken to compare the AQoL-8D with other instruments. The 2.4 million items of data (15 instruments, 6 countries, 7 chronic disease areas plus healthy respondents) are freely available on the AQoL website (www.aqol.com.au). The site provides population norms for each of the instruments and transformations from AQoL-4D and AQoL-8D to other preference-based measurement systems (Chen et al., 2016). These transformations rescale AQoL-4D and AQoL-8D scores while (largely) retaining the sensitivity of the initial data. Publications confirm that, relative to other preference-based measurement systems, AQoL-8D is particularly sensitive to the psychosocial domains of health and the diseases where these are important (Richardson et al., 2015b, 2016). The AQoL-8D largely explains variation in subjective well-being (Richardson et al., 2015a) and capabilities as measured by the ICECAP A (Mitchell et al., 2015).

APPLICATION

The national Dutch guidelines on pharmacoeconomics recommend preference-based health measures to estimate QALYs in cost-effectiveness analyses. For many years, in the absence of a Dutch EQ-5D value function, the UK value function was used. However, there was evidence that EQ-5D value functions may differ across countries (Badia et al., 2001; Tsuchiya et al., 2002; Johnson et al., 2005). This was the reason to develop a Dutch EQ-5D value function.

A company for marketing research recruited the respondents for the Dutch EQ-5D (three-level version) valuation study. Respondents aged 18–75 years were selected. Quota sampling was used to achieve a sample that is representative for the Dutch population with regard to age and gender. Although a simulation study showed that 200 respondents should suffice to estimate a value function, the sample size was increased to 300 respondents to allow for the application of poststratification weights in case of deviations from the Dutch adult population on characteristics other than age and gender. Interviewers, who were trained by the researchers, conducted the face-to-face interviews at the office of the research company during the summer of 2003.

The Dutch EQ-5D valuation study is a replication of the Measurement and Valuation of Health (MVH) study (Dolan et al., 1996) conducted in the United Kingdom, and it also uses the MVH study protocol (Dolan, 1997). This protocol describes a face-to-face interview that can be separated into several sections. After the respondents filled in a questionnaire with socioeconomic and background questions, they described their own health using the standard EQ-5D. Subsequently, the respondents ranked 17 EQ-5D health states plus state 11111 and immediate death, printed on cards, by putting the "best" health state on top and the "worst" at the bottom. After the ranking exercise, the respondents were asked to place the 19 cards with health states on a VAS, often referred to as the EuroQol "thermometer" with endpoints of 100 for the best imaginable health state and 0 for the worst. This was followed by the TTO task for a set of 17 EQ-5D states. State 11111 (i.e., full health) and death cannot be directly valued, as in TTO their values are preassigned to 1.00 and 0.00, respectively.

In this study, the TTO valuation part of the interview was done on the computer, which is a deviation from the original MVH protocol. A graphic computer program replaced the TTO boards. This program integrated the MVH TTO protocol, the scoring administration, and the visual TTO aids. The interviewer operated the computer program, just as the TTO board was operated by interviewers in the original MVH study. The program presented the health states in random order. Respondents were led by a process of outward titration to select a length of time t in state 11111 that they regarded as equivalent to 10 years in the target state. The shorter the "equivalent" length of time in full health, the worse the target state (the health state for which we want to derive a value). In case a respondent preferred death over the target state, the choice was between dying immediately and spending a length of time $(10-t)$ in the target state followed by t years in state 11111. The more time required in state 11111 to compensate for a shorter time in the target state, the worse the target state.

Modeling

For states regarded as better than death, the TTO value v is $t/10$, where t is the time in full health. For states worse than death, v is $-t/(10-t)$. The negative values for states worse than death (v') were transformed so that they were bounded by a maximum negative value of -1 using:

$$v' = v/(1 - v) = -t/10 \tag{12.2}$$

Regression analyses were performed to estimate a statistical model that can be used to interpolate TTO values for all possible EQ-5D health states from the direct valuations of the subset of 17 states. In the regression, the dependent variable is 1 minus the TTO value for the health states, and the independent variables describe the health states.

Two models were estimated: a main-effects model and an N3 model. The latter gives additional weight to extreme problems by adding a variable that indicates that at least one EQ-5D attribute is at the worst level. The former comprised 10 dummy variables that indicate the presence of either a level two or three in a given attribute of the evaluated state. The regression equation for the main-effects model is as follows:

$$y = \alpha + \sum_a \sum_l \beta_{al} x_{al} + e \qquad 12.3$$

where y is 1 minus value; x_{al} represents the 10 dummy variables, which indicate the presence of either a level two or three in a given attribute; a stands for the attributes; and l for either level two or three. The five attributes are mobility, self-care, usual activities, pain/discomfort, and anxiety/depression. For example, the estimated equation for state 11312 is $y = a + b_{UA3} x_{UA3} + b_{AD2} x_{AD2} + e$. The N3 model extends the main-effects model with a dummy variable indicating that at least one attribute is at level three, i.e., the problems are extreme. The regression equation for the N3 model is:

$$y = \alpha + \sum_a \sum_l \beta_{al} x_{al} + \gamma N3 + e \tag{12.4}$$

where y is 1 minus value; x_{al} represents the 10 dummy variables, which indicate the presence of either a level two or three in a given attribute; and N3 is a dummy that indicates whether at least one attribute is at level three. Using the N3 model, the estimated equation for state 11312 is: $y = a + b_{UA3} x_{UA3} + b_{AD2} x_{AD2} + \gamma N3 + e$.

Since each respondent was expected to have a different response pattern, a random-effects model was estimated. In this model the error term e is divided into two components: a traditional error term unique to each observation and an error term representing the extent to which the intercept of an individual respondent differs from the overall intercept.

Results

In total, 309 respondents were interviewed. Two of them broke off the interview before the TTO valuation task, so their TTO responses were completely missing. Because of technical problems, the TTO values of two other respondents were lost. The same exclusion criteria as those adopted in the MVH study were used. Seven respondents were excluded because they did not want to trade off any lifetime for quality improvement, resulting in a value of 1.00 for each state. TTO responses from 298 respondents were available for analysis.

The TTO data are complete for all respondents, mainly because that part of the interview was computer assisted. Comparison with the Dutch population showed that the age by gender distribution of the sample corresponded with the population data. The sample showed a small underrepresentation of married persons and an overrepresentation of respondents with a high educational level. Among the background characteristics, the only one that showed a statistically significant difference was whether or not a respondent believed in life after death. Those who did gave higher health-state values.

The mean values for the 17 health states that were directly valued ranged from 0.91 for state 21111 to −0.30 for state 33333; the median values from 0.99 to −0.38 (Table 12.5 and Fig. 12.8). State 33333, the worst possible EQ-5D state, was the only one that was rated as worse than death by a majority of the respondents. The coefficients for the two estimated models are shown in Table 12.6. The coefficient for mobility at level two of both models did not show a statistically significant difference from zero. The coefficient for usual activities at level two showed no statistically significant difference from zero for the N3 model. The proportion of explained variance of the two models was about the same: 0.37 for the main-effects model and 0.38 for the N3 model. The use of the random-effects model demonstrated that there were significant differences between respondents regarding their overall level of valuing health states ($p < .001$). This same model resulted in coefficients similar to the coefficients of a standard regression model. However, the standard errors are smaller for the random-effects model.

The N3 model had the best predictive performance in terms of both the mean absolute error and the correlation between observed and predicted values over the 17 states. Therefore, the set of coefficients of the N3 model (last column of Table 12.6) was chosen as the Dutch EQ-5D-3L value function. To assess the effects of applying the Dutch value function instead of the UK value function, both value sets were compared (Fig. 12.8).

TABLE 12.5 Mean, Median, and Standard Deviation for Observed Value, and Percentages of Negative Values per Health State ($n = 298$)

State	Observed Value			Negative Value (%)
	Mean	Median	Standard Deviation	
21111	0.91	0.99	0.21	1
11211	0.88	0.97	0.21	1
12111	0.88	0.97	0.24	2
11121	0.86	0.95	0.22	1
11112	0.78	0.90	0.34	3
11312	0.56	0.68	0.45	11
22222	0.54	0.68	0.47	12
13311	0.47	0.63	0.52	15
32211	0.42	0.55	0.53	16
11113	0.37	0.53	0.57	22
11131	0.32	0.50	0.59	22
11133	0.04	0.03	0.59	41
32223	0.04	0.06	0.56	37
32313	0.04	0.03	0.59	39
23232	0.03	0.09	0.58	38
33323	−0.11	0.0	0.58	46
33333	−0.30	−0.38	0.48	62

With permission from Lamers, L.M., McDonnell, J., Stalmeier, P.F.M., Krabbe, P.F.M., Busschbach, J.J.V., 2006. The Dutch tariff: results and arguments for an effective design for national EQ-5D valuation studies. Health Economics 15 (10), 1121–1132.

For the more severe health states, the Dutch values were higher than the MVH values. As one could expect from these higher mean values, the Dutch value function reflected the tendency of Dutch respondents to ascribe less weight than the UK respondents to most attributes. The most striking difference was for severe problems in mobility, with a coefficient that was 0.143 lower in the Dutch study. The anxiety and depression attribute was the only one to which Dutch respondents ascribed more weight than respondents in the

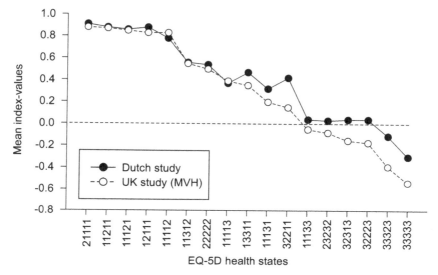

FIGURE 12.8 Mean Dutch and UK time trade-off values for 17 health states. *MVH*, Measurement and Valuation of Health. *With permission from Lamers, L.M., McDonnell, J., Stalmeier, P.F.M., Krabbe, P.F.M., Busschbach, J.J.V., 2006. The Dutch tariff: results and arguments for an effective design for national EQ-5D valuation studies. Health Economics 15, 1121–1132.*

United Kingdom. The implication of these differences is that comparisons of QALYs calculated using different value functions should be made with caution, especially for patient groups with severe conditions or mental-health problems.

What could have caused the higher Dutch values? The Dutch EQ-5D valuation study deviated in several respects from the original protocol. First, the researchers used the Dutch version of EQ-5D and translated the valuation procedure into the Dutch language, a process that could have introduced some noise. An important deviation from the MVH protocol was the use of a smaller set of health states for direct valuation: 17 states in the Dutch study versus 42 in the MVH study. There are several other points that may explain these differences. Differences could also be attributed to interviewer effects, mode of data collection, sample selection, and other elements of the study design that are difficult to control or replicate. In the MVH valuation study, each respondent had to evaluate a set of 13 states (selected from a set of 42) consisting of two very mild, three mild, three moderate, and three severe states, state 33333, and unconsciousness. When using 17 health states, all respondents evaluated the complete set, comprising five very mild, four mild, four moderate, and three severe states, and state 33333. This set was relatively less severe than the set of 13 states evaluated in the MVH study.

TABLE 12.6 Parameter Estimated for Main-Effects and N3 Random-Effects Regression Model

Parameters	Main Effects	N3
Constant	0.135	0.071
M02	−0.024[a]	0.036[a]
M03	0.173	0.161
SC2	0.085	0.082
SC3	0.166	0.152
UA2	0.103	0.032[a]
UA3	0.151	0.057
PD2	0.041	0.086
PD3	0.419	0.329
AD2	0.133	0.124
AD3	0.408	0.325
N3		0.234
R^2-value	0.37	0.38

[a]*Not statistically significant different from 0 (p > .05).*
With permission from Lamers, L.M., McDonnell, J., Stalmeier, P.F.M., Krabbe, P.F.M., Busschbach, J.J.V., 2006. The Dutch tariff: results and arguments for an effective design for national EQ-5D valuation studies. Health Economics 15 (10), 1121−1132.

Another possible explanation for the differences in observed values for health states lies in the differences in the two populations' health preferences. Since factors such as beliefs about life after death, which might vary between countries, affect people's valuation of health states, differences in the populations' health-related preferences form a plausible explanation for at least part of the differences between the values found in the Netherlands and the United Kingdom.

DEAD END FOR TIME TRADE-OFF?

I have spent years doing methodological research on various methods to measure health, and the TTO was one of the techniques that attracted much of my attention. That was also noticed by colleagues (Streiner et al., 2015, p. 64) who cited my work. From their perspective, one of my publications (Krabbe et al., 1997) was the final nail in the coffin of econometric methods, as I compared the SG and the TTO with the VAS. In that 1997 publication, one of my conclusions was that the VAS was the most feasible and reliable and that, with a simple power transformation, it is possible to go from scores derived from the VAS to those from other techniques. Based on these findings, Streiner et al. no longer saw a need for the more complex and cognitively demanding econometric methods.

The conclusions drawn by my colleagues, however, could hardly have been based on my earlier study, since the conclusions in my 1997 publication, where I compared different methods to derive values for health states, were much more modest. Although it was a small experimental study, it was methodologically thorough. At that time, I was pleasantly surprised that I was referred to in their prominent book about the development of health measurement scales but also amazed that my work was placed in such a context. Nevertheless, after years of experience with the TTO, I have more empirical evidence of the limitations of this technique. The most noticeable pieces of evidence will be briefly discussed below.

Compelling arguments against using TTO techniques have been raised by several authors, most of them being health economists themselves (Gafni and Torrance, 1984; Richardson, 1994; Johannesson et al., 1994; Bleichrodt and Johannesson, 1997a,b; Bleichrodt, 2002; Dolan and Stalmeier, 2003; Drummond et al., 2005; Attema et al., 2013). In fact, the TTO seems to be associated with many problems: practical (e.g., difficult for people to perform, trained interviewer assistance required); theoretical (e.g., axiomatic violations, problems in dealing with states worse than dead); and bias related (e.g., time preference, mode of administration). From a measurement perspective, the TTO technique has been criticized for its susceptibility to framing issues (e.g., duration of the time frame, indifference procedure).

Most valuation techniques produce values that reflect not only judgments about health but other elements as well (Krabbe and Bonsel, 1998; Essink-Bot and Bonsel, 2002; Salomon and Murray, 2004). For example, people with an aversion to risk are less willing to accept hypothetical risks, so their SG values will be systematically higher. In the TTO technique, time preference will reduce the differences between the trade-off options and therefore inflate estimated values. One consequence of incorporating distinct elements may be that it may lead to measures that cannot be placed on a unidimensional scale. The biases introduced by extraneous elements can be avoided to some extent in the design, or they can be adjusted for. But they cannot be eliminated completely, because the essential characteristic of these techniques is the trade-off between distinctive elements (commodities).

A new and even more complicated valuation technique was recently introduced: the lead-time TTO (Attema et al., 2013). It is designed to compensate to some extent for the problems of the original TTO, particularly for the assessment of states considered worse than dead. Apart from the fact that states worse than dead pose methodological problems, merely confronting people with "death" may introduce all kind of biases in their responses. States worse than death are assessed with lead-time TTO in the new five-level version of the EQ-5D instrument (Devlin et al., 2016). The introduction of the lead-time TTO valuation technique is an example of a seemingly pragmatic solution to a serious methodological problem. No convincing (measurement) theory is presented to explain and justify this extension of the conventional TTO technique.

Nevertheless, despite all the known biases and other disturbing factors, it is probably fair to say that overall, the TTO technique provides somewhat granular measures of the quality of health states that in many, if not most, applications can provide sufficient information.

DISCUSSION

Researchers who acknowledge the importance of health-outcome measures agree that health attributes should be weighted if we are interested in the relative importance that individuals place on health. Health states must be described in terms of their seriousness and then assigned meaningful values. To assign values to health states, a preference-based approach to measurement is mandatory. A key element of any preference-based measurement approach is that respondents have to compare two or more health states and express a preference for one of them. Another characteristic of most preference-based approaches is that the health-state descriptions should capture relevant attributes of health that have to be assessed by respondents all at once.

The choice-based models discussed in Chapter 11 are also preference based and produce values that express the quality of health. However, these values are on a relative interval scale, where the lowest value represents the worst health state (based on the specific classification system) and the highest value the best health state or full health. Yet these lower values have an arbitrary metric. The information provided by such values allows us to compare the seriousness of different health states. However, we cannot express in percentages how much better or worse one health state is than another.

The values derived with the valuation techniques discussed in this chapter are somewhat different. Index values are absolute measures. With these we can express the quality of health in percentages (a value of 0.78 means that the quality of health is 78% of full health). The location of "death" has been determined, and the values for the health states are scaled on this death = 0.00 through 1.00 range. Of course, the assumption made by making death equal to zero is that there may be health states worse than death and that such states are considered of no value (even negative). Another assumption is that "death" is a natural or logical lower point (0.00) and is understood thus by (most) people. Such index values can be incorporated into health summary measures such as QALYs and the Quality-adjusted Time Without Symptoms of disease and Toxicity of treatment (Q-TWiST).

What is crucial in the preference-based methods (choice models and valuation techniques) is not the instruction given to individuals or the task that they have to perform. The really important aspect is the underlying response mechanism. Both the choice models and the valuation techniques can be classified as trade-off techniques. Respondents have to compare different attributes for at least two different health-state descriptions and make trade-offs to arrive at a response. In choice methods, respondents compare two health states and have to express which one they prefer (which one is better). For the valuation techniques, the response task is a bit more difficult. Valuation techniques involve reaching an indifferent situation between two different health states. Indifference means that respondents consider both conditions as equal in value. Respondents may reach this equilibrium in one step (in practice, this is done with an iterative procedure to make the task more manageable). Respondents have to make explicit the value of a health state by providing an absolute scaling response. That is different from the choice models, where respondents only have to express which health state is better.

Eliciting individuals' values is not an easy exercise, irrespective of the technique. It can be difficult to judge whether or not the respondents understood the task. With the possible exception of the VAS, valuation techniques put a large cognitive burden on the respondents by demanding a relatively high degree of abstract reasoning (Green et al., 2000). Qualitative research, using study designs such as "think-aloud" interviews that can be conducted alongside valuation tasks, can provide valuable insights into how respondents interpret and complete these tasks. Previous qualitative research has shown that respondents often found it difficult to distinguish between the health states to be valued, to understand the hypothetical nature of health states, and to conceptualize death, or that they held their own spiritual beliefs about death (Shah et al., 2015).

A serious limitation of existing preference-based measurement systems, such as the TTO-based EQ-5D, is that these are not patient centered. Instead, the five health attributes of the EQ-5D are generated by expert opinion. In addition, the derived values for health states do not come from patients but are based on assessments from the general population. Thus, overtly healthy people are asked to judge hypothetical health states. But surely, patients are the best judges of their own health. Therefore, it would be sensible to elicit patients' values. Lately, the EuroQol Research Foundation started to explore (discrete) choice methods (Chapter 11). In their research program, pairs of health states are judged in terms of their quality, and extensions are introduced by including duration statements (Chapter 14). However, the content of the health concept is still based on expert opinion, and the valuation responses do not come from experienced patients.

REFERENCES

Aletras, V.H., Kontodimopoulos, N., Niakas, D.A., Vagia, M.G., Pelteki, H.J., Karatzoglou, G.I., Sintonen, H., Yfantopoulos, J.N., 2009. Valuation and preliminary validation of the Greek 15D in a sample of patients with coronary artery disease. Value in Health 12 (4), 574–579.

Abellán Perpiñán, J.M., Sánchez Martínez, F.I., Martínez Pérez, J.E., Méndez, I., 2012. Lowering the "floor" of the SF-6D scoring algorithm using a lottery equivalent method. Health Economics 21 (11), 1271—1285.

Anderson, N.H., 1979. How functional measurement can yield validated interval scales of mental quantities. Journal of Applied Psychology 61, 677—692.

Arons, A.M.M., Schölzel-Dorenbos, C.J.M., Olde Rikkert, M.G.M., Krabbe, P.F.M., 2016. A simple and practical index to measure Dementia-related quality of life. Value in Health 19 (1), 60—65.

Attema, A.E., Versteegh, M.M., Oppe, M., Brouwer, W.B., Stolk, E.A., 2013. Lead time TTO: leading to better health state valuations? Health Economics 22 (4), 376—392.

van Agt, H.M.E., Essink-Bot, M.L., Krabbe, P.F.M., Bonsel, G.J., 1994. Test—retest reliability of health state valuations collected with the EuroQol questionnaire. Social Science and Medicine 39 (11), 1537—1544.

Badia, X., Roset, M., Herdman, M., Kind, P., 2001. A comparison of UK and Spanish general population time trade-off values for EQ-5D health states. Medical Decision Making 21 (1), 7—16.

Bansback, N., Brazier, J., Tsuchiya, A., Anis, A., 2012. Using a discrete choice experiment to estimate health state utility values. Journal of Health Economics 31 (1), 306—318.

Bleichrodt, H., 2002. A new explanation for the difference between time trade-off utilities and standard gamble utilities. Health Economics 11 (5), 447—456.

Bleichrodt, H., Johannesson, M., 1997a. An experimental test of a theoretical foundation for rating-scale valuations. Medical Decision Making 17 (2), 208—216.

Bleichrodt, H., Johannesson, M., 1997b. The validity of QALYs: an experimental test of constant proportional trade-off and utility independence. Medical Decision Making 17 (1), 21—32.

Boye, K.S., Matza, L.S., Feeny, D.H., Johnston, J.A., Bowman, L., Jordan, J.B., 2014. Challenges to time trade-off utility assessment methods: when should you consider alternative approaches? Expert Review of Pharmacoeconomics and Outcomes Research 14 (3), 437—450.

Brazier, J., Deverill, M., 1999. A checklist for judging preference-based measures of health related quality of life: learning from psychometrics. Health Economics 8 (1), 41—51.

Brazier, J., Ratcliffe, J., Salomon, J.A., Tsuchiya, A., 2007. Measuring and Valuing Health Benefits for Economic Evaluation. Oxford University Press, New York.

Brazier, J.E., Dixon, S., Ratcliffe, J., 2009. The role of patient preferences in cost-effectiveness analysis: a conflict of values? PharmacoEconomics 27 (9), 705—712.

Brazier, J.E., Roberts, J., 2004. Estimating a preference-based index from the SF-12. Medical Care 42 (9), 851—859.

Brazier, J.E., Roberts, J., Deverill, M., 2002. The estimation of a preference-based single index measure for health from the SF-36. Journal of Health Economics 21 (2), 271—292.

Brazier, J.E., Usherwood, T.P., Harper, R., Jones, N.M.B., Thomas, K., 1998. Deriving a preference based single index measure for health from the SF-36. Journal of Clinical Epidemiology 51 (11), 1115—1129.

Brooks, R.G., 1995. Health Status Measurement: A Perspective on Change. Macmillan Press, London.

Brooks, R., 1996. EuroQol: the current state of play. Health Policy 37 (1), 53—72.

Brooks, R., 2012. The EuroQol Group After 25 Years. Springer, Dordrecht.

Chatterji, S., Ustün, B.L., Sadana, R., Salomon, J.A., Mathers, C.D., Murray, C.J.L., 2002. The Conceptual Basis for Measuring and Reporting on Health. Global Programme on Evidence for Health Policy Discussion Paper No. 45. World Health Organization, Geneva.

Chen, G., Khan, M.A., Iezzi, A., Ratcliffe, J., Richardson, J., 2016. Mapping between six multi-attribute utility instruments. Medical Decision Making 36 (2), 160−175.

Chestnut, L.G., Kellet, L.R., Lambert, W.E., Rowe, R.D., 1996. Measuring heart patients willingness to pay for changes in angina symptoms. Some methodological issues. Medical Decision Making 16 (1), 65−77.

Coon, C.D., McLeod, L.D., 2013. Patient-reported outcomes: current perspectives and future directions. Clinical Therapeutics 35 (4), 399−401.

Coretti, S., Ruggeri, M., McNamee, P., 2014. The minimum clinically important difference for EQ-5D index: a critical review. Expert Review of Pharmacoeconomics and Outcomes Research 14 (2), 221−233.

Devlin, N., Shah, K., Feng, Y., Mulhern, B., van Hout, B., 2016. Valuing Health-Related Quality of Life: An EQ-5D-5L Value Set for England (Research Paper 16/01). Office of Health Economics, London. https://www.ohe.org/publications/valuing-health-related-quality-life-eq-5d-5l-value-set-england.

Devlin, N.J., Krabbe, P.F.M., 2013. The development of new research methods for the valuation of EQ-5D-5L. European Journal of Health Economics 14 (Suppl. 1), 1−3.

Dolan, P., 1997. Modeling valuations for EuroQol health states. Medical Care 35 (11), 1095−1108.

Dolan, P., Gudex, C., Kind, P., Williams, A., 1996. The time trade-off method: results from a general population study. Health Economics 5 (2), 141−154.

Dolan, P., Kahneman, D., 2008. Interpretations of utility and their implications for the valuation of health. Economic Journal 118 (525), 215−234.

Dolan, P., Stalmeier, P., 2003. The validity of time trade-off values in calculating QALYs: constant proportional time trade-off versus the proportional heuristic. Journal of Health Economics 22 (3), 445−458.

Drummond, M.F., Sculpher, M.J., Torrance, G.W., O'Brien, B.J., Stoddart, G.L., 2005. Methods for the Economic Evaluation of Health Care Programmes, third ed. Oxford University Press, Oxford.

van Dalen, H., Williams, A., Gudex, C., 1994. Lay people's evaluations of health: are there variations between different subgroups? Journal of Epidemiology and Community Health 48 (3), 248−253.

Essink-Bot, M.L., Bonsel, G.J., 2002. How to derive disability weights. In: Murray, C.J.L., Salomon, J.A., Mathers, C.D., Lopez, A.D. (Eds.), Summary Measures of Population Health: Concepts, Ethics, Measurement and Applications. World Health Organization, Geneva, pp. 449−465.

Fanshel, S., Bush, J.W., 1970. A health-status index and its application to health-services outcomes. Operations Research 18 (6), 1021−1066.

Fallowfield, L., 1990. The Quality of Life: The Missing Measurement in Health Care. Souvenir Press, London.

Feeny, D., Furlong, W., Torrance, G.W., Goldsmith, C.H., Zhu, Z., DePauw, S., Denton, M., Boyle, M., 2002. Multiattribute and single-attribute utility functions for the health utilities index mark 3 system. Medical Care 40 (2), 113−128.

Food and Drug Administration, 2009. Guidance for Industry Use in Medical Product Development to Support Labeling Claims Guidance for Industry, Guidance for Industry Patient-reported Outcome Measures: Use in Medical Product Development to Support Labeling Claims. Silver Spring.

Froberg, D.G., Kane, R.L., 1989. Methodology for measuring health-state preferences − I: measurement strategies. Journal of Clinical Epidemiology 42 (5), 345−354.

Fryback, D.G., Palta, M., Cherepanov, D., Bolt, D., Kim, J., 2010. Comparison of 5 health related quality of life indexes using item response theory analysis. Medical Decision Making 30 (1), 5–15.

Furlong, W., Feeny, D., Torrance, G.W., Goldsmith, C.H., DePauw, S., Zhu, Z., Denton, M., Boyle, M., 1998. Multiplicative Multi-attribute Utility Function for the Health Utilities Index Mark 3 (HUI3) System: A Technical Report. McMaster University Centre for Health Economics and Policy Analysis Working Paper 98-11. McMaster University, Ontario.

Gafni, A., Torrance, G.W., 1984. Risk attitude and time preference in health. Management Science 30 (4), 440–451.

Gafni, A.G., 1994. The standard gamble method: what is being measured and how it is interpreted. Health Services Research 29 (2), 207–224.

Gandjour, A., 2010. Theoretical foundation of patient v. population preferences in calculating QALYs. Medical Decision Making 30, E57–E63.

Ginsberg, A.S., Offensend, F.L., 1968. An application of decision theory to a medical diagnosis-treatment problem. IEEE Transactions on Systems Science and Cybernetics 4 (3), 355–362.

Gold, M.R., Siegel, J.E., Russel, L.B., Weinstein, M.C., 1996. Cost-Effectiveness in Health and Medicine. Oxford University Press, New York.

Green, C., Brazier, J., Deverill, M., 2000. Valuing health-related quality of life. A review of health state valuation techniques. PharmacoEconomics 17 (2), 151–165.

Grewal, I., Lewis, J., Flynn, T., Brown, J., Bond, J., Coast, J., 2006. Developing attributes for a generic quality of life measure for older people: preferences or capabilities? Social Science and Medicine 62 (8), 1891–1901.

Gudex, C., 2005. The descriptive system of the EuroQol instrument. In: Kind, P., Brooks, R., Rabin, R. (Eds.), EQ-5D Concepts and Methods: A Developmental History. Springer, Dordrecht, pp. 19–27.

Gustafson, D.H., Holloway, D.C., 1975. A decision theory approach to measuring severity in illness. Health Services Research 10 (1), 97–106.

Hakim, Z., Pathak, D.S., 1999. Modelling the Euroqol data: a comparison of discrete choice conjoint and conditional preference modelling. Health Economics 8 (2), 103–116.

Hawthorne, G., Richardson, J., Day, N.A., 2001. A comparison of the Assessment of Quality of Life (AQoL) with four other generic utility instruments. Annals of Medicine 33 (5), 358–370.

Hawthorne, G., Richardson, J., Osborne, R., 1999. The Assessment of Quality of Life (AQoL) instrument: a psychometric measure of health related quality of life. Quality of Life Research 8 (3), 209–224.

Heintz, E., Krol, M., Levin, L.A., 2013. The impact of patients' subjective life expectancy on time tradeoff valuations. Medical Decision Making 33, 261–270.

Hodgkin, P., Taylor, J., 2013. Power to the people: what will bring about the patient centred revolution? BMJ 347, 1–2.

Hunink, M.M., Weinstein, M.C., Wittenberg, E., Drummond, M.F., Pliskin, J.S., Wong, J.B., Glasziou, P.P., 2014. Decision Making in Health and Medicine: Integrating Evidence and Values. Cambridge University Press.

Hurst, N.P., Kind, P., Ruta, D., Hunter, M., Stubbings, A., 1997. Measuring health-related quality of life in rheumatoid arthritis: validity, responsiveness and reliability of EuroQol (EQ-5D). British Journal of Rheumatology 36 (5), 551–559.

Johannesson, M., Jönsson, B., Borgquist, L., 1991. Willingness to pay for antihypertensive therapy – results of a Swedish pilot study. Journal of Health Economics 10 (4), 461–473.

Johannesson, M., Pliskin, J.S., Weinstein, M.C., 1994. A note on QALYs, time tradeoff, and discounting. Medical Decision Making 14 (2), 188−193.

Johnson, J.A., Luo, N., Shaw, J.W., Kind, P., Coons, S.J., 2005. Valuations of EQ-5D health states: are the United States and United Kingdom different? Medical Care 43 (3), 221−228.

Kahneman, D., Sugden, R., 2005. Experienced utility as a standard of policy evaluation. Environmental and Resource Economics 32 (1), 161−181.

Kaplan, R., 1994. Value judgment in the Oregon Medicaid experiment. Medical Care 32 (10), 975−988.

Kaplan, R., 2005. Kaplan about J.W. Bush. In: Lenderking, W., Revicki, D. (Eds.), Advancing Health Outcomes Research Methods and Clinical Applications. Degnon Associates, McLean, VA.

Kaplan, R., Anderson, J., 1988. A general health policy model: update and applications. Health Services Research 23 (2), 203−235.

Kaplan, R., Bush, J.W., Berry, C., 1976. Health status: types of validity and the index of well-being. Health Services Research 11 (4), 478−507.

Kaplan, R.M., Anderson, J.P., 1990. Quality of life assessments in clinical trials. In: Spilker, B. (Ed.), Quality of Life Assessments in Clinical Trials. Raven Press, New York, pp. 131−149.

Keeney, R.L., Raiffa, H., 1976. Decision with Multiple Objectives: Preferences and Value Trade-Offs. John Wiley, New York.

Konerding, U., 2013. What does Cronbach's alpha tell us about the EQ-5D? A methodological commentary to "psychometric properties of the EuroQol Five-Dimensional Questionnaire (EQ-5D-3L) in caregivers of autistic children". Quality of Life Research 22 (10), 2939−2940.

Krabbe, P.F.M., 2014. Measurement properties of valuation techniques. In: Culyer, A.J. (Ed.), Encyclopedia of Health Economics. Elsevier, San Diego, pp. 228−233.

Krabbe, P.F.M., 2006. Valuation structures of health states revealed with singular value decomposition. Medical Decision Making 26 (1), 30−37.

Krabbe, P.F.M., Bonsel, G.J., 1998. Sequence effects, health profiles and the QALY model: in search of realistic modeling. Medical Decision Making 18 (2), 178−186.

Krabbe, P.F.M., Essink-Bot, M.L., Bonsel, G.J., 1996. On the equivalence of collectively and individually collected responses: standard-gamble and time-tradeoff judgments of health states. Medical Decision Making 16 (2), 120−132.

Krabbe, P.F.M., Essink-Bot, M.L., Bonsel, G.J., 1997. The comparability and reliability of five health-state valuation methods. Social Science & Medicine 45 (11), 1641−1652.

Krabbe, P.F.M., Salomon, J.A., Murray, C.J.L., 2007. Quantification of health states with rank-based nonmetric multidimensional scaling. Medical Decision Making 27 (4), 395−405.

Krabbe, P.F.M., Stalmeier, P.F.M., Lamers, L.M., van Busschbach, J.J., 2006. Testing the interval-level measurement property of multi-item visual analogue scales. Quality of Life Research 15 (10), 1651−1661.

Krabbe, P.F.M., Tromp, N., Ruers, T.J.M., van Riel, P.L.C.M., 2011. Are patients' judgments of health status really different from the general population? Health and Quality of Life Outcomes 9 (1), 31−40.

Lamers, L.M., McDonnell, J., Stalmeier, P.F.M., Krabbe, P.F.M., Busschbach, J.J.V., 2006. The Dutch tariff: results and arguments for an effective design for national EQ-5D valuation studies. Health Economics 15 (10), 1121−1132.

Matza, L.S., Boye, K.S., Feeny, D.H., Bowman, L., Johnston, J.A., Stewart, K.D., McDaniel, K., Jordan, J., 2015. The time horizon matters: results of an exploratory study varying the timeframe in time trade-off and standard gamble utility elicitation. European Journal of Health Economics 1−12.

McNeil, B.J., Pauker, S.G., Sox, H.C., Tversky, A., 1982. On the elicitation of preferences for alternative therapies. New England Journal of Medicine 306 (21), 1259−1262.

McNeil, B.J., Weichselbaum, R., Pauker, S.G., 1981. Speech and survival: tradeoffs between quality and quantity of life in laryngeal cancer. New England Journal of Medicine 305 (17), 982−987.

McPhail, S., Beller, E., Haines, T., 2010. Reference bias: presentation of extreme health states prior to EQ-VAS improves health-related quality of life scores. A randomised cross-over trial. Health and Quality of Life Outcomes 8 (146), 1−11.

Misajon, R., Hawthorne, G., Richardson, J., Barton, J., Peacock, S., Iezzi, A., Keeffe, J., 2005. Vision and quality of life: the development of a utility measure. Investigative Ophthalmology & Visual Science 46 (11), 4007−4015.

Mitchell, P.M., Al-Janabi, H., Richardson, J., Iezzi, A., Coast, J., 2015. The relative impacts of disease on health status and capability wellbeing: a multi-country study. PLoS One 10, e0143590.

Miyamoto, J.M., Eraker, S.A., 1985. Parameter estimates for a QALY utility model. Medical Decision Making 5 (2), 191−213.

Moock, J., Kohlmann, T., 2008. Comparing preference-based quality-of-life measures: results from rehabilitation patients with musculoskeletal, cardiovascular, or psychosomatic disorders. Quality of Life Research 17 (3), 485−495.

Mulhern, B., Bansback, N., Brazier, J., Buckingham, K., Cairns, J., Devlin, N., Dolan, P., Hole, A.R., Kavetsos, G., Longworth, L., Rowen, D., Tsuchiya, A., 2014. Preparatory study for the revaluation of the EQ-5D tariff: methodology report. Health Technology Assessment 18 (12), 222.

Murray, C.J.L., 1996. Rethinking DALYs. In: Murray, C.J.L., Lopez, A.D. (Eds.), The Global Burden of Disease: A Comprehensive Assessment of Mortality and Disability from Diseases, Injuries, and Risk Factors in 1990 and Projected to 2020. Harvard University Press, Cambridge.

Murray, C.J.L., Lopez, A.D., 1996a. Evidence-based health policy: lessons from the global burden of disease study. Science 274 (5288), 740−743.

Murray, C.J.L., Lopez, A.D., 1996b. The Global Burden of Disease: A Comprehensive Assessment of Mortality and Disability from Diseases, Injuries, and Risk Factors in 1990 and Projected to 2020. Harvard School of Public Health/World Health Organization, Geneva.

Nord, E., 1992. Methods for quality adjustment of life years. Social Science & Medicine 34 (5), 559−569.

Nord, E., 1995. The person-trade-off approach to valuing health care programs. Medical Decision Making 15 (3), 201−208.

Norman, R., Viney, R., Brazier, J.E., Cronin, P., King, M.T., Ratcliffe, J., Street, D., 2013. Valuing SF-6D health states using a discrete choice experiment. Medical Decision Making 34 (6), 773−786.

von Neumann, J., Morgenstern, O., 2004. Theory of Games and Economic Behavior, sixteenth ed. Princeton University Press, Princeton.

Oppe, M., Devlin, N.J., van Hout, B., Krabbe, P.F.M., de Charro, F., 2014. A program of methodological research to arrive at the new international EQ-5D-5L valuation protocol. Value in Health 17 (4), 445−453.

Parducci, A., 1968. The relativism of absolute judgments. Scientific American 219 (6), 84−90.

Parkin, D., Devlin, N., 2006. Is there a case for using visual analogue scale valuations in cost-utility analysis? Health Economics 15 (7), 653−664.

Patrick, D.L., Bush, J.W., Chen, M.M., 1973. Methods for measuring levels of wellbeing for a health status index. Health Services Research 8 (3), 228–245.

Pliskin, J.S., Shepard, D.S., Weinstein, M.C., 1980. Utility functions for life years and health status. Operations Research 28 (1), 206–224.

Pohjolainen, V., Räsänen, P., Roine, R.P., Sintonen, H., Wahlbeck, K., Karlsson, H., 2010. Cost-utility of treatment of bulimia nervosa. The International Journal of Eating Disorders 43 (7), 596–602.

Richardson, J., 2002. Evaluating summary measures of population health. In: Murray, C.J.L., Salomon, J.A., Mathers, C.D., Lopez, A.D. (Eds.), Summary Measures of Population Health Concepts, Ethics, Measurement and Applications. World Health Organization, Geneva.

Richardson, J., 1994. Cost utility analysis: what should be measured? Social Science & Medicine 39 (1), 7–21.

Richardson, J., Chen, G., Khan, M.A., Iezzi, A., 2015a. Can multi attribute utility instruments adequately account for subjective well-being? Medical Decision Making 35 (3), 292–304.

Richardson, J., Iezzi, A., Khan, M.A., 2015b. Why do multi-attribute utility instruments produce different utilities: the relative importance of the descriptive systems, scale and "micro-utility" effects. Quality of Life Research 24 (8), 2045–2053.

Richardson, J., Khan, M.A., Iezzi, A., Maxwell, A., 2014a. Comparing and explaining differences in the content, sensitivity and magnitude of incremental utilities predicted by the EQ-5D, SF-6D, HUI 3, 15D, QWB and AQoL-8D multi attribute utility instruments. Medical Decision Making 35 (3), 276–291.

Richardson, J., Khan, M.A., Iezzi, A., Maxwell, A., 2016. Measuring the sensitivity and construct validity of six utility instruments in seven disease states. Medical Decision Making 36 (2), 147–159.

Richardson, J., McKie, J., Bariola, E., 2014b. Multi attribute utility instruments and their use. In: Culyer, A.J. (Ed.), Encyclopedia of Health Economics. Elsevier, San Diego, pp. 341–357.

Richardson, J., Sinha, K., Iezzi, A., Khan, M.A., 2014c. Modelling utility weights for the assessment of quality of life (AQoL) 8D. Quality of Life Research 23 (8), 2395–2404.

Roest, F.H., Eijkemans, M.J., van der Donk, D.J., Levendag, P.C., Meeuwis, C.A., Schmitz, P.I., Habbema, J.D., 1997. The use of confidence intervals for individual utilities: limits to formal decision analysis for treatment choice. Medical Decision Making 17 (3), 285–291.

Rosser, R., Kind, P., 1978. A scale of valuations of states of illness: is there a social consensus? International Journal of Epidemiology 7 (4), 347–358.

Salomon, J.A., Murray, C.J., 2004. A multi-method approach to measuring health-state valuations. Health Economics 13 (3), 281–290.

Shah, K., Mulhern, B., Longworth, L., Janssen, M.F., 2015. An empirical study of two alternative comparators for use in time-trade off studies. Value in Health 19, 53–59.

Sintonen, H., 1981. An approach to measuring and valuing health states. Social Science & Medicine 15 (2), 55–65.

Sintonen, H., 1994. The 15D-Measure of Health-Related Quality of Life. I. Reliability, Validity and Sensitivity of Its Health State Descriptive System. Centre for Health Program Evaluation, Working Paper 41, Melbourne. Downloaded on 6 February 2016 from:http://business.monash.edu/__data/assets/pdf_file/0009/391374/wp41-1.pdf.

Sintonen, H., 1995. The 15D-Measure of Health-related Quality of Life. II. Feasibility, Reliability and Validity of Its Valuation System. Centre for Health Program Evaluation, Working Paper 42, Melbourne. Downloaded on 6 February 2016 from:http://business.monash.edu/__data/assets/pdf_file/0003/391422/wp42.pdf.

Sintonen, H., Pekurinen, M., 1993. A fifteen-dimensional measure of health-related quality of life (15D) and its applications. In: Walker, S.R., Rosser, R.M. (Eds.), Quality of Life Assessment: Key Issues in the 1990s. Kluwer Academic Publishers, Dordrecht, pp. 185−195.

Stevens, S.S., 1951. Mathematics, measurement and psychophysics. In: Stevens, S.S. (Ed.), Handbook of Experimental Psychology. Wiley, New-York, pp. 1−49.

Stevens, S.S., 1957. On the psychophysical law. The Psychological Review 64 (3), 153−181.

Streiner, D.L., Norman, G.R., Cairney, J., 2015. In: Health Measurement Scales: A Practical Guide to Their Development and Use, fifth ed. Oxford University Press, Oxford.

Torrance, G., 2006. Utility measurement in healthcare: the things I never got to. PharmacoEconomics 24 (11), 1069−1078.

Torrance, G.W., Feeny, D., Furlong, W., 2001. Visual analog scales: do they have a role in the measurement of preferences for health states? Medical Decision Making 21 (4), 329−334.

Torrance, G., Furlong, W., Feeny, D., Boyle, M., 1995. Multi-attribute preference functions: health utilities index. PharmacoEconomics 7 (6), 503−520.

Torrance, G.W., 1976. Social preferences for health states: an empirical evaluation of three measurement techniques. Socio-Economic Planning Sciences 10 (3), 129−136.

Torrance, G.W., Boyle, M.H., Horwood, S.P., 1982. Application of multi-attribute utility theory to measure social preferences for health states. Operations Research 30 (6), 1043−1069.

Torrance, G.W., Feeny, D.H., Furlong, W.J., Barr, R.D., Zhang, Y., Wang, Q., 1996. Multiattribute utility function for a comprehensive health status classification system. Health Utilities Index Mark 2. Medical Care 34 (7), 702−722.

Torrance, G.W., Sackett, D.L., Thomas, W.H., 1973. Utility maximization model for program evaluation: a demonstration application. In: Berg, R.L. (Ed.), Health Status Indexes: Proceedings of a Conference Conducted by Health Services Research. Hospital Research and Education Trust, Chicago.

Torrance, G.W., Thomas, W.H., Sackett, D.L., 1972. A utility maximization model for evaluation of health care programs. Health Services Research 7 (2), 118−133.

Tsuchiya, A., Ikeda, S., Ikegami, N., Nishimura, S., Sakai, I., Fukuda, T., Hamashima, C., Hisashige, A., Tamura, M., 2002. Estimating an EQ-5D population value set: the case of Japan. Health Economics 11 (4), 341−353.

Ubel, P.A., Jankovic, A., Smith, D., Langa, K.M., Fagerlin, A., 2005. What is perfect health to an 85-year-old?: evidence for scale recalibration in subjective health ratings. Medical Care 43 (10), 1054−1057.

Ubel, P.A., Loewenstein, G., Jepson, C., 2003. Whose quality of life? A commentary exploring discrepancies between health state evaluations of patients and the general public. Quality of Life Research 12 (6), 599−607.

de Wit, G.A., Busschbach, J.J.V., De Charro, F.T., 2000. Sensitivity and perspective in the valuation of health status: whose values count? Health Economics 9 (2), 109−126.

von Winterfeldt, D., Edwards, W., 1986. Decision Analysis and Behavioral Research. Cambridge University Press, Cambridge.

Wegener, B., 1982. Social Attitudes and Psychophysical Measurement. Lawrence Erlbaum Associates, Hillsdale.

Whiteside, R., 1984. A conversation with James W. Bush, M.D. Simulation 43 (2), 102−106.

WHO, 2001. International Classification of Functioning, Disability and Health (ICF). World Health Organization, Geneva.

Wittrup-Jensen, K.U., Pedersen, K.M., 2008. Modelling Danish Weights for the 15D Quality of Life Questionnaire by Applying Multi-Attribute Utility Theory (MAUT). Health Economics Papers 7. University of Southern Denmark. Downloaded on 6 February 2016 from:http://static. sdu.dk/mediafiles/Files/Om_SDU/Centre/c_ist_sundoke/Forskningsdokumenter/publications/ Working%20papers/20087.pdf.

Xu, R., Insinga, R.P., Golden, W., Hu, X.H., 2011. EuroQol (EQ-5D) health utility scores for patients with migraine. Quality of Life Research 20 (4), 601–608.

Yang, Z., 2015. Inconsistency in the valuations of EuroQol EQ-5D-5L health states in China was more related to interviewer and to interview process than to respondents' characteristics. Value in Health 18 (7), A737–A738.

Chapter 13

Clinimetrics

Chapter Outline

INTRODUCTION

Clinimetrics is the term introduced by Alvan Feinstein in the early 1980s to indicate a research domain concerned with indices, rating scales, and other expressions that are used to describe or measure symptoms, physical signs, and other clinical phenomena (Feinstein, 1983). Feinstein is widely known for advancing the field of measurement in medicine and may be even considered as the founding father of clinical epidemiology. He developed a special fondness for Boolean algebra and Venn Diagrams (Feinstein, 1987, p. 37/64/133), which seemed to him the best way to summarize the complex phenomena he had observed (Fletcher, 2001).

In 1987 he published his influential book *Clinimetrics*. In this work, Feinstein stressed the importance of subjective data in the setting of medical evaluation. He wrote, "The biomedical paradigm of explanation doesn't work because what is needed for clinical work is 'soft data' describing human life, based in judgments of clinical practice." While he called these data "soft," he clearly argued that maintaining consistency and accuracy in the measurement of disease and treatment effects on patients is not only possible but also necessary. Clinimetrics was designed to provide reproducible and quantitative information to support clinical decisions where there, simply, was not enough guidance from pathophysiology (Sullivan, 2003).

The term clinimetrics was introduced to describe an approach to scale development in the area of health that ostensibly differs from the traditional "psychometrics" approach, i.e., classical test theory (Chapter 9). Clinimetrics differs from standard psychometrics, both from a conceptual and a methodological viewpoint. It has a set of rules that govern the structure of indexes, the choice of component variables, and the evaluation of consistency and validity. However, many of the methods to assess the level of reliability that are widely applied in classical test theory are also used in the framework of clinimetrics.

One may say that clinimetrics originated mainly in classical test theory (Chapter 9) and has less to do with the other three approaches of subjective measurement (Chapters 10–12). It seems largely to deal with clinical measures and multiitem profile measures (classical test theory). Far less attention is given to more advanced measurement frameworks such as item-response theory. Index measures play a substantial role in this field, but only in the form of unweighed or nonanalytical weighed index measures. The class of sophisticated, preference-based methods (Chapters 11 and 12) seems to play no role whatsoever within clinimetrics (Fig. 13.1).

BOX 13.1 Alvan Feinstein

Alvan Feinstein was a clinician, a researcher, and an epidemiologist who made a significant impact on clinical investigation, especially in the field of clinical epidemiology. Feinstein received his medical degree at the University of Chicago School of Medicine, but previously he had also earned a degree in mathematics. In 1962, Feinstein joined the Yale University School of Medicine faculty. He published his first paper as a medical student in 1951 and more than 400 throughout his career. He wrote six major textbooks, two of which, *Clinical Judgment* (1967) and *Clinical Epidemiology* (1985), are among the most widely referenced books in clinical epidemiology. He founded and was editor of the *Journal of Clinical Epidemiology* (1988–2001).

Box 13.1 Alvan Feinstein—cont'd

FIGURE 13.1   Alvan Feinstein (1925–2001). *With permission from Yale Office of Public Affairs & Communications.*

Indices and Rating Scales

Feinstein (1987) has argued that many clinical scales possess fundamentally different characteristics from psychometric scales, and that their development and validation should therefore proceed along separate paths. He proposed the name clinimetrics for the domain concerned with the construction of clinical indexes and scales. A "good" and useful clinimetric scale may consist of items comprising a variety of symptoms and other clinical variables, and it does not necessarily need to satisfy the same requirements that are demanded by conventional psychometric scales. In this sense, psychometricians try to measure a single construct with multiple items and use the validation methods described in Chapter 7 to demonstrate that the items are all measuring (more or less) the same single construct. Clinicians try to measure single constructs (index) based on multiple attributes. They aim their strategies at choosing the most important attributes to be included in the index. This distinction between the psychometric scales based on classical test theory and the index approach from clinimetrics is largely connected—although not recognized as such at the onset of clinimetrics—to the two different ways of measurement discussed in Chapter 6: reflective versus formative measurement. These are two measurement approaches that deal with different data structures and assume different causality pathways of the variables captured in a health measurement instrument.

A simple example of a clinimetric index measurement approach is the way frailty is measured. As individuals age, their vulnerability to adverse outcomes (including death) increases. Some individuals experience a state of increased vulnerability, known as frailty. Frailty is often based on the concept of deficit accumulation. Deficits in health may include a range of symptoms, morbidities, or functional limitations that can be quantified (summed up) using an index. Several frailty indices have been developed (Jones et al., 2004; Peters et al., 2012; Peña et al., 2014). Clinicians have developed index measures not only for frailty but for numerous other health outcomes as well.

Indices

In clinical studies global judgments are often based on summarizing several aspects (attributes) related to a specific health concept (e.g., pain, frailty, depression). However, if patients improve on some attributes and worsen on others, a judgment is required on the relative importance of the various attributes to decide if the patient has improved or worsened overall. This is often expressed in the form of indices, where the weighing and combining of the individual attributes (variables) is done according to defined rules. Indices are useful to summarize clinical information on individuals or groups.

In the beginning, it was thought that such indices may be derived using statistical methods or by asking physicians to provide weights. Several statistical techniques (e.g., discriminant analysis, regression analysis, factor analysis) were proposed to find the "best" combination of attributes and their weights (Bombardier and Tugwell, 1982). This idea may rest on a simplification of the problem, because in general these statistical techniques are not suitable to indicate most attributes.

However, most often no weights are given to the different domains or attributes, nor to the levels or categories of the domains or attributes. In the case that weights are attached to the attributes and/or their levels, these weights are almost always based on expert opinion. In clinical settings, weighting is not worked out in a formal measurement (statistical) framework. As such, index instruments developed in the setting of clinimetrics cannot be considered preference-based instruments (Chapters 10–12).

Scales

One problem with psychometric instruments from a clinimetric point of view is that they contain a large set of multiple items. Depending on which psychometric analytical techniques are applied, the resulting item selection may obscure (or may not even include) the particular symptom (e.g., dyspnea or joint pain) that is the focal concern to be relieved by treatment. Another problem with the classical psychometric scales is what Feinstein (1994) calls "the unfamiliar methods used to support the efficacy of the psychometric scores with a statistical analysis of 'reliability' and 'validity.'" Feinstein proceeds by saying, "Awed or baffled by the arcane vocabulary and statistics, clinical investigators may have neglected the appraisal of 'face validity'—a statistically unmeasurable attribute that refers to the measurement's clinical 'sensibility' or 'common sense' in doing its intended job." Feinstein noted a third problem: an instrument receiving high statistical results for "reliability" and "validity" in one clinical setting may be poorly suited for some other clinical setting. For example, measurements of activities of daily living often have excellent statistical credentials when applied to impaired patients in nursing homes but do not focus on the appropriate problems when used for ambulatory patients elsewhere. A fourth problem he noted is that "multidimensional" indexes are seldom satisfactory for measuring change. As Feinstein continues, "This problem was pointed out many years ago by Nunally (PK: Nunnally), a psychometric titan who corresponds to R.A. Fisher in the statistical pantheon, but the warning has often been ignored. The clinical-psychometric investigators have then been surprised or dismayed to find that the multiple-item indexes were not effective in discerning clinical response changes after therapy. Although clinicians can readily construct special 'transition indexes' that identify changes easily and effectively, the approach is not a standard part of psychometric strategy, and use of the transition method is often discouraged by consultants who cannot conveniently fit the results into a repeated-measures analysis of variance."

THREE CATEGORIES

In the same period that Feinstein advanced his ideas, researchers from McMaster University (Canada) proposed that for a better understanding, health-status measures should be divided into three broad categories based on their purpose (Kirshner and Guyatt, 1985). These three broad categories of potential applications of health-status measures were discrimination (differentiating between people who have a better health status and those who have a worse health status); prediction; and evaluation (how much the health status has changed over time). They stated that a health-status instrument to monitor patient groups should be differently devised and constructed than a health-status instrument to discriminate among patients. To support this conclusion,

TEST - RETEST RELIABILITY		
Patients	Time 1	Time 2
1	15*	14
2	14	15
3	15	14
4	14	15
5	15	14
6	14	15
7	15	14
8	14	15

*Score is out of 20.

FIGURE 13.2 Hypothetical example used in the paper of Kirshner and Guyatt (1985) to explain their thoughts about different requirements needed for health instruments. *With permission from Elsevier Limited, UK.*

they showed fictitious data to represent results obtained from a test–retest procedure (Fig. 13.2). However, artificial examples do not always reflect reality. As we can see from their example, there is almost no variance. One general rule in statistics is if there is no variance, then statistics fails.

Some years later, Norman (1989) responded to this by saying: "Kirshner and Guyatt (Kirshner and Guyatt, 1985) claim that the goals of discrimination (detecting differences between individuals) and evaluation (detecting changes within individuals over time) are different, may require different instruments, and that 'the requirements for maximizing one of the functions .... may actually impede the others'" (p. 28). The reliability coefficient is a generally accepted measure of the ability of an instrument to discriminate among individuals. If Kirshner and Guyatt are correct, we cannot assume that an instrument with acceptable reliability can be appropriately applied to an experimental situation where the goal is to change or improve function within an individual. In fact, in the extreme case, high reliability might be viewed as a reason for rejecting an instrument. Norman went on to conclude "From the analyses in the paper, it is evident that the concept of reliability, assessing stable differences between individuals, and responsiveness to change, assessing overall effects of treatment, are related." Later, this organization in categories of health-status measures has been criticized further (Norman, 1989; Williams and Naylor, 1992; Hays and Hadorn, 1992).

Fayers and Machin (2000) come up with another explanation to support the idea of different instrument development for different purposes. "If an instrument is intended to be discriminative, it may be less important to include symptoms that are common to all patients and unlikely to differ between the various treatment groups. For example, fatigue is not only common for patients with thyroid disease but is also common amongst patients without the disease, and hence it might be considered an unimportant item in a purely discriminative instrument." This explanation does make sense. Yet another way of reasoning could lead to the opposite conclusion: that fatigue should definitely be part of the instrument. Why should we develop an instrument that does not comprise one or more of the most salient characteristics of the phenomenon that we try to measure?

In the years that the influence of clinimetrics was rising, the idea was gaining ground that, depending on how a measurement instrument was to be used, it should be differently developed, because different measurement properties are required. For the case of prediction, this view seems to be changing, as a recognized group of clinimetricians (de Vet et al., 2011) recently stated that nowadays, prediction models are used to define a set of variables that best predict the future course. These prediction tools are usually referred to as prediction models or prediction rules, rather than measurement instruments (Steyerberg, 2009). They continue, "In our opinion, it is better to speak of discriminative, evaluative, or predictive applications, than of instruments, because the same instrument can be used for different purposes."

RESPONSIVENESS

As Kirshner and Guyatt also stated in their 1985 paper, "while demonstration of reliability and validity is sufficient for concluding that an instrument is useful for descriptive (or predictive) purposes, we must also have information about an evaluative instrument's responsiveness before we can confidently use it as an outcome measure in a clinical trial." Responsiveness is the usefulness of clinical (evaluative) instruments to measure changes within persons over time. To be of value, an instrument should be stable when no change occurs, although it should reveal differences in case of improvement or deterioration of a person's health status.

The concept of responsiveness has drawn considerable attention, mainly from the users of multiitem (multidomain) health instruments (Chapter 9). During the 1970s and 1980s, the concept of responsiveness first received attention in the medical literature (Deyo and Centor, 1986; Guyatt et al., 1987). Most of those who were applying the framework of clinimetrics were working in the field of medicine. One reason for Guyatt to introduce the concept of responsiveness was that, in this field, "responsiveness" to health interventions is central to the clinical framework (Guyatt et al., 1987; Jaeschke et al., 1989).

Responsiveness is considered by clinimetric researchers to be only relevant for measurement instruments that are used in evaluative applications. That is, it is relevant when the instrument is used in a longitudinal study to measure change. To be of value, the measure should be stable when no change occurs but reveal changes in cases of improvement or deterioration (Deyo and Inui,

1984; Muraweski and Miederhoff, 1998; Kirshner and Guyatt, 1985). Even though many within the field of descriptive health measurement and patient-reported outcomes research argue that responsiveness is important, little consensus seems to exist regarding how to express it (Norman et al., 2007; Revicki et al., 2008; Angst, 2011).

In theory, if a health instrument is responsive, the change it measures should represent the effect caused by a treatment, intervention, or natural course. In practice, we often do not know what this effect is because there is no gold standard to define and quantify changes in health. Consequently, it may be difficult or even impossible to unravel whether the instrument is under-estimating, overestimating, or correctly estimating the change in health.

Responsiveness or Sensitivity

The question of whether to use the term "responsiveness" or "sensitivity" has also attracted attention. It seems that "responsiveness" is used to avoid confusion with "sensitivity," which appears frequently in the diagnostic literature. According to Guyatt et al. (1992), sensitivity entails two aspects. The first is the ability to distinguish between individuals and groups in different health states cross-sectionally (discrimination); the second is to detect changes in individuals or groups over time (responsiveness to change) (Hays and Hadorn, 1992).

Separate Measurement Properties

There have been major discussions about whether responsiveness should be treated as a separate measurement property, or as an aspect of reliability or validity. Many researchers consider responsiveness as an aspect of validity. Their reasoning is that the only difference between validity and responsiveness is that validity refers to the validity of a single score, and responsiveness refers to the validity of a change score. This definition implies that if we want to measure change, a valid instrument should truly measure changes in the construct(s) it purports to measure. However, the COSMIN panel (Mokkink et al., 2010a,b,c) treated responsiveness as a separate measurement property to emphasize this distinction between the validity of a single score and the validity of a change score. The panel concluded that both are important and may lead to different results.

COMPUTATION OF RESPONSIVENESS

Much has remained unclear about this newly introduced measurement concept. There is no consensus in the literature on the concept and the interpretation of responsiveness, nor on how it should be measured (Norman,

1989; Hays and Hadorn, 1992; Guyatt et al., 1998; Beaton et al., 2001). Various analytical methods have been put forward to assess responsiveness, and several different responsiveness indices have been introduced (Kazis et al., 1989; Beaton et al., 1997, 2001; Wright and Young, 1997). In a review of the medical literature, Terwee et al. (2003) encountered numerous definitions and 31 different ways to measure responsiveness. This lack of consensus has led to a great proliferation of statistics, and it is not uncommon for some researchers to employ several simultaneously in the same article. The various statistics to compute responsiveness (to change) seem to have different objectives, prompting different conclusions.

The standard error of measurement has been suggested as an approach to assessing responsiveness (Wyrwich et al., 1999; Guyatt et al., 2002). Other ways to express responsiveness are by calculating the effect size and standardized mean response. However, Streiner et al. (2015) presented an alternative that seems more satisfactory. Their way of dealing with responsiveness is based on a combination of study design and appropriate statistics (analysis of variance). Another less systematic approach would be to use other instruments as comparatives under identical conditions to assess their relative responsiveness (see Application section). It may turn out that the latter two strategies provide the best information about responsiveness and are, in fact, all we may have (Stockler et al., 1998).

Distribution Based

Distribution-based approaches compare the change in patient-reported outcome scores to some measure of variability such as the standard error of measurement, the standard deviation, or the effect size. The distribution-based indices provide no direct information about the minimal important difference (MID) (see below). They are simply a way of expressing the observed change in a standardized metric (Revicki et al., 2008). Other methods are sensitivity and specificity, limits of agreement, and intraclass correlation coefficients.

Anchor Based

Anchor-based methods examine the relationship between a health measure and an independent measure (or anchor) to elucidate the meaning of a particular degree of change. Thus, anchor-based approaches require an independent standard or anchor that is in itself interpretable and at least moderately correlated with the instrument being explored. The obvious way of testing responsiveness is to divide the scores for differences between the patients who had improved and those who had worsened compared to scores in patients whose health status remain unaffected. For this approach, another (global) measurement must be available (Jaeschke et al., 1989). It is clear that because the anchoring approach depends on another (subjective) health measure, a definite responsiveness statistic based on this approach cannot be estimated.

The anchor-based approaches use an external indicator to assign subjects to classes reflecting, for example, no change, small positive changes, large positive changes, small negative changes, or large negative changes. The anchors can be clinical (e.g., laboratory measures, physiological measures, and clinician ratings) or patient based, such as global ratings of change or actual changes in health measures. The approach based on patient-reported anchor measures seems to be the most suitable approach to estimate the level of responsiveness of an instrument, but also to estimate its minimal importance difference (see below).

Effect Sizes

Many studies use effect sizes as a measure of responsiveness, calculated as the mean change score in a group of patients divided by the standard deviation of the baseline scores or the standard deviation of the change scores. The most popular effect-size approach uses Cohen's (Cohen, 1977; Walsh et al., 2003) standardized effect size, the mean change divided by a standard deviation. Cohen suggested that standardized effect sizes of 0.2−0.5 should be regarded as "small," 0.5−0.8 as "moderate," and those above 0.8 as "large." Researchers then conclude that their instrument is responsive when the effect size is large. However, a high magnitude of change gives little indication of the ability of the instrument to detect change over time on the construct to be measured. That is because effect sizes are highly dependent on the standard deviation and will then be higher in a relatively homogeneous population or when the variation in treatment effect is small. Therefore, effect size in itself provides limited evidence of responsiveness.

Minimal Important Differences

The ability of an instrument to detect small but important clinical changes has been suggested as one criterion on which to base a choice among instruments (Kirshner and Guyatt, 1985; Guyatt et al., 1987). One year before Kirshner and Guyatt presented their idea in their 1985 publication, Deyo and Inui (1984) came forward with similar considerations: "If these instruments are to be used for serial assessments of individual patients, it is important to know how much score variability may occur when the patient is actually clinically stable, and whether the instrument will reflect changes that would be considered clinically important." Realizing the importance of detecting small but important differences also got attention in later years (Jaeschke et al., 1989; Lydick and Epstein, 1993).

MID was originally defined as "the smallest difference in score in the domain of interest which patients perceive as beneficial and which would mandate, in the absence of troublesome side effects and excessive cost, a change in the patient's management" (Jaeschke et al., 1989). This definition was later simplified to "the smallest change that is important to patients."

Responsiveness represents the instrument's ability to detect changes, whereas the MID denotes the smallest score or change in score that would likely be important from the patient's or clinician's perspective (Revicki et al., 2008). Demonstrating the ability to detect responsiveness is necessary but not sufficient for estimating the smallest change in score that can be regarded as important. The amount of change score has been referred to as the MID, and when connected to clinical anchors, sometimes as the minimal clinically important difference. The MID concept was defined to help with the interpretation of observed differences obtained in longitudinal studies. Just as for responsiveness, two general approaches have been used to determine MID: distribution-based methods and anchor-based methods. Both measure a quantifiable change in outcomes, but the specific approach selected will determine the type of change measured (Copay et al., 2007). The best method for determining the MID of an instrument is still under debate. However, anchor-based methods are recommended by numerous authors, as they compare observed score differences with external criteria that have clinical relevance. The area under the receiver operating curves is often used. It refers to the ability of an instrument to discriminate between patients who are considered to be improved and patients who are not considered to be improved according to the gold standard (de Vet et al., 2011). Because receiver operating curves require a dichotomous variable, arbitrary selection or grouping of scale levels will be necessary if categorical or rating scales are used to measure clinical change.

More importantly, distribution-based approaches do not address the question of clinical importance. They seem to ignore the purpose of MID, which is to distinctly separate clinical importance from statistical significance. Another limitation of distribution-based approaches is that they are sample specific; in the sense that the MID value depends on the variability of the scores in the studied samples.

Significant Versus Importance

To keep the discussion about responsiveness and MID clear, we have to distinguish "differences" from "changes." Two groups can exhibit differences in cross-sectional and longitudinal end points, whereas only individual patients can experience change (Gagnon, 2007).

Occasionally dispute arises over the relationship between the MID and the statistical significance of treatment effects. Is it possible to find a statistically significant effect without detecting an MID? Is it also possible that no statistically significant effect can be detected in a given situation although an MID is reported?

Some authors use the p-value obtained from a paired t-test as a measure of responsiveness and conclude that their instrument is responsive if the p-value is small. This does not have to be true, because the p-value is a measure of the statistical significance of the change scores rather than of the magnitude of the change scores. Statistical significance depends on the magnitude of change, the standard deviation of the change scores, and the sample size. Therefore, the paired t-test cannot be considered a good parameter of responsiveness.

Furthermore, statistical significance is something different than meaning and importance. No amount of statistical or analytical methodology will ever be able to interpret end points that were meaningless in their definitions (Gagnon, 2007).

Choice of Anchor

One may choose clinical end points, global improvement scales, changes in other patient-reported outcome measures, or a combination of patient- and clinical-based outcomes as an anchor. An anchor needs to have a nontrivial association with change in the outcome of interest. In other words, if the correlation between the anchor and the patient-reported outcome is zero, then the anchor is not suitable. Such an association is usually investigated, in terms of strength and direction, with Spearman's correlation coefficient (Revicki et al., 2008).

While various methods have been proposed for estimating the MID, empirical studies to date have generally relied on patients' global ratings of change in their own health. In this approach, the subjects are asked to retrospectively judge whether a particular aspect of their health has improved, stayed the same, or worsened, using a single item (the anchor) with several response options across a range from "much worse" through "no change" to "much better." The degree of change on the global rating scale is then related back to the magnitude of change in the health instrument under study to obtain a numerical value denoting the MID. Of course, a serious drawback of this approach is that the reporting on the anchor item can be unsound due to recall bias.

Although the use of an external criterion is common to all anchor-based approaches, three variations may be identified. The first anchor-based approach (Jaeschke et al., 1989; Juniper et al., 1994) defines MID as the change in health-outcome scores of a group of patients selected according to their answers on a global assessment scale. A second approach is to compare the health-outcome scores (Kulkarni, 2006) or the health-change scores of groups of patients (Hägg et al., 2003) with different responses on a global assessment scale. A third but not widely used approach lets patients compare themselves with other patients. In this procedure, patients are paired with other patients to discuss their health situation. After the discussion, patients rate themselves as the same or, to varying degrees, worse or better than the patient with whom they spoke (Copay et al., 2007).

WHAT IS WHAT?

For traditional clinical measures, such as the forced expiratory volume (FEV), clinicians and researchers typically have an understanding of what constitutes a minimal (clinically) important difference. They would likely interpret an improvement of 50 mL in FEV as clinically important, even though it might not be statistically significant in a study. A similar understanding of the MID is apparent when dealing with scores on health instruments. Clinimetricians assert that this number (MID) is specific to each instrument and can draw upon logical analysis of the properties of the instrument as well as upon data from a variety of experimental designs.

There is an ongoing discussion about which methods should be used to determine the MID. The developers of the COSMIN checklist (see below) also recognized that the literature contains a lively discussion about which methods should be used to determine the MID of a health instrument (Mokkink et al., 2010a,b,c). Consequently, the opinions of the COSMIN panel members differed, and within the COSMIN study no consensus on standards for assessing the MID could be reached (Mokkink et al., 2010b).

The thrust of MID studies is the search for a unique threshold value, whereas, ironically, the different methods produce a variety of MID values. In addition, the distinction between responsiveness and MID seems ambiguous, given that the same method (anchor-based) can be applied for both metrics. Anchor-based methods will produce a different MID depending on the anchor scale and the arbitrary selection or grouping of scale levels; and that is a common procedure in MID studies. Conceptually, a MID is a difference between two adjacent levels on a scale, such as "unchanged" and "slightly better." The MID would then depend on the number of levels on a scale: the larger the number of levels, the smaller the difference between two adjacent levels, and the smaller the MID.

INSTRUMENTS

The following section presents some examples of typical clinical tests, indices, and other scores. These instruments include the 6-min walk test, the

Apgar score, a frailty index, and the Patient Health Questionnaire (PHQ). Interestingly, the Apgar score is formally an index whereas the PHQ is better described as a diagnostic or screening instrument.

Six-Minute Walk Test

The original purpose of the 6-min walk was to test exercise tolerance in chronic respiratory disease and heart failure. The test has since been used as a performance-based measure of functional exercise capacity in other populations, including healthy older adults, people undergoing knee or hip arthroplasty, fibromyalgia, and scleroderma. It has also been used with children. The 6-min walk test (6MWT) measures the distance an individual is able to walk over a total of 6 min on a hard, flat surface. The goal is for the individual to walk as far as possible in 6 min. The individual is allowed to self-pace and rest as needed while traversing back and forth along a marked walkway. The 6-min walk test was developed in 1963 by Balke. Several variations of the 6-min walk test have been tested; the 6-min timed walk was recommended, given its reproducibility and ease of administration compared to longer timed tests. Its test—retest reliability has been reported as high, with an ICC of 0.90 at baseline, 0.88 at 18 weeks, and 0.91 at 43 weeks in a cohort of patients with heart failure.

	0 Points	1 Point	2 Points	Points totaled
Activity (muscle tone)	Flaccid	Some flexion of extremities	Active movement	
Pulse (heart rate)	Absent	Below 100 bpm	Over 100 bpm	
Grimace (reflex irritability)	Absent	Grimace	Active motion (sneeze, cough, pull away)	
Appearance (skin color)	Bleu, pale	Body pink, extremities blue	Completely pink	
Respiratory effort	Absent	weak, irregular, gasping	Vigorous cry	

Severely depressed	0-3
Moderately depressed	4-6
Excellent condition	7-10

FIGURE 13.3 Apgar score. *bpm*, beats per minute.

Apgar Score

The Apgar score (Apgar, 1953) is used to assess the clinical condition of a newborn baby. It provides a nice illustration of an explicitly pragmatic index measure in medicine. The Apgar score is calculated based on an infant's condition at 1 min and 5 min after birth. If the 5-min Apgar score is low, additional scores may be assigned every 5 min. It consists of five variables or indicators: skin color (appearance), heart rate (pulse), reflex response (grimace), muscle tone (activity), and respiratory rate, leading to the acronym Apgar (Fig. 13.3). Its scale combines heterogeneous elements (e.g., variables, attributes) by taking values 0, 1, or 2 for each of the observations (color of complexion, heart rate, respiration rate, reflex response to nose catheter, and muscle tone) and sums these to yield a total (index) score. These five elements are not intended to be highly interrelated indicators of a baby's condition; therefore, these can be considered causal indicators (Chapter 5). Scores in the range of 7–10 are considered normal.

However, is the Apgar scale a measurement instrument as well a useful clinical information tool? For this to be the case, the numbers generated by the Apgar score need to satisfy criteria as measurements of an explicit construct. Neither Apgar nor Feinstein provide evidence to support the claim that the numbers generated are measurements, nor did Apgar define the construct she sought to evaluate very clearly. In the introduction she stated its objective as the "… reestablishment of simple, clear classification or 'grading' of newborn infants which can be used as a basis for discussion and comparison of the results of obstetric practices, types of maternal pain relief and the effects of resuscitation." Further in the article she referred to the "evaluation of the condition of the newborn infant one minute after birth" and she mentioned "prognosis." Nevertheless, the Apgar score is a lucid example of a simple and attractive clinical (clinimetric) index instrument. Yet, at a conceptual level, questions arise about the suitability of the Apgar score as a measurement instrument, as the scores seem not applicable for statistical analysis (Hobart, 2007) (Fig. 13.4).

BOX 13.2 Virginia Apgar

Virginia Apgar (1909–74) was an American-Armenian obstetrical anesthesiologist. She was a leader in the field of anesthesiology and introduced obstetrical considerations to the established field of neonatology. To the public, however, she is best known as the inventor of the Apgar score (1953), a way to quickly assess the health of newborn infants immediately after birth. Apgar was also a lecturer and then clinical professor of pediatrics at Cornell University School of Medicine, where she taught teratology (the study of birth defects).

Continued

Box 13.2 Virginia Apgar—cont'd

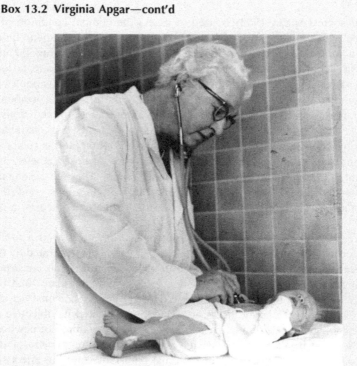

FIGURE 13.4 Virginia Apgar examining a newborn baby in 1966. *With permission from Library of Congress Prints and Photographs Division, New York World-Telegram and the Sun Newspaper Photograph Collection.*

Frailty Index

Some individuals experience a state of heightened vulnerability, known as frailty, which can be quantified using a frailty index. Following a standard procedure, a frailty index can be constructed as the proportion of age-related health deficits an individual has accumulated. Deficits can be any diseases, signs, symptoms, laboratory abnormalities, or functional or cognitive impairments. Since the introduction of the concept, several instruments have been developed to determine the level of frailty among elderly people (Jones et al., 2004; Lutomski et al., 2013; Peña et al., 2014). Although researchers disagree on a precise definition, it is commonly accepted that frailty indicates a state of vulnerability regarding the future occurrence of poor health outcomes, such as mortality, hospitalization, institutionalization, chronic conditions, and/

or loss of function in one or more domains (i.e., the physical, psychological, cognitive, and social domains) (Peters et al., 2012).

Patient Health Questionnaire

A typical example of a diagnostic index is the PHQ. The PHQ-9 incorporates DSM-IV depression diagnostic criteria with other leading major depressive symptoms into a brief self-report tool (Kroenke et al., 2001). This instrument is based on nine items and has been developed under the supervision of the pharmaceutical company Pfizer, which holds the copyright. It offers clinicians a concise, self-administered screening and diagnostic tool for mental health disorders, using questions that have been field tested in office practice. The PHQ-9 is derived from the self-report version of the Primary Care Evaluation of Mental Disorders (PRIME-MD), a diagnostic tool developed in the mid-1990s by Pfizer Inc. The PRIME-MD consisted of 26 items (Spitzer et al., 1999). There is even a two-item PHQ version (Whooley et al., 1997). The PHQ-2 is a shorter version of the PHQ-9. It consists of two screening questions to assess the presence of a depressed mood and a loss of interest or pleasure in routine activities; a positive response to either question indicates that further testing is required.

COSMIN CHECKLIST

A recent initiative is the development of a checklist for evaluating the methodological quality of studies on measurement properties for health instruments (Mokkink et al., 2010a). The developers invited 91 experts to participate in a Delphi panel, 57 of whom (63%) agreed to take part. The main reason for nonparticipation was lack of time. Twenty panel members (35%) completed all four rounds. One of the panel members (not me) decided to withdraw after the second round because of strongly divergent opinions (Mokkink et al., 2010b).

Multiplicity of statistics for measuring characteristics such as responsiveness and MID are available. There are several parameters proposed in the literature to assess responsiveness that the Delphi panel considers inappropriate. The panel reached consensus about the use of effect sizes and related measures, such as standardized response mean, Norman's responsiveness coefficient, and relative efficacy statistic. As for the verdict passed by the COSMIN group: all these are inappropriate measures of responsiveness (Mokkink et al., 2010a,b,c).

The truism that consensus does not come easy is illustrated in the publications of the COSMIN group. In some cases, changes were made by the Steering Committee. For example, the Steering Committee advocated using the term "reliability" instead of the proposed term "reproducibility" (Mokkink et al., 2010a, p. 742).

Unfortunately, no mention of preference-based health measurement approaches (Chapters 11 and 12) was made in the COSMIN checklist, and item-response theory is addressed in a somewhat limited way.

GUIDELINES, CHECKLISTS, AND STATEMENTS

The development the COSMIN checklist does not stand on its own. Numerous guidelines, checklists, and consensus statements are formulated in the field of health. However, guidelines, checklists, and consensus statements are not synonymous with evidence and sound reasoning. Many situations require sound arguments, logical theories, and a mathematically proven coherent framework instead of fixed rules. Delicate combinations of different measurement approaches, different settings, and different ways to develop health instruments may contradict the use of checklists. There is a real danger that scientific deliberation in the development of health-status measurement instruments and the evaluation of such instruments will be overlooked (Mokkink et al., 2010c).

Measurement standards have been established by the Committee on the Standards of the American Psychological Association, the American Educational Research Association, and the National Council on Measurement in Education. Relevant properties addressed in these standards include the measurement framework, the reliability of the measurement, and how one establishes the validity of particular interpretations of test scores. But these standards are directed toward tests that measure ability. Measuring (perceived) health is different. Whereas ability has to do with correct or wrong answers to questions or tasks, health has to do with attitudes and perceptions (Chapter 5). Because measurement methods and their related measurement properties were often developed within different disciplines, no single analytical approach can be regarded as standard. Every analytical method is also based on assumptions concerning the type of data (nominal, ordinal, continuous), the sampling (random, consecutive, convenience), and on the treatment of random and systematic error. Therefore, it is not possible to be too prescriptive regarding the "best" statistical method; the choice depends on the purpose and the design of the study (Kottner et al., 2011).

APPLICATION

This section presents an example of the assessment of responsiveness. It is based on an earlier publication in which two widely applied health measurement instruments were compared head-to-head (Krabbe et al., 2004). The main aim of that study was to evaluate the responsiveness of the EQ-5D index. For this purpose, data were obtained in a study on the health of patients who had undergone surgical management for colorectal liver metastases. Health was measured using the EQ-5D and the disease-specific EORTC QLQ C-30.

The EQ-5D (3 level version) generic health instrument, which provides an index value (utility or preference score), has been widely used in clinical studies, especially to conduct economic evaluations. It is considered reliable and valid. Far less is known about the responsiveness of the index value of the EQ-5D. Some authors have suggested that targeted disease-specific or condition-specific health instruments are likely to be more responsive than generic instruments, whose strengths include breadth and applicability across conditions and interventions (Wiebe et al., 2003).

The EQ-5D consists of two parts: a classification system of five health domains (attributes) and a visual analogue scale (VAS). Based on the responses to the classification, an index is estimated by applying an algorithm (Dolan, 1997). The EQ VAS and EORTC QLQ C-30 were used as reference instruments. The EQ-5D index can be regarded as a societal measure, whereas the EQ VAS is a "holistic" assessment from the patient's perspective. The EQ VAS task leaves patients free to merge all relevant health attributes, so this holistic health concept may be broader than the five EQ-5D attributes. Therefore, it forms a good basis for comparison with the EQ-5D index. The EORTC QLQ C-30 instrument focuses on disease-specific problems, and its items (questions) are computed into single EORTC QLQ C-30 scales. The responsiveness of the EQ-5D index was investigated in three ways: (1) The EQ-5D index was compared with the EQ VAS. (2) EQ-5D attributes were compared with corresponding EORTC QLQ C-30 scales. (3) The EQ-5D index and the EQ VAS were compared with the EORTC QLQ C-30 global health-status scale. It was assumed that the EQ-5D attributes would be less responsive than the parallel EORTC QLQ C-30 scales. However, it was also expected that the composite index of the EQ-5D and the EQ VAS would be equally responsive, or even show higher responsiveness than the global health-status scale of the EORTC QLQ C-30.

Between June 1999 and January 2003, 75 consecutive patients with colorectal liver metastases were asked to participate in this prospective study the day before undergoing a premeditated liver surgery to exterminate their metastatic disease. At exploratory laparotomy, three interventions could be distinguished: group I, surgical liver resection with or without additional local ablative therapy (cryosurgery/freezing or radiofrequency ablation/heating of the tumor); group II, local ablative therapy alone; and group III, no treatment (unresectable disease due to the extension of disease, the localization near the vessels or extrahepatic metastases, implying that no surgery could be performed). Liver resection is considered to be a far more severe operation than local tumor ablation alone. Patients only undergoing an explorative laparotomy (group III) have no blood loss, no long period of anesthesia, no intensive care stay, and less trauma from surgery. However, also after an explorative laparotomy, patients have postoperative pain and discomfort due to the subcostal incision and the forced retraction of the rib cage.

The self-report health instruments were completed at baseline (1 day before intervention), 0.5 month later (discharge from hospital was generally within 10 days), and 3 and 6 months after intervention. At baseline, the instruments were administered in hospital, while after intervention the instruments were sent to the patients at home.

Effect size (ES) was chosen as the metric of responsiveness. The method of calculation was $ES = (M_1 - M_2)/S_1$, where M_1 is the preassessment, M_2 is the postassessment, and S_1 is the standard deviation of the preassessment. An effect size of 0.2–0.5 was considered to be small, 0.5–0.8 moderate, and above 0.8 large (Cohen, 1977). Effect sizes greater than 0.5 were considered to be clinically relevant. Mean outcomes with standard deviations were computed for all EQ-5D attributes, the EQ-5D index, and EQ VAS at baseline, 0.5, 3, and 6 months after intervention.

A total of 75 patients were enrolled in the study. The EQ-5D (index and VAS) and EORTC QLQ C-30 global health-status scales discriminated well among the three intervention groups in our study population and showed similar patterns. Postintervention, all groups showed a decrease in outcome, with the exception of the EQ-5D anxiety/depression attribute, on which all of the patients reported fewer problems. Obvious trends were seen for all groups: deterioration in mean outcomes 0.5 month after intervention and improvement at 3 months after intervention. The finding that the outcomes of the three intervention groups were not comparable at baseline was remarkable. The unresectable group had the poorest outcomes throughout follow-up on the EQ-5D index. Various explanations can be given for the steady deterioration in the unresectable patients. However, for the head-to-head comparison of the responsiveness of two different health instruments, such clinical and methodological issues are less relevant. More importantly, with regard to responsiveness, all three health outcomes showed similar differences among the three intervention groups over time.

The distribution of the outcomes on the EQ-5D classification showed that level-three responses (confined to bed, unable to, extreme) were seldom reported after intervention. All outcome measures (means) indicated that patients had recovered fairly well within 3 months and completely within 6 months. Effect sizes between outcomes obtained at baseline and 0.5 month after intervention were moderate to large on all EQ-5D attributes with the exception of the anxiety/depression attribute (Table 13.1), on which particularly the respectable and the local ablation group showed small effect sizes (data not shown). The only effect size that showed improvement in outcome was seen in the anxiety/depression attribute. Comparison of outcomes obtained at baseline and 3 months after intervention mainly revealed small effect sizes on the EQ-5D attributes. Effect sizes on the anxiety/depression attribute were still positive, again showing amelioration in outcome. Effect sizes between outcomes obtained at baseline and 6 months after intervention were small on the majority of EQ-5D attributes.

TABLE 13.1 Effect Sizes (Baseline vs Postintervention) on EQ-5D Attributes, EQ-5D Index, and EQ Visual Analogue Scale Compared to Parallel EORTC QLQ C-30 Scales (Negative Effect Sizes Correspond With Deterioration)

EQ-5D	EORTC Scale	Number of EORTC Items	Number of EORTC Responses	Baseline vs ½ Month Post		Baseline vs 3 Months Post		Baseline vs 6 Months Post	
				EQ-5D	EORTC QLQ C-30	EQ-5D	EORTC QLQ C-30	EQ-5D	EORTC QLQ C-30
Mobility	Physical functioning	5	4	−1.26	−1.82	−0.52	−0.42	−0.26	−0.11
Self-care	n/a	n/a	n/a	−[a]	n/a	−[a]	n/a	−[a]	n/a
Usual activities	Role functioning	2	4	−1.86	−1.64	−0.64	−0.48	−0.22	−0.18
Pain/discomfort	Pain	2	4	−1.24	−1.47	−0.61	−0.69	−0.24	−0.28
Anxiety/depression	Emotional functioning	4	4	0.18	0.12	0.22	0.28	0.22	0.28
Index	Global health status	2	7	−1.33	−1.07	−0.58	−0.30	0.00	0.03
Visual analogue scale	Global health status	2	7	−1.25	−1.07	−0.48	−0.30	−0.04	0.03

[a]Could not be estimated because of standard deviation of zero.

Comparison of the EQ-5D index with the EQ VAS showed effect sizes of similar magnitude (Table 13.1). Comparison of EQ-5D attributes with parallel EORTC QLQ C-30 scales showed that effect sizes were of comparable magnitude. For instance, the effect sizes calculated for means obtained at baseline and 0.5 month after intervention on the EQ-5D usual activities attribute and the EORTC QLQ C-30 role functioning scale were -1.86 and -1.64, respectively.

Effect sizes of the EQ-5D index, EQ VAS, and EORTC QLQ C-30 global health-status scales were of comparable magnitude. The assumption that the EQ VAS would be more responsive than the EQ-5D index was not confirmed. Moreover, in the majority of cases the EQ-5D index and EQ VAS proved to be more responsive than the EORTC QLQ C-30 global health-status scales. The EQ-5D attributes proved to be equally responsive in comparison with corresponding EORTC QLQ C-30 scales on a number of occasions. The latter result was somewhat surprising. It was expected that the EORTC QLQ C-30 scales would be more responsive than the EQ-5D attributes for the following reasons. The EQ-5D classification offers patients three response levels for each attribute, whereas the EORTC QLQ C-30 offers four. Furthermore, the EORTC QLQ C-30 scales are based on aggregated data (multiple items). These two aspects, i.e., number of response categories and number of items, at least in this study, did not affect the responsiveness of the instrument.

DISCUSSION

Readers are cautioned that this is a somewhat lengthy discussion. A major reason for this extensive treatment is that, from its inception, clinimetrics has touched upon some interesting but difficult concepts, such as responsiveness and MID. Physicians and other practitioners in medicine have day-to-day involvement with patient treatment and consultation. Within this practical setting, they are confronted with the challenges of monitoring (measuring) the health condition of their patients. That may explain why many concepts from clinimetrics are loosely grounded in measurement theory but worked out much more along the lines of clinical reasoning. Clinimetrics has evolved from the practical clinical situation. It did not start from basic general rules (e.g., data theory, statistics, information theory), which may also explain the origin of the terminology used.

Some aspects of the clinimetric framework seem to be inconsistent or to overlook key measurement concepts. Normally, any mechanism developed or detected in one field of science should be transferable to any other field. For example, the principles of Shannon's information theory should be equally valid in health sciences and in social sciences. Therefore, the claim that clinimetrics is a specialized discipline with its own specific concepts and methods has been questioned (Williams and Naylor, 1992; Streiner, 2003).

Three Categories

Many clinimetricians distinguish three broad categories of potential applications of health-status measures, namely for discrimination, prediction, and evaluation purposes. The assumption is that a health instrument to monitor patient groups should be differently devised and constructed than a health instrument to discriminate among patients. This argument has been repeated over and over again. As clinical practice makes a clear distinction between diagnosis and prognosis, the idea that different applications would require different health-status instruments seems quite appealing. Yet, in the absence of a clear theoretical underpinning of the claim that health-status instruments should be differently constructed, this resonates with one of the lines in the movie about the discovery of the double helix: "They're all wrong [the textbooks]. Take it from me. It's one of those guesses that's been repeated so often it's got the status of fact" (Wikipedia, 2016).

Responsiveness

So far, I have not come across any literature that offers a clear, overarching explanation and theoretical framework for how to deal with responsiveness, nor with the concept of MID. In one of the textbooks about clinimetrics, the authors take the position that different approaches can be used, and the evidence from these approaches should be combined to draw conclusions about the degree of responsiveness of the instrument in a specific population and context (de Vet et al., 2011). Other authors have recommended that the estimation of the MID for a specific patient-reported health measure should be based on multiple approaches and a triangulation of methods (Revicki et al., 2008). To some extent this opinion seems extraordinary. Science is characterized by the use of standardized and well-defined methods. Therefore, the lack of agreement on how to formulate and evaluate a measurement property such as responsiveness or MID may indicate the following: (1) One of the proposed methods is correct and the others are not; (2) They are all derivatives of each other and are therefore (more or less) equal; (3) All methods are wrong; there is no such thing as responsiveness or MID that can be quantified.

Interestingly, the notion of responsiveness as a critical component of instrument development appears to be unique to the measurement of health status as performed under the measurement framework of clinimetrics. Formal areas of research in the social sciences (e.g., psychometrics, mathematical psychology, and measurement theory) offer no empirical, theoretical, or mathematical support for the notion of responsiveness. Highly regarded standards such as those of the American Psychological Association, the American Educational Research Association, and the National Council on Measurement in Education make no mention of responsiveness at all.

Responsiveness seems to be more of a normative than an analytical issue. If we are interested in developing a metric that expresses in one way or another the relative (or even absolute) "responsiveness" of a health instrument, then we probably should develop other types of instruments. If, for example, a health-status instrument could be devised that has ratio-scale measurement properties, it would be possible to express its responsiveness in a more straightforward manner. So far, only some of the preference-based measures (Chapter 12) claim to have such qualities. Item-response theory methods may also contribute to the endeavor to get a better handle on the assessment of the responsiveness of health scales. Another characteristic that is related to the level of responsiveness is the granularity of an instrument. This has to do with the scale range of the instrument and its precision. A thermometer for checking the temperature of wine has a different granularity than an oven thermometer. The "responsiveness" of an instrument is largely affected by the "range" or metric of a measurement instrument together with its reliability (precision). However, in conventional profile instruments (multiple aggregated items for separate domains) as well as in index instruments, measurement aspects such as scale range, metric qualities, level of information, granularity, and other characteristics cannot be addressed. Maybe another engineer is needed to solve this issue of responsiveness in the field of health measurement; it might just take another Thurstone to solve this persisting problem. For the time being, it seems that responsiveness in itself can only be examined by means of relative comparison with other measurement instruments.

Minimal Important Difference

As others have noted, many issues are still unresolved in the determination of MID, issues that render its new metric status decidedly premature (Copay et al., 2007). An interesting observation was made at a conference by the philosopher/economist Gagnon (2001). He stated that clinical importance is what we bring to the analysis rather than what emerges from the data. No data-driven criterion signifies clinical importance. The evaluation of clinical importance is imposed upon the data not derived from it. As such, the evaluation of clinical importance goes well beyond the mechanical application of a criterion to the data. Each method of clinical importance can aid our interpretation of effects in clinical studies. But whether taken separately or as a group, these methods do not provide criteria for determining clinical importance. Thus, the search for a particular MID is probably misguided. Similar thoughts in a somewhat different context have been put forward by Bland and Altman (1986) and can probably also be found in the work of Kant (1787). Much of this reflection has to do with the fact that MID is, apart from all the problems that we have with our health instrument measures, largely a normative aspect.

The Asset of Clinimetrics

The good thing about clinimetrics is that it recognizes that meaningful concepts and constructs are important in science and in practice. These concepts and constructs are developed largely on the basis of theories, experience, knowledge, and understanding. Therefore, when developing a new instrument, substantial investment in the early phase of research is crucial. Clinical doctors, nurses, psychologists, psychiatrists—and especially the patients themselves—are among the many whose opinions should be elicited and who should be included in qualitative studies preceding any numerical or statistical investigations. It is this consideration that is better understood by researchers and practitioners operating in the field of clinimetrics. Health measurement instruments may be developed by traditional psychometric methods, as these still have much to offer. But a specific variable or health attribute could be dropped and then reintroduced into the final instrument because practical experience tells us that the variable that had initially been dropped on psychometric grounds is nevertheless important, perhaps even crucial to the content.

REFERENCES

Angst, F., 2011. The new COSMIN guidelines confront traditional concepts of responsiveness. BMC Medical Research Methodology 11, 152.

Apgar, V., 1953. A proposal for a new method of evaluation of the newborn infant. Current Researches in Anesthesia & Analgesia 32 (4), 260–267.

Balke, B., 1963. A simple field test for the assessment of physical fitness. REP 63-3. [Report]. Civil Aeromedical Research Institute (U.S.) 53, 1–8.

Beaton, D.E., Bombardier, C., Katz, J.N., Wright, J.G., 2001. A taxonomy for responsiveness. Journal of Clinical Epidemiology 54 (12), 1204–1217.

Beaton, D.E., Hogg-Johnson, S., Bombardier, C., 1997. Evaluating change in health status: reliability and responsiveness of five generic health status measures in workers with musculoskeletal disorders. Journal of Clinical Epidemiology 50 (1), 79–93.

Bland, J.M., Altman, D.G., 1986. Statistical methods for assessing agreement between two methods of clinical measurement. Lancet 1 (8476), 307–310.

Bombardier, C., Tugwell, P., 1982. A methodological framework to develop and select indices for clinical trials: statistical and judgmental approaches. The Journal of Rheumatology 9 (5), 753–757.

Cohen, J., 1977. Statistical Power Analysis for the Behavioral Sciences. Academic Press, New York.

Copay, A.A.G., Subach, B.R., Glassman, S.S.D., Polly, D.W.D., Schuler, T.T.C., 2007. Understanding the minimum clinically important difference: a review of concepts and methods. Spine Journal 7 (5), 541–546.

de Vet, H.C.W., Terwee, C.B., Mokkink, L.B., Knol, D.L., 2011. Measurement in Medicine: A Practical Guide. Cambridge University Press, Cambridge.

Deyo, R.A., Inui, T.S., 1984. Toward clinical applications of health status measures: sensitivity of scales to clinically important changes. Health Services Research 19 (3), 275–289.

Deyo, R.A., Centor, R.M., 1986. Assessing the responsiveness of functional scales to clinical change: an analogy to diagnostic test performance. Journal of Chronic Diseases 39 (11), 897–906.

Dolan, P., 1997. Modeling valuations for EuroQol health states. Medical Care 35 (11), 1095–1108.

Fayers, P.M., Machin, D., 2000. Quality of Life. Assessment, Analysis and Interpretation. Wiley, Chichester.

Feinstein, A., 1994. Clinical judgment revisited: the distraction of quantitative models. Annals of Internal Medicine 120 (9), 799–805.

Feinstein, A.R., 1983. An additional basic science for clinical medicine: IV. The development of clinimetrics. Annals of Internal Medicine 99 (6), 834–848.

Feinstein, A.R., 1987. Clinimetrics. Yale University Press, New Haven.

Fletcher, R.H., 2001. Alvan Feinstein, the father of clinical epidemiology, 1925–2001. Journal of Clinical Epidemiology 54 (12), 1188–1190.

Gagnon, D.D., 2001. Some logical problems with tests for clinically important changes in HRQoL. In: Poster at the 8th Annual Conference of the International Society for Quality of Life Research, Amsterdam, The Netherlands.

Gagnon, D.D., 2007. Clinical significance of patient-reported questionnaire data: two more steps? Journal of Clinical Epidemiology 60 (1), 103–104.

Guyatt, G., Walter, S., Norman, G., 1987. Measuring change over time: assessing the usefulness of evaluative instruments. Journal of Chronic Diseases 40 (2), 171–178.

Guyatt, G.H., Juniper, E.F., Walter, S.D., Griffith, L.E., Goldstein, R.S., 1998. Interpreting treatment effects in randomised trials. BMJ 316 (7132), 690–693.

Guyatt, G.H., Kirshner, B., Jaeschke, R., 1992. Measuring health status: what are the necessary measurement properties? Journal of Clinical Epidemiology 45 (12), 1341–1345.

Guyatt, G.H., Osoba, D., Wu, A.W., Wyrwich, K.W., Norman, G.R., 2002. Methods to explain the clinical significance of health status measures. Mayo Clinic Proceedings 77 (4), 371–383.

Hägg, O., Fritzell, P., Nordwall, A., 2003. The clinical importance of changes in outcome scores after treatment for chronic low back pain. European Spine Journal 12 (1), 12–20.

Hays, R.D., Hadorn, D., 1992. Responsiveness to change: an aspect of validity, not a separate dimension. Quality of Life Research 1 (1), 73–75.

Hobart, J., 2007. A brief critique of clinimetrics. The Lancet Neurology.

Jaeschke, R., Singer, J., Guyatt, G.H., 1989. Measurement of health status: ascertaining the minimal clinically important difference. Controlled Clinical Trials 10 (4), 407–415.

Jones, D.M., Song, X., Rockwood, K., 2004. Operationalizing a frailty index from a standardized comprehensive geriatric assessment. Journal of the American Geriatrics Society 52 (11), 1929–1933.

Juniper, E.F., Guyatt, G.H., Willan, A., Griffith, L., 1994. Determining a minimal important change in a disease-specific quality of life questionnaire. Journal of Clinical Epidemiology 47, 81–87.

Kant, I., 1787. Kritik der reinen Vernunft. Hartknoch, Riga (translated version: Kritiek van de zuivere rede. Boom, Amsterdam, 2004).

Kazis, L.E., Anderson, J.J., Meenan, R.F., 1989. Effect sizes for interpreting changes in health status. Medical Care 27 (3), S178–S189.

Kirshner, B., Guyatt, G., 1985. A methodological framework for assessing health indices. Journal of Chronic Diseases 38 (1), 27–36.

Kottner, J., Audige, L., Brorson, S., Donner, A., Gajewski, B.J., Hróbjartsson, A., Roberts, C., Shoukri, M., Streiner, D.L., 2011. Guidelines for reporting reliability and agreement studies (GRRAS) were proposed. Journal of Clinical Epidemiology 64 (1), 96–106.

Krabbe, P.F.M., Peerenboom, L., Langenhoff, B.S., Ruers, T.J.M., 2004. Responsiveness of the generic EQ-5D summary measure compared to the disease-specific EORTC QLQ C-30. Quality of Life Research 13 (7), 1247–1253.

Kroenke, K., Spitzer, R.L., Williams, J.B.W., 2001. The PHQ-9. Journal of General Internal Medicine 16 (9), 605–613.

Kulkarni, A.V., 2006. Distribution-based and anchor-based approaches provided different interpretability estimates for the Hydrocephalus Outcome Questionnaire. Journal of Clinical Epidemiology 59 (2), 176–184.

Lutomski, J.E., Baars, M.A.E., van Kempen, J.A., Buurman, B.M., den Elzen, W.P.J., Jansen, A.P.D., Kempen, G.I.J.M., Krabbe, P.F.M., Steunenberg, B., Steyerberg, E.W., Olde Rikkert, M.G.M., Melis, R.J.F., 2013. The validation of a frailty index from the older persons and informal caregivers survey Minimal DataSet (TOPICS-MDS). Journal of the American Geriatrics Society 61 (9), 1625–1627.

Lydick, E., Epstein, R.S., 1993. Interpretation of quality of life changes. Quality of Life Research 2 (3), 221–226.

Mokkink, L.B., Terwee, C.B., Patrick, D.L., Alonso, J., Stratford, P.W., Knol, D.L., Bouter, L.M., de Vet, H.V.W., 2010a. The COSMIN study reached international consensus on taxonomy, terminology, and definitions of measurement properties for health-related patient-reported outcomes. Journal of Clinical Epidemiology 63 (7), 737–745.

Mokkink, L.B., Terwee, C.B., Knol, D.L., Stratford, P.W., Alonso, J., Patrick, D.L., Bouter, L.M., de Vet, H.C., 2010b. The COSMIN checklist for evaluating the methodological quality of studies on measurement properties: a clarification of its content. BMC Medical Research Methodology 10 (22), 1–8.

Mokkink, L.B., Terwee, C.B., Patrick, D.L., Alonso, J., Stratford, P.W., Knol, D.L., Bouter, L.M., de Vet, H.C.W., 2010c. The COSMIN checklist for assessing the methodological quality of studies on measurement properties of health status measurement instruments: an international Delphi study. Quality of Life Research 19 (4), 539–549.

Muraweski, M.M., Miederhoff, P.A., 1998. On the generalizability of statistical expressions of health related quality of life instrument responsiveness: a data synthesis. Quality of Life Research 7 (1), 11–22.

Norman, G.R., 1989. Issues in the use of change scores in randomized trials. Journal of Clinical Epidemiology 42 (11), 1097–1105.

Norman, G.R., Wyrwich, K.W., Patrick, D.L., 2007. The mathematical relationship among different forms of responsiveness coefficients. Quality of Life Research 16 (5), 815–822.

Peña, F.G., Theou, O., Wallace, L., Brothers, T.D., Gill, T.M., Gahbauer, E.A., Kirkland, S., Mitnitski, A., Rockwood, K., 2014. Comparison of alternate scoring of variables on the performance of the frailty index. BMC Geriatrics 14 (1), 25.

Peters, L.L., Boter, H., Buskens, E., Slaets, J.P.J., 2012. Measurement properties of the Groningen Frailty Indicator in home-dwelling and institutionalized elderly people. Journal of the American Medical Directors Association 13 (6), 546–551.

Revicki, D., Hays, R.D., Cella, D., Sloan, J., 2008. Recommended methods for determining responsiveness and minimally important differences for patient-reported outcomes. Journal of Clinical Epidemiology 61 (2), 102–109.

Spitzer, R.L., Kroenke, K., Williams, J.B.W., 1999. Validation and utility of a self-report version of PRIME-MD: the PHQ primary care study. Primary care evaluation of mental disorders. Patient health questionnaire. JAMA 282 (18), 1737–1744.

Steyerberg, E.W., 2009. Clinical Prediction Models: A Practical Approach to Development, Validations, and Updating. Springer, New York.

Stockler, M.R., Osoba, D., Goodwin, P., Corey, P., Tannock, I.F., 1998. Responsiveness to change in health-related quality of life in a randomized clinical trial: a comparison of the prostate cancer specific quality of life instrument (PROSQOLI) with analogous scales from the EORTC QLQ-C30 and a trial specific module. Journal of Clinical Epidemiology 51 (2), 137–145.

Streiner, D.L., 2003. Clinimetrics vs. psychometrics: an unnecessary distinction. Journal of Clinical Epidemiology 56 (12), 1142–1145.

Streiner, D.L., Norman, G.R., Cairney, J., 2015. Health Measurement Scales: A Practical Guide to Their Development and Use, fifth ed. Oxford University Press, Oxford.

Sullivan, M., 2003. The new subjective medicine: taking the patient's point of view on health care and health. Social Science and Medicine 56 (7), 1595–1604.

Terwee, C.B., Dekker, F.W., Wiersinga, W.M., Prummel, M.F., Bossuyt, P.M.M., 2003. On assessing responsiveness of health-related quality of life instruments: guidelines for instrument evaluation. Quality of Life Research 12 (4), 349–362.

Walsh, T.L., Hanscom, B., Lurie, J.D., Weinstein, J.N., 2003. Is a condition- specific instrument for patients with low back pain/leg symptoms really necessary? The responsiveness of the Owestry Disability Index, MODEMS, and the SF-36. Spine 28 (6), 607–615.

Whooley, M.A., Avins, A.L., Miranda, J., Browner, W.S., 1997. Case-finding instruments for depression. Two questions are as good as many. Journal General Internal Medicine 12 (7), 439–445.

Wiebe, S., Guyatt, G., Weaver, B., Matijevic, S., Sidwell, C., 2003. Comparative responsiveness of generic and specific quality-of-life instruments. Journal of Clinical Epidemiology 56 (1), 52–60.

Wikipedia, 2016. Jerry Donohue in the Movie Life Story (1987): About the Discovery by Watson & Crick of the Double Helix. Retrieved on 04 April 2016 from: https://en.wikipedia.org/wiki/Jerry_Donohue, http://www.downtownmagnets.org/ourpages/portnoff/CS1_CPRWE/Units/DNA/Script LifeStory.pdf.

Williams, J.I., Naylor, C.D., 1992. How should health status measures be assessed? Cautionary notes on procrustean frameworks. Journal of Clinical Epidemiology 45 (12), 1347–1351.

Wright, J.G., Young, N.L., 1997. A comparison of different indices of responsiveness. Journal of Clinical Epidemiology 50 (3), 239–246.

Wyrwich, K.W., Tierney, W.M., Wolinsky, F.D., 1999. Further evidence supporting an SEM-based criterion for identifying meaningful intra-individual changes in health-related quality of life. Journal of Clinical Epidemiology 52 (9), 861–873.

Part IV

Chapter 14

New Developments

Chapter Outline

INTRODUCTION

This chapter is about some new developments in the five research traditions for subjective health measurement that are treated separately in this book. It should come as no surprise that nothing presented in this chapter is related to classical test theory. That theory is really classical; it had already reached the limits of its possibilities years ago. The prospects for item-response theory (IRT) are different, however. Development is still going on in this area, and some of the advances will be discussed in this chapter. However, most development seems to be taking place in the area of preference-based measurement.

An emerging insight is that patient involvement is important, particularly if researchers want to know which aspects of health are salient from the perspective of the patients themselves. This recognition may explain the increasing interest in devoting more attention to qualitative research at an early stage in the process of developing health measurement instruments. But we will start with a brief overview of the opportunities that computers and information and communication technology can offer to develop health measurement instruments that have unprecedented potential.

COMPUTERS AND INFORMATION AND COMMUNICATION TECHNOLOGY

It is perfectly reasonable to develop computer programs that facilitate health measurement. The most straightforward way to measure health electronically would be to use software that simulates a traditional pencil-and-paper "questionnaire." This can be readily accomplished using computers with software that mimics the traditional act of filling in questionnaires. However, this is only a very limited use of the many options that are currently available. What began with using computers to automatically collect and store responses from traditional paper questionnaires has progressed to computerized adaptive measurement. The next step may be to deploy the full spectrum of options offered by information and communication technology, as we now are entering an epoch beyond heuristically programmed algorithmic computer (Fig. 14.1).

The most efficient use of computers nowadays is in the application of IRT. We can exercise the full flexibility of IRT measurement models when these are combined with computerized adaptive testing (CAT). For example, in a CAT administration of an IRT-based measure of physical function, a person who has already indicated that he or she is able to walk a mile would not be asked to respond to items about his or her ability to walk a block. Instead,

FIGURE 14.1 HAL 9000 (heuristically programmed algorithmic computer) is a sentient (or artificial general intelligence) computer that controls the systems of the Discovery One spacecraft and interacts with the ship's astronaut crew in the movie "2001: A Space Odyssey." The movie (1968) is based on collaboration between the American film director Stanley Kubrick (1928–99) and science fiction writer Arthur C. Clarke (1917–2008). HAL is capable of speech, speech recognition, facial recognition, natural language processing, lip reading, art appreciation, interpreting emotional behavior, automated reasoning, and playing chess.

the computer would select items about higher levels of physical functioning (such as ability to run a mile) for that subject. Such tailored "testing" is more efficient as it allows for more precise estimates of physical functioning with fewer items per respondent. The respondents receive different sets of items, yet each person's scores can be combined or compared with those of others because the items have been calibrated by IRT models. This tailored approach contrasts with the paper-and-pencil approach, where all items must be presented collectively in a fixed format and all items have to be assessed. CAT offers the respondents a smaller set of items, yet its measures are more reliable. For fixed-length (traditional) patient-reported instruments, IRT models may provide the greatest benefit for those scales that contain many items assessing a single domain. This applies to instruments about pain, fatigue, depression, and other distinct domains of health (Fayers, 2007).

All choice models that allow researchers to derive measures for health are basically versions of the Thurstonian paired comparison choice model. It has been shown that the paired comparison choice model and its extensions can even be embedded in the contemporary and very general framework of structural equation modeling (Maydeu-Olivares and Böckenholt, 2005). In the early days, not all choice models could be estimated because of mathematical problems and a lack of computing power. However, modern computational estimation techniques have overcome many of the earlier restrictions, so choice models can be estimated in their full generality. Analytical complexity is, unfortunately, typical of most modern choice models. A few software packages have addressed the issue by building in analytical procedures that, to varying degrees, simplify the analysis of data derived from choice models. Nowadays, complex Bayesian estimation procedures and Monte Carlo estimation methods are available. To conclude, many augmentations of conventional measurement methodology would simply not be possible without advanced information and communication technology and computers.

At present, computer technology is used for many of the preference-based measurement methods to deal more efficiently with the manifold tasks that respondents have to perform. For instance, the EuroQol Research Foundation has enhanced the newest five-level version of its instrument by developing an online system to carry out the survey. That system, called EuroQol Valuation Technology (EQ-VT), randomly assigns respondents to blocks of tasks, then presents the pairs for the paired comparison tasks in random order, and randomizes the location of the states within the pair (i.e., left and right). Apart from running the time trade-off and the paired comparison tasks, the system also captures the response data (Oppe et al., 2014).

In the setting of choice methods to derive health-state values, many interesting developments are going on. One in particular concerns controlling the presentation of the health-state descriptions. Traditional procedures to devise "discrete choice experiments (DCEs)" are based on main effect models and efficient designs. But the latter generates sets of health states that are quite difficult to compare. The respondents have to assess and compare all combinations of the various health attributes, as all of these may be different. Newer approaches present pairs of health states for which only a limited number of attributes (say, two) differ. The selection and control process for such designs, as well as their graphical representation (colors), can be carried out easily with computers.

Even more interesting opportunities are likely to arise in the coming years. For instance, instead of verbal descriptions of health conditions, we may expect to encounter animations or movies. Visual media will largely circumvent the problems that arise when health descriptions are expressed and communicated in language. To some extent, the linguistic medium complicates and obstructs the measurement of perceived health (Barofsky, 2012). Another development may allow instruments to be automatically translated into other languages with context recognition. To illustrate how sensitive the issue of translation can be, consider an early translation of the English version of the EQ-5D into German. Level 3 of the attribute "mobility," phrased "I am confined to bed," was translated as "Ich bin ans Bett gebunden." The latter means literally "tied up in bed" (Claes et al., 1998).

SELECTING HEALTH ITEMS OR ATTRIBUTES

A top-down approach has been used when developing the majority of the existing health instruments. That is, their content has been derived from the existing body of literature, instruments, and health surveys. Patient or public involvement, if any, occurred in the later stages, when an instrument was being validated. In general, there is a lack of awareness that careful selection of key health items (or attributes) is paramount in developing a sound health measurement instrument. That said, a nice example of an exception to this rule is the CHU9D (Chen et al., 2015; Stevens, 2016).

Developers of health-outcome instruments have realized that content validity cannot be established through quantitative psychometric analyses. This point was highlighted in the FDA PRO guidance (2009). Face and content validity are important aspects of health-state instrument development, yet they have seldom been appraised. An appropriate selection of health items can be enhanced by acknowledging the importance of clinical insight and patient participation to the development process. Since the phenomena to be measured are often complex and intangible, it is likely that salient health aspects will be missed if scientists with little or no experience of illness and disease are in charge of deciding which items are important and should be incorporated in an instrument.

At an early stage, patient input is of utmost importance to identify the concepts and craft the wording for the items of a health instrument. After drafting the instrument, the (often required) second step is to conduct cognitive interviews. At that point, the patients fill in the instrument and then comment on its relevance, clarity, and comprehensiveness. This qualitative part of the research should be conducted in samples of patients from the target population to establish content validity for a particular use. The rationale is that while an instrument may have content validity for one disease, it might omit concepts that are important when assessing another disease (Patrick et al., 2011a,b; Stevens and Palfreyman, 2012).

Top-down or theoretically driven development is more typical of how measures were previously derived. Today's approaches are much more bottom-up, relying on targeted patient samples to determine a conceptual framework of what is important to assess in a given context. As the methods used are often qualitative (interviews and focus groups), the resulting content is variable and can be idiosyncratic (Fig. 14.2). Today, instruments are designed in ways that "measure what matters to patients," regardless of whether they fit into some previously articulated theoretical structure.

Until now, most content for health-status instruments has been generated in consultation with expert panels or by a synthesis of existing instruments. Probably a more valid and certainly a more patient-centered approach would be to organize focus groups. All aspects of health that are important to the patients may be uncovered in these sessions until saturation has been reached. It is reached at the point when no additional aspects are raised in a subsequent focus group. Another approach is to present a large patient target group with a "complete" overview of relevant health attributes that have been selected after theoretical reflection and in-depth study of existing health-outcomes

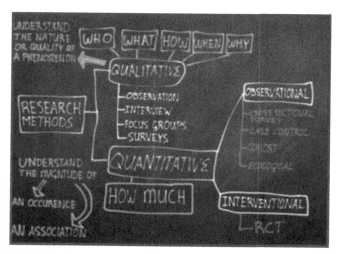

FIGURE 14.2 Qualitative versus quantitative research goals. *With permission from Global Health with Greg Martin, Research Methods − Introduction. https://www.youtube.com/watch? v=PDjS20kic54.*

instruments, followed by sessions with focus groups, when the participants are asked to select the most important attributes based on their health experience.

Such a systematic approach is different from quantitative methods. In fact, most of this book deals with quantitative research. However, many quantitative methods are inappropriate for qualitative issues. Sometimes researchers apply statistical techniques to determine which items are important. For example, factor analysis is often used for this purpose. However, factor analysis is directed by relationships (i.e., correlations) between variables (i.e., items). This means that two items with more or less equal distributions of responses (i.e., frequencies) on the response categories will load on the same factor. Therefore, the results of factor analysis will not tell us anything about the importance of these items. To find out what the important items are, we have to conduct qualitative research or studies with specific response tasks to generate data that can be properly analyzed. Response data collected on the basis of patient exercises can be analyzed by special procedures to highlight specific relationships and clusters. One such procedure is Q methodology (Stenner et al., 2003).

Since the phenomena are often complex and intangible, important aspects are likely to be missed by the blind application of the statistical or psychometric approach. An essential ingredient is lacking in the psychometric approach (and certainly in the preference-based approach): clinical and patient insight, which even sophisticated statistics cannot provide. Neither a quantitative nor a qualitative approach alone seems sufficient (Zyzanski and Perloff, 1999; Fava et al., 2004).

DYNAMIC CONTENT

The perception of well-being may vary between individuals and within an individual over time. This is especially pertinent to the field of health, where different values are ascribed to elements of health at different ages, and where the meaning of activities and health conditions may change with time. The prominence of particular health aspects may shift along a continuum from typical mid-life (chronically) ill patients to the relatively healthy elderly and onto the elderly confronted with comorbidity.

Elderly people use both conventional health care (pursuing a cure) and social care or long-term care. The aim of conventional health care is to achieve health gains or at least to halt further deterioration. The focus of social care is to improve well-being. A life-span perspective takes into account how a person grows, develops, and declines. It places people in different frameworks for understanding their situation. With increasing life expectancy, more people receive "care" instead of "cure." Also, more people are suffering from chronic diseases. In many patient populations, health conditions, even when chronic, are neither life-threatening nor indications for surgery or other invasive treatments.

Consider dementia. Attributes such as "independence," "intimacy," and "problem solving/planning" can all be related to factors other than health (Schölzen-Dorenbos et al., 2012). Thus, it seems inappropriate to measure health outcomes in old age by the same standards that apply to middle age. If we want to measure the perceived health-status in elderly people, then it may be better to speak of population-centered health outcomes instead of patient-centered health outcomes.

In that light, we may need different health measurement instruments not only for different patient populations but also for different stages in life. Therefore, it is of utmost importance to develop measurement instruments that offer the respondents the flexibility to select their own set of health items.

PROMIS

PROMIS stands for patient-reported outcomes measurement information system, and its instruments are based on IRT. The purpose of PROMIS is "to create valid, reliable, and generalizable measures of clinical outcomes of interest to patients." This initiative includes a series of item banks (sets of items or questions measuring a specific concept) to measure patient-reported aspects of health (www.nihpromis.org). The items were derived from existing instruments. IRT modeling is applied on previously collected data from multiple studies. Consequently, the samples are large enough to estimate item parameters. Content experts identify the "best" representation of their area; supporting face and content validity and items are subsequently statistically tested for their item characteristics. The PROMIS instruments can be administered through CAT.

Even though this large-scale initiative is innovative, the methodology is quite standard, and the measures aligned with it are less advantageous than generally recognized. All PROMIS instruments are based on items from existing health instruments. Thus, most items are not provided by experienced patients but are largely based on expert opinion. PROMIS instruments are capable of measuring, for individual patients, the severity of a health condition on specific health domains. However, these instruments are not preference based. Therefore, they do not quantify the impact that limitations and complaints have on a health condition as experienced by the patient. Patients may turn out to be quite limited on a certain health domain, but that measure does not reveal to what extent this limitation affects the experienced well-being for patients on that domain. In addition, the measures on the distinct domains cannot be integrated to arrive at a single general health measure.

An attempt has been made to transform the PROMIS instruments that are based on IRT into a preference-based instrument. This has been done for the 29-item instrument (PROMIS-29; Craig et al., 2014a,b). However, as the PROMIS-29 includes four five-level items on seven domains (physical functioning, anxiety, depression, fatigue, sleep disturbance, social functioning, pain) as well as a single 11-level pain-intensity scale, the preference study could become imbalanced. To overcome any cognitive overload in this supplementary preference-based study, not all of the items from the original PROMIS-29 were presented to respondents. Instead, only one item (attribute) was presented during each judgmental task (paired comparisons), assuming that the respondents considered the other attributes to be optimal. The element of time (number of years living in one of the two health conditions) was also specified in the paired comparison tasks. By doing so, instead of measuring quality of health the computational procedure measured quality-adjusted life years (see Chapter 12). Overall, the method of deriving values for the PROMIS-29 makes several challenging assumptions about people's cognitive and judgmental capacities. Moreover, these people were not experienced patients but a sample drawn from the general population. Overall, patient centeredness seems not to be a cornerstone of PROMIS.

FROM RELATIVE TO ABSOLUTE VALUES

Choice models (Chapter 11) are attractive for the measurement of health, particularly for expressing it in a single measure. However, these type of preference-based models produce relative measures or values, and it is somewhat difficult to ascribe meaning to these. That is largely because most health instruments measure on an interval level. While this is typical for all choice models, the same restriction is associated with multiitem profile instruments (Chapter 9). Differences on an interval-level scale may be interpretable in the sense that an increase from 2.69 to 2.78 is more (or better) than an increase from 2.69 to 2.74. But it is difficult and sometimes impossible to interpret these measures in an absolute sense. The reason is that values derived with choice models lack an absolute upper and lower end point or anchor. Health states are positioned on a scale that uses the definition of the best and worst health states that were included in the choice tasks as upper and lower end points. However, we do not know how good or bad these end point states are in absolute terms.

What we can do quite well with such relative values is compare the effect among patients who have filled in the instrument. So, with a pre—post design, we can infer how much each patient has improved in comparison with other patients. The choice models are attractive for several reasons.

Nonetheless, everything in life comes at a price, and these choice models have a downside. For cost-effectiveness analyses—not for clinical applications—separate exercises are needed to estimate the position where states are considered to become worse than death (position of death = 0.0). This position must be known before quality-adjusted life years (Chapter 3) can be computed with current procedures (van Hoorn et al., 2014). We need to calibrate the scales derived with choice models to a standardized scale with fixed and comparable upper and lower end points to obtain an absolute index scale. Some attempts have been made to develop methods to transform relative values into absolute index values.

A simple approach is to conduct a choice study along with a complementary study based on valuation techniques. In that way, the values derived from the choice model can be calibrated with time trade-off values for the worst health states (mostly a negative value: worse than death). The downside in this case is that researchers still have to conduct complicated time trade-off studies, whereby biases from the two different methods might be merged. This approach also has a theoretical disadvantage: values for health are not derived within a single (coherent) measurement framework.

Other strategies eliminate the need to use external time trade-off values to generate calibrated values from choice models. They extend the choice task beyond a simple paired comparison between two multiattribute descriptions of health states. Two extensions are possible. A simple approach is to include "death" as a separate third option in the choice task, allowing all health-state values to be estimated relative to death. McCabe et al. (2006) and Salomon (2003) have proposed solutions where the state "death" is mixed in with the choice set. In this way, a parameter for the state "death" is estimated as part of the model. However, Flynn et al. (2008) noted that estimated values are likely to be biased when certain assumptions about the choices between health states and death are not satisfied. When a significant proportion of the sample regards all life as worth living (e.g., because of religious beliefs), this bias is likely to arise.

An alternative solution is to extend the information presented in the choice task with a life-span attribute (number of years living in a health state). The latter strategy has become popular among scientists exploring the use of choice models to derive index values. The approach to include a life-span attribute mimics the conventional time trade-off valuation technique, allowing trade-offs between quality of life and length of life (Mulhern et al., 2014). When a life-span attribute is included, a regression model incorporates a parameter for years and interaction terms between each level of the health-state attributes and the life-span attribute. With this interaction term, life span acts as a multiplier on the amount of value that is attributed to a health state for a given time period (Box 14.1). The proposition underlying these multiplicative models is that the derived values based on such a model can be positioned on a 0.0–1.0 scale (Bansback et al., 2012), because the value for a health state will be zero when there is no life span left. The developers of this approach refer to it as a DCE framed as a time trade-off task (DCE_{TTO}). The results obtained with this novel approach warrant further investigation. Specifically, proof of concept studies adopting DCE_{TTO} produced lower values and classified more states as worse than death in comparison to time trade-off techniques. To explain these contradictory results, we should recall that the TTO task explicitly addresses whether or not a health state is considered better or worse than death, which is not the case in a DCE_{TTO} choice task. However,

> **BOX 14.1**
> To determine the coefficients for the DCE_{TTO} a conditional logit model is used as described by Bansback and colleagues. Briefly, the value function v of each respondent i is defined to be a multiplicative between a vector of levels for each EQ-5D attribute x life years t in each health scenario j. The full model can be written as:
>
> $$v_{ij} = \alpha + \beta t_{ij} + \lambda x_{ij} \cdot t_{ij} + \varepsilon_{ij} \qquad (14.1)$$
>
> Of these, α is a constant, expected to be equal to zero; β represents the value of living in full health for the specified duration and is expected to be positive; λ represents the disutility of living with the specified set of EQ-5D-5L health problems for the same duration and thus is expected to be negative; and ε_{ij} is a random term. The anchoring of the utility function for death at 0.0 is achieved through the relative size of the two regression coefficients β and λ in Eq. (14.1) above. In this model t is considered to be linear.

DCE_{TTO} allows values for states worse than death to be predicted using the same methodology and modeling process as states better than death, and with no arbitrary transformations (Mulhern et al., 2014).

Norman et al. (2016) even go a step further. They introduce a choice model with pairs of health descriptions, each based on a fixed set of health attributes with varying levels. Accompanying these multiattribute health descriptions is a separate life-span attribute (DCE_{TTO}). In addition to the two health-state descriptions including a life-span statement, a third option is presented: "death" (Fig. 14.3). Within the EuroQol Research Foundation, there is a debate on the normative interpretation of findings with these multiplex measurement methods.

BETTER THAN DEAD

Better than dead (BTD) is the name of a relatively simple method to derive index values for health states (Stalmeier et al., 2007). It may be adequate to solve a critical problem that is associated with current economic valuation techniques, namely how to measure states considered "worse than dead." There is disagreement about the assessment and computation of the values for such health states (Lamers, 2007). The current techniques employ different tasks to elicit values for states that are judged as being better or worse than dead (Chapter 12). Various transformations have been proposed to position better and worse than dead states on a single metric scale. However, there is no agreed method for doing so, and all methods probably introduce biases. Moreover, they all assume that health-state values are independent of the duration of such states, despite clear indications that this assumption does not hold (Sutherland et al., 1982).

If you had to choose between the following scenarios:

Scenario A	Scenario B	Scenario C
• You have no problem in walking about • You have some problems washing and dressing yourself • You have no problems with performing your usual activities • You have moderate pain or discomfort • You are extremely anxious or depressed	• You have some problems in walking about • You have no problems with self-care • You are unable to perform your usual activities • You have extreme pain or discomfort • You are not anxious or depressed	Death
You will live in this state for **4 years**, than die.	You will live in this state for **4 years**, than die.	

Which is **best**?	○	○	○
Which is **worst**?	○	○	○

FIGURE 14.3 Example of the response task with a life-span statement (on the bottom of the box) and "death" as a third option. *With permission from Springer.*

The BTD method involves comparisons of two health episodes, where an episode consists of a health state with a specified duration. One of the health episodes is always equal to dead. For example, a respondent is asked to make a choice between living 8 years in a moderate health state or being dead. Not surprisingly, good episodes are often preferred to being dead, and bad episodes are less often preferred to being dead. By collecting choices for many health states of different durations, information is gained about the value of these episodes.

The computation of index values by the BTD method is relatively straightforward. If a particular episode—for example, 20 years in a mild condition—is preferred to dead, it is assigned a score of 1.0. Conversely, if 20 years in a mild condition is not preferred to dead, that episode receives a score of −1.0. These 1.0 or −1.0 scores are aggregated over the respondents to arrive at an index value for this specific health profile (20 years, mild). Using this scheme, better and worse than dead health states are positioned on a single index-value scale.

In the first empirical study based on the BTD method, preferences were obtained for three different durations combined with five health-state descriptions, based on the EQ-5D-3L health instrument (Stalmeier et al., 2007). The results of this study showed a logical ordering of health states (Fig. 14.4).

In a subsequent study, the BTD method was administered as a stand-alone web-based survey. The main goal was to explore its feasibility and robustness when estimating weights for the levels of the different attributes (van Hoorn et al., 2014). A representative sample of 291 persons responded to 108 choices, taken from a total of 50 EQ-5D-3L health states and six durations, ranging

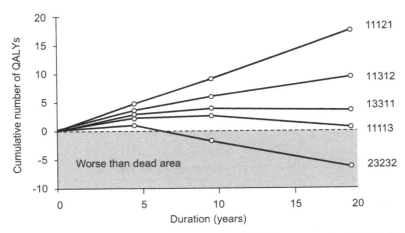

FIGURE 14.4 Cumulative quality-adjusted life years (QALYs) as measured with Better-Than-Dead preferences as a function of the duration (in years) of a health state and the value of this health state (for the specified duration).

from 1 to 40 years. Regression analysis (a logistic random effects model) was used to estimate the weights, and a logical ordering of health states was found. The weights for the levels were rather similar to the Dutch valuation study weights, which employed a standard protocol (Lamers et al., 2006). The relative position of the health states and being dead was also determined. An episode equals dead if the respondents show a 50/50 split in BTD preferences. In this case, the values for the episode and for dead were equal, so both values were set at zero.

Fig. 14.4 shows a phenomenon specific to severe health states. As mentioned above, the value of a health state can be time dependent. For severe health states, this implies that their value may be low but still positive (BTD). However, if such health states persist, they may be perceived as worse than dead (i.e., dead is preferred over the health state). Due to this relationship, a severe health state with a short duration may be preferred to the same health state with a long duration. The literature refers in this context to "maximal endurable time" states (Sutherland et al., 1982). In the studies conducted so far, about 15—25% of the respondents showed maximal endurable time preferences.

EYE TRACKING

Most preference-based methods involve judgmental tasks for which it is assumed that respondents pay attention to the instructions, the full description of the health states, and other elements such as visual cues. However, there is little direct verification of whether respondents pay (equal) attention to the various elements.

Eye tracking may be an attractive and informative tool to explore the attendance process during judgmental tasks. By monitoring eye movements, it is possible to identify which information is attended. The implicit assumption is that attendance to certain elements of the task means that such elements are perceived, i.e., that the respondents are paying attention to it. Eye tracking is used in marketing, cognitive science, human computer interaction, and in psychology to study visual perception and language processing.

Eye tracking implies the process of capturing the eye movements or the points of gaze, which is the area of focus at a particular time on the visual scene. Generally, to see objects, the eyes need to fixate the gaze for a certain period of time, typically between 200 and 600 ms. The process of vision refers to scanning the object with rapid eye movements, which are called saccades. The mechanism of many eye trackers involves recording (on video) the eye movements and mapping them to the computer screen for further analysis. The videos are the basis for the focus position analysis that detects changes in the eye and pupil location.

Once the respondents' eye movements are recorded, they can be analyzed using heat maps (Fig. 14.5), fixation paths, areas-of-interest statistics, and other graphs. Heat maps and fixation paths present the overall pattern of the respondents' focus of attention: the areas and intensity of focus, length of focus, and direction path of focus. The areas-of-interest approach is based on the observation that some objects are more interesting and attract more attention, so the eyes focus longer on these specific objects (Selivanova and Krabbe, 2016).

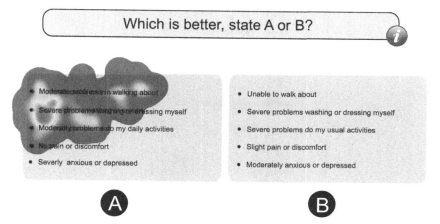

FIGURE 14.5   A heat-map visualization from eye tracking an EQ-5D health-state description in a paired comparison task.

PREFERENCE-BASED, MULTIATTRIBUTE RASCH MODEL

So far, models based on IRT have been used to measure distinct domains or aspects of health but not the overall quality of health as perceived by individual patients. However, an application to measure the latter does seem feasible. A straightforward measurement system based on a combination of IRT and discrete choice methods was recently worked out. It is perfectly suited to deal with responses from patients themselves (Krabbe, 2013). The core of this new system is the Rasch model (Rasch, 1960/1980; Fisher, 1983; De Boeck and Wilson, 2004; Groothuis-Oudshoorn et al., forthcoming).

Respondents first indicate their own health condition in a specified health-status classification system. Subsequently, they are confronted with paired comparison choice tasks that offer one or more comparator health states (based on the same classification system; Fig. 14.6).

The comparator health states are closely related to the patient's own condition. One of the attractive qualities of this novel framework is that derived values are fully patient reported. In other words, both the classification

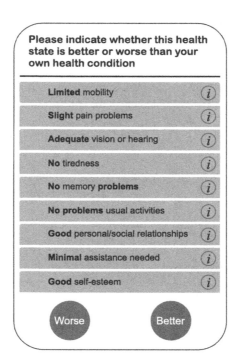

FIGURE 14.6 Basic comparator task in a multiattribute preference response (MAPR) model based on a health-status classification system (prototype Chateau Santé—Base ©): 9 attributes, each response category has 4 levels.

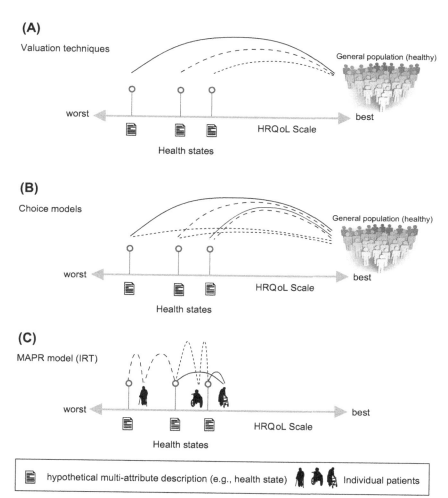

FIGURE 14.7 Schematic representation of the judgmental task for three health states using A = conventional monadic measurement (valuation techniques) by a sample of the general population; B = conventional discrete choice task (paired comparison) by a sample of the general population; C = multiattribute preference response (MAPR) model based on item-response theory (IRT) for individual patients (3 patients in this example, each assessing 2 health states located nearby).

of the health attributes and the subsequent preference-based comparison task are based on patient input. Fig. 14.7 is a schematic rendering of the current patient-reported, preference-based, multiattribute health measurement system compared with more traditional preference-based measurement approaches.

Development is ongoing; the next step is to design a protocol to obtain personalized responses, whereby individual patients select the health attributes they consider important out of a larger set of candidate attributes. The multiattribute preference response (MAPR) model produces relative values for health states, and these are perfectly suitable for clinical applications. For application in DALYs and QALYs, however, MAPR-derived values need to be rescaled around the position where states are considered to become worse than death (position of death = 0.0). Supplementary studies can be conducted to localize the juncture where health states are considered worse than death (van Hoorn et al., 2014; Arons et al., forthcoming-b).

MULTIDIMENSIONAL ITEM-RESPONSE THEORY

Unidimensional models based on IRT are often simpler and have some interesting measurement properties. Nonetheless, many constructs are unavoidably multidimensional in nature. For instance, a subjective phenomenon such as health may be understood as a construct with subscale components nested within or alongside a more general construct. One drawback of multidimensional item-response theory (MIRT) models is that the estimation of the item parameters in higher dimensional space is computationally difficult when using standard numerical estimation procedures. However, with recent advances in estimation theory, coupled with the expanding computational power of personal computers, MIRT has become recognized as a feasible methodology for statistical analysis (Reckase, 2009).

Substantial research on MIRT models has been done over the last two decades. There is a bundle of such models available: multidimensional two-parameter logistic, multidimensional Rasch, multidimensional normal-ogive, multidimensional generalized partial credit, multidimensional partial credit, and multidimensional graded response. Unlike unidimensional item-response theory, an item in MIRT has direction in addition to location on a scale; item information at a certain score point is not a single number but a matrix that gives different values in different directions; and item or test characteristic functions are not curves but surfaces. For the estimation of these models, Bayesian Markov chain Monte Carlo estimation procedures are often used (Edwards, 2010; Fox, 2010).

MULTIDIMENSIONAL SCALING

Multidimensional scaling refers to a family of mathematical (not statistical) models that can be used to analyze distances between objects (e.g., health states). Information contained in a set of data is represented by a set of points in space. These points are arranged in such a way that the geometrical distance between them will reflect empirical relationships in the data. The geometrical relationships can be situated in multidimensional space but can also encompass the one-dimensional mode. They may be interval (metric) or rank distances (nonmetric). For example, the "psychological distance" between health states denotes the perceived similarity between them and would result in a scale from the worst to the best health state (Borg and Groenen, 2005). Multidimensional scaling is based on similarity or dissimilarity data.

In multidimensional scaling analyses, "proximities" refer to observed differences between objects. These are described as either similarities or dissimilarities. For similarities, larger numbers indicate that objects are nearer, whereas the opposite applies to dissimilarities. Multidimensional scaling originated in psychometrics but has become a general data analytical technique, now used in a wide variety of fields (Tenenbaum et al., 2000).

Shepard (also known for the Shepard tone, 1964) developed a major extension of classical metric multidimensional scaling in 1962. Early work demonstrated that it was possible to derive metric multidimensional scaling solutions assuming only ordinal relationships between the objects (Shepard, 1962a,b).

Similarity data can be modeled as distances among pairs of health states in geometric space by means of multidimensional scaling. This is illustrated in Fig. 14.8, which displays four health states. The interval distance between them is represented by the length of the arrows. If we approximate the distances between the pairs of health states, we can use these rough estimates to infer the actual distances, which is done with metric multidimensional scaling. Conversely, when the distances are elicited as or converted into rank distances (the numbers in Fig. 14.8), we can use nonmetric multidimensional scaling. A benefit of nonmetric multidimensional scaling is that it allows responses that are less precise. As a consequence, easier response tasks can be used to obtain this type of similarity data.

The purpose of multidimensional scaling is to map the relative location of objects using data that show how the objects differ. Seminal work on this method was undertaken by Torgerson (1958). A reduced version is one-dimensional scaling. It is applied when we have good reason to believe

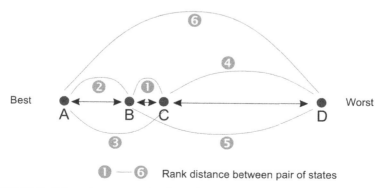

FIGURE 14.8 Schematic representation (input for nonmetric multidimensional scaling) of the psychological distances and ranks between pairs of health states (A—D are health states).

there is only one underlying dimension of interest. Computational tests are then performed to verify that more than one dimension does not add (substantially) to the solution of the scaling.

A classic example of the application of multidimensional scaling concerns airline mileages between 10 big cities in the United States. The problem is to construct a map showing city locations based only on the intercity distances. We may be able to do so easily for distances between three or four cities, but adding cities makes the task impossible without recourse to mathematical procedures. Multidimensional scaling is one example of this procedure. With 10 cities, it turns out that a two-dimensional (flat-Earth) multidimensional scaling solution almost perfectly recovers the locations. (Theoretically, a three-dimensional solution would be perfect, as it allows for the curvature of the Earth's surface.) For our purposes, it is even more interesting that an almost identical map of the 10 cities can be made by using ranks of the distances instead of actual mileage (Fig. 14.9).

In theory, multidimensional scaling is an elegant and robust method (Kruskal, 1964a,b; Shepard, 1962a,b, 1966). In practice, however, it might be very demanding at the data-collection stage. For the time being, multidimensional scaling seems more suited for exploring and deriving distances (quantification) in specific areas (Krabbe et al., 2007) or situations where conventional preference-based methods are not feasible or fail.

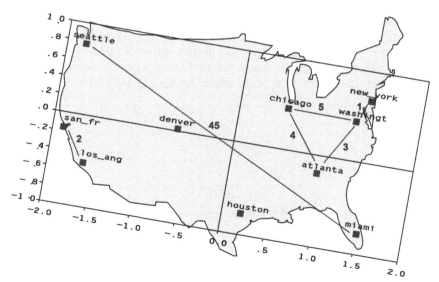

FIGURE 14.9 The two-dimensional solution (PROXSCAL, SPSS) of nonmetric multidimensional scaling on the ranks of the 45 airline mileages (distance ranks 1—5 and 45 are depicted) between 10 cities in the United States.

Together with a PhD student of mine, I conducted a multidimensional scaling study, though with disappointing results (Arons and Krabbe, 2014). No matter how much we tried, we could not get interpretable and logical results, nor were the diagnostic statistics satisfying. This was one of those studies that taught us a great deal (no failures, no science). As it turned out, the small set of health states that we assessed had been perceived by the respondents as almost equal in regard to the quality (value) attached to each one. So, in our empirical study, the health states that we wanted to scale by applying multidimensional scaling could not be represented as nicely and logically as in the artificial example of Fig. 14.8. Nevertheless, that study elucidated the MDS procedures.

BEST–WORST SCALING

Another approach that might be useful to derive health-state values is best–worst scaling (BWS), also known as maximum difference scaling (Marley and Louviere, 2005). There are three types of BWS, namely attribute BWS, profile BWS, and attribute-level BWS. In attribute BWS, respondents choose the best and worst attributes from those available. An example is the study by Finn and Louviere (1992), where respondents select the best and worst attributes. In the field of health-state measurement, such a task might provide information relevant to the selection of attributes for a classification system.

In profile BWS, the respondents are shown two or more profiles (for our purposes, health-state descriptions with varying attribute levels). They are asked to indicate which profile they consider best and which one they consider worst. By asking the respondents to choose the best and worst—or the smallest and largest, most and least liked, etc.—profile BWS provides more information than paired comparison methods and requires fewer responses from the subjects. In the field of health-state measurement, such a response task might be useful. However, if the classification system that is being valued consists of many attributes and levels, the respondents might find this task too difficult to complete if more than two profiles are presented (Xie et al., 2012).

The final type is the attribute-level BWS, in which the respondents are presented with a single profile (health-state description). They have to indicate which attribute (with a specific level) they consider the best and which the worst (Fig. 14.10). This type of BWS might prove to be very useful in the context of health-state measurement, since the amount of information respondents need to process seems less than in pair-wise comparison.

Attribute-level BWS has been applied successfully in the field of health-state measurement (Ratcliffe et al., 2011; Brazier et al., 2012; Arons et al., forthcoming-a). For a discussion of some methodological concerns, the reader is referred to Flynn et al. (2007, 2008; Flynn, 2010; see also Louviere et al., 2015). Marley and Louviere (2005) have shown that the analysis of nominal best–worst choice data leads to a common interval or ratio scale for all the

Which attribute is the best, and which is the worst?

Best   Worst

○   • Severe problems in walking about   ○
○   • Moderate problems washing or dressing myself   ○
○   • Unable to do my usual activities   ○
○   • Moderate pain or discomfort   ○
○   • Severely anxious or depressed   ○

FIGURE 14.10 The profile best—worst scaling version (here for EQ-5D-5L health-state description). *HRQoL*, Health-related quality of life.

factors, depending on how the data are analyzed. Although BWS has been used to examine preferences for complex attitudinal dimensions, recent studies have applied it to measure food- and meal-related properties and liking. Interestingly, this research area is where I started my endeavor years ago in search of new methods and methodologies to improve subjective measurement.

REFERENCES

Arons, A.M.M., Groothuis-Oudshoorn, C.G.M., Krabbe, P.F.M. Comparing two measurement models based on ordinal responses to derive multi-attribute health-state values: choices between states or attributes (forthcoming-a).

Arons, A.M.M., Krabbe, P.F.M., 2014. Quantification of health by scaling similarity judgments. PLoS One 9, e89091.

Arons, A., Selivanova, C.G.M., Groothuis-Oudshoorn, P.F.M. Krabbe Calibrating the position of "dead" among health states in the choice modeling framework (forthcoming-b).

Bansback, N., Brazier, J., Tsuchiya, A., Anis, A., 2012. Using a discrete choice experiment to estimate health state utility values. Journal of Health Economics 31 (1), 306—318.

Barofsky, I., 2012. Quality: It's Definition and Measurement as Applied to the Medically Ill. Springer, New York.

Borg, I., Groenen, P.J.F., 2005. In: Modern Multidimensional Scaling: Theory and Applications, second ed. Springer, New York.

Brazier, J., Rowen, D., Yang, Y., Tsuchiya, A., 2012. Comparison of health state utility values derived using time trade-off, rank and discrete choice data anchored on the full health—dead scale. European Journal of Health Economics 13 (5), 575—587.

Chen, G., Flynn, T., Stevens, K., Brazier, J., Huynh, E., Sawyer, M., Roberts, R., Ratcliffe, J., 2015. Assessing the health-related quality of life of Australian Adolescents: an empirical comparison of the Child Health Utility 9D and EQ-5D-Y instruments. Value in Health 18 (4), 432—438.

Claes, C., Greiner, W., Uber, A., Graf von der Schulenburg, J.M., 1998. In: Rabin, R.E., Busschbach, J.J.V., de Charro, F.T.H., Essink-Bot, M.L., Bonsel, G.J. (Eds.), The New German Version of the EuroQol Quality of Life Questionnaire. EuroQol Plenary Meeting Discussion Papers 1997, Rotterdam.

Craig, B.M., Reeve, B.B., Brown, P.M., Cella, D., Hays, R.D., Lipscomb, J., Pickard, S.A., Revicki, D.A., 2014a. US valuation of health outcomes measured using the PROMIS-29. Value in Health 17 (8), 846−853.

Craig, B.M., Reeve, B.B., Cella, D., Hays, R.D., Pickard, A.S., Revicki, D.A., 2014b. Demographic differences in health preferences in the United States. Medical Care 52 (4), 307−313.

De Boeck, P., Wilson, M., 2004. Explanatory Item Response Models: A Generalized Linear and Nonlinear Approach. Springer, New York.

Edwards, M.C., 2010. A Markov chain Monte Carlo approach to conformatory item factor analysis. Psychometrika 75 (3), 474−497.

Fava, G.A., Ruini, C., Rafanelli, C., 2004. Psychometric theory is an obstacle to the progress of clinical research. Psychotherapy and Psychosomatics 73 (3), 145−148.

Fayers, P.M., 2007. Applying item response theory and computer adaptive testing: the challenges for health outcomes assessment. Quality of Life Research 16 (Suppl. 1), 187−194.

Finn, A., Louviere, J.J., 1992. Determining the appropriate response to evidence of public concern: the case of food safety. Journal of Public Policy Market 11, 12−25.

Fisher, G.H., 1983. Logistic latent trait models with linear constraints. Psychometrika 48 (1), 3−26.

Flynn, T.N., 2010. Valuing citizen and patient preferences in health: recent developments in three types of best-worst scaling. Expert Review of Pharmacoeconomics & Outcomes Research 10 (3), 259−267.

Flynn, T.N., Louviere, J.J., Marley, A.A., Coast, J., Peters, T.J., 2008. Rescaling quality of life values from discrete choice experiments for use as QALYs: a cautionary tale. Population Health Metrics 6 (6), 1−11.

Flynn, T.N., Louviere, J.J., Peters, T.J., Coast, J., 2007. Best−worst scaling: what it can do for health care research and how to do it. Journal of Health Economics 26 (1), 171−189.

Food and Drug Administration, 2009. Guidance for Industry Use in Medical Product Development to Support Labeling Claims Guidance for Industry, Guidance for Industry Patient-Reported Outcome Measures: Use in Medical Product Development to Support Labeling Claims. Silver Spring.

Fox, J.P., 2010. Bayesian Item Response Modeling: Theory and Applications. Springer, New York.

Groothuis-Oudshoorn, C.G.M., van der Heuvel, E., Krabbe, P.F.M., 2016. An Item Response Theory Model to Measure Health: The Multi-attribute Preference Response Model (forthcoming).

Krabbe, P.F.M., 2013. A generalized measurement model to quantify health: the multi-attribute preference response model. PLoS One 8 (11), e79494.

Krabbe, P.F.M., Salomon, J.A., Murray, C.J.L., 2007. Quantification of health states with rank-based nonmetric multidimensional scaling. Medical Decision Making 27 (4), 395−405.

Kruskal, J.B., 1964a. Multidimensional scaling by optimizing goodness of fit to a nonmetric hypothesis. Psychometrika 29 (1), 1−27.

Kruskal, J.B., 1964b. Nonmetric multidimensional scaling: a numerical method. Psychometrika 29 (2), 115−129.

Lamers, L.M., 2007. The transformation of utilities for health states worse than death − consequences for the estimation of EQ-5D value sets. Medical Care 45 (3), 238−244.

Lamers, L.M., McDonnell, J., Stalmeier, P.F.M., Krabbe, P.F.M., Busschbach, J.J.V., 2006. The Dutch tariff: results and arguments for an effective design for national EQ-5D valuation studies. Health Economics 15 (10), 1121−1132.

Louviere, J.J., Flynn, T.N., Marley, A.A.J., 2015. Best-Worst Scaling: Theory, Methods and Applications. Cambridge University Press, Cambridge.

Marley, A.A.J., Louviere, J.J., 2005. Some probabilistic models of best, worst, and best−worst choices. Journal of Mathematic Psychology 49 (6), 464−480.

Maydeu-Olivares, A., Böckenholt, U., 2005. Structural equation modeling of paired comparisons and ranking data. Psychological Methods 10 (3), 285−304.

McCabe, C., Brazier, J., Gilks, P., Tsuchiya, A., Roberts, J., O'Hagan, A., Stevens, K., 2006. Using rank data to estimate health state utility models. Journal of Health Economics 25 (3), 418−431.

Mulhern, B., Bansback, N., Brazier, J., Buckingham, K., Cairns, J., Devlin, N., Dolan, P., Hole, A.R., Kavetsos, G., Longworth, L., Rowen, D., Tsuchiya, A., 2014. Preparatory study for the revaluation of the EQ-5D tariff: methodology report. Health Technology Assessment 18 (12), 1−222.

Norman, R., Mulhern, B., Viney, R., 2016. The impact of different DCE-based approaches when anchoring utility scores. PharmacoEconomics 34 (8), 805−814.

Oppe, M., Devlin, N.J., van Hout, B., Krabbe, P.F., de Charro, F., 2014. A program of methodological research to arrive at the new international EQ-5D-5L valuation protocol. Value in Health 17 (4), 445−453.

Patrick, D.L., Burke, L.B., Gwaltney, C.J., Leidy, N.K., Martin, M.L., Molsen, E., Ring, L., 2011a. Content validity − establishing and reporting the evidence in newly developed patient-reported outcomes (PRO) instruments for medical product evaluation: ISPOR PRO good research practices task force report: Part 1-Eliciting concepts for a new PRO instrument. Value in Health 14 (8), 967−977.

Patrick, D.L., Burke, L.B., Gwaltney, C.J., Leidy, N.K., Martin, M.L., Molsen, E., Ring, L., 2011b. Content validity − establishing and reporting the evidence in newly developed patient-reported outcomes (PRO) instruments for medical product evaluation: ISPOR PRO good research practices task force report: Part 2-Assessing respondent understanding. Value in Health 14 (8), 978−988.

Rasch, G., 1960/1980. Probabilistic Models for Some Intelligence and Attainment Tests. (Copenhagen, Danish Institute for Educational Research), Expanded Edition (1980) With Foreword and Afterword by B.D. Wright. University of Chicago Press, Chicago.

Ratcliffe, J., Couzner, L., Flynn, T., Sawyer, M., Stevens, K., Brazier, J., Burgess, L., 2011. Valuing Child Health Utility 9D health states with a young adolescent sample: a feasibility study to compare best−worst scaling discrete-choice experiment, standard gamble and time trade-off methods. Applied Health Economics and Health Policy 9 (1), 15−27.

Reckase, M.D., 2009. Multidimensional Item Response Theory. Springer, New York.

Salomon, J.A., 2003. Reconsidering the use of rankings in the valuation of health states: a model for estimating cardinal values from ordinal data. Population Health Metrics 1, 1−12.

Schölzel-Dorenbos, C.J.M., Arons, A.M.M., Wammes, J.J.G., Olde Rikkert, M.G.M., Krabbe, P.F.M., 2012. Validation study of the prototype of a disease-specific index measure for health-related quality of life in dementia. Health and Quality of Life Outcomes 10 (118), 1−11.

Selivanova, A., Krabbe, P.F.M., 2016. Eye tracking to detect the attendance process in assessing health-state descriptions. EuroQol Proceedings 2016 (forthcoming).

Shepard, R.N., 1962a. The analysis of proximities: multidimensional scaling with an unknown distance function. I. Psychometrika 27 (2), 125−140.

Shepard, R.N., 1962b. The analysis of proximities: multidimensional scaling with an unknown distance function. II. Psychometrika 27 (3), 219−246.

Shepard, R.N., 1964. Circularity in judgements of relative pitch. Journal of the Acoustical Society of America 36 (12), 2346–2353.

Shepard, R.N., 1966. Metric structures in ordinal data. Journal of Mathematical Psychology 3 (2), 287–315.

Stalmeier, P.F.M., Lamers, L.M., Busschbach, J.J.V., Krabbe, P.F.M., 2007. On the assessment of preferences for health and duration: maximal endurable time and better than dead preferences. Medical Care 45 (9), 835–841.

Stenner, P.H.D., Cooper, D., Skevington, S.M., 2003. Putting the Q into quality of life: the identification of subjective constructions of health-related quality of life using Q methodology. Social Science and Medicine 57 (11), 2161–2172.

Stevens, K.J., 2016. How well do the generic multi-attribute instruments incorporate patient and public views into their descriptive systems? Patient 9 (1), 5–13.

Stevens, K., Palfreyman, S., 2012. The use of qualitative methods in developing the descriptive systems of preference-based measures of health-related quality of life for use in economic evaluation. Value in Health 15 (8), 991–998.

Sutherland, H.J., Llewellyn-Thomas, H., Boyd, N.F., Till, J.E., 1982. Attitudes toward quality of survival. The concept of "maximal endurable time". Medical Decision Making 2 (3), 299–309.

Tenenbaum, J.B., de Silva, V., Langform, J.C., 2000. A global geometric framework for nonlinear dimensionality reduction. Science 290 (5500), 2319–2323.

Torgerson, W.S., 1958. Theory and Methods of Scaling. John Wiley & Sons, New York.

van Hoorn, R.A., Donders, A.R.T., Oppe, M., Stalmeier, P.F.M., 2014. The better than dead method: feasibility and interpretation of a valuation study. PharmacoEconomics 32 (8), 789–799.

Xie, F., Pullenayegum, E., Gaebel, K., Oppe, M., Krabbe, P.F.M., 2012. Eliciting preferences to the EQ-5D-5L health states: discrete choice experiment or multiprofile case of best–worst scaling. Value in Health 15 (4), A198–A199.

Zyzanski, S.J., Perloff, E., 1999. Clinimetrics and psychometrics work hand in hand. Archives of Internal Medicine 159 (15), 1816.

Chapter 15

Perspectives

Chapter Outline

INTRODUCTION

It is fair to say that in the medical setting, the emphasis in measuring the "quality of health" is on pragmatic approaches. From the outset, there has been interest in the content of instruments for measuring patients' health status, yet far less attention has been directed toward the theoretical considerations underpinning those instruments. Moreover, it seems that many of the researchers who are developing and applying health-outcome instruments are largely unaware of certain crucial assumptions and conditions of measurement.

In the absence of a systematic and thorough implementation of modern measurement theory, a situation has arisen whereby contemporary health measures are only moderately informative. What is needed for an adequate (i.e., valid and precise) measurement of the health condition of patient groups and individual patients is an array of sophisticated measurement models.

To arrive at measures for health that have cardinal- or interval-level measurement properties, certain basic conditions must be met. For instance, the invariance principle must be applied when collecting response data; another prerequisite is unidimensionality of the measurement scale. We may even go on to develop measurement instruments that express the overall quality of health conditions instead of the reported frequencies or intensities of complaints. For this step, preference-based measurement procedures are required.

Almost all instruments used in clinical settings were developed within the framework of classical test theory, so they do not have an absolute scale (ratio level, Chapter 4); that is, these scales lack interpretable and fixed lower/upper end points. One of the most important but neglected topics in clinical assessment is the so-called metric question. It is important because the ability to draw meaningful inferences from health measures is directly proportionate to our understanding of the metric of our scales (Reise and Waller, 2009).

Human cognition lies at the core of psychology, and this field has developed measurement models with response mechanisms (e.g., trace lines) that are connected to cognitive theories. Economics has shown far less interest in basic response mechanisms. Instead, economic theory is mainly grounded in assumptions on how people should respond. This is in line with the old adage that psychology is descriptive, whereas economics is prescriptive. A similar distinction can be found between proponents of the Rasch model (descriptive) and those who favor item-response theory (prescriptive).

OUR WISH LIST

It is clear how the ideal health measurement instrument should look. The "tool" should be sufficiently sensitive and specific to evaluate health-care provision for almost all special groups. It should contain items that are phrased at a reading level appropriate to the great majority of the population; that is, the statements should be short and easy to answer or process, and their meaning should be commonly understood. In particular, the instrument should be easy to complete and easy to administer. Scoring must be relatively easy, so the response categories must be unambiguous. The measurement tool must be short and simple enough to be self-administered; it should be inexpensive; and, above all, it must be valid and reliable. In addition, the process must be acceptable and valuable to patients, clinicians, and administrators; it should ensure timely communication among all who provide care; and it should allow efficient data collection, analysis, and reporting (Hunt and McEwen, 1980). Moreover, an attractive, user-friendly format that allows patients to score on an electronic medium at home would be highly advisable (Table 15.1).

SIMPLICITY

Simplicity is a highly prized good in science. It comes in several shapes and is achieved through diverse operations. Some procedures are not simple in themselves but can be made to appear so. The famous and elegant formula, $E = mc^2$ is so plain and simple that one might get the idea that the work on which it is grounded is also simple (Einstein, 1905). This is definitely not the case. Behind this formula lies a whole world of complex mechanisms and consequences. Not surprisingly, Einstein himself is credited with the intriguing idea that "Everything should be made as simple as possible, but not simpler."

Another analogy is drawn from the Rasch model. According to Ben Wright (a major proponent of the Rasch mechanism), "The Rasch model is so simple that its immediate relevance to contemporary measurement practice and its

TABLE 15.1 Overview of the Main Features of Health Measurement Instruments Developed in the Five Research Traditions (the More Dots, the Better)

Feature	Classical Test Theory	Item-Response Theory	Choice Models	Valuation Techniques	Clinimetrics
Sensitive	●●●	●●●	●●●	●●●	●●●
Understandable	●●●	●●●	●●●	●	●●●
Easy to administer	●●	●●●	●	●	●●
Inexpensive	●●	●●	●●●	●	●●
Reliability	●●●	●●●●●	●●●	●●	●●●
Validity	●●	●●●	●●	●●	●●●
Efficient data collection	●●	●●●●	●	●●●	●●
Attractive	●●	●●●	●●	●	●●
User friendly	●●●	●●●	●●●	●	●●●

extensive possibilities for solving measurement problems may not be fully apparent" (1977). The observation made by Wright in regard to the simplicity (and beauty) of the Rasch model resonates with a cryptic remark made by the famous Dutch football player Johan Cruijff in a very different setting: "You only see it once you get it." In other words, a mental shift is necessary to understand new or "unknown" approaches.

Simplicity is also important for the development of instruments to measure (patient reported) health outcomes, in particular health and health status. What we want are procedures that are easy to administer and easy for the respondents to understand. For example, some "questionnaires" merely ask people "to tick the box." To the respondent, this may look like a simple response task, but in fact it may form part of a state-of-the-art measurement system in which the user only sees the "tick-the-box" question. In the background, a refined and complex machinery may be operating, but the user gets the impression that it is too simple to be true. Interactive applications on mobile devices in combination with wireless connections to a central server may prompt seemingly simple questions or response tasks, but such systems can have strengths that are not present in the bulk of existing health instruments.

The use of paired comparisons to record human judgments dates back to the late 19th century. Regarding comparative responses, Titchener (1901, p. 92) noted that "We have no absolute measure of the amount of pleasantness or unpleasantness that corresponds to a given stimulus; but we are able to say, when two stimuli are presented, which of them is the more or which the less pleasant" (cited in: Brown and Peterson, 2009). Similarly, Nunnally and Bernstein (1994, p. 51) assert that "people are almost invariably better (more consistent and/or accurate) at making comparative responses than absolute responses. (…) People rarely make absolute judgments in daily life, since most judgments are inherently comparative." In conclusion, response tasks should be simple.

Other considerations with a connection to simplicity underpin measurement theory. For instance, it may be better to describe the topic of interest (e.g., health states) as uniformly and distinctively as possible (Glynn, 2010). As Nunnally and Bernstein (1994) explicitly state on one of the first pages of their massive book, "a measure should generally concern some one thing—some distinct, unitary attribute" (p. 4).

There is another reason why we have to strive for simplicity in measuring such a complex phenomenon as health. There is a limit to what people can process and the amount of information that certain measurement procedures can produce. Coombs (1964) raised this topic in terms of the amount of information that can be obtained by a given method (its channel capacity) and the amount of implied or repeated information (redundancy) it contains, which can be used to check the subject's consistency. His summary is well-worth repeating: "On a priori grounds we would expect that the higher the channel capacity the better, but this is certainly not true. A price is paid for data, not only in financial terms but in wear and tear on the organism at source. A method with too high channel capacity may, through boredom and fatigue, result in a decrease of information transmitted, through stereotype behavior. Furthermore, the potential variety of messages from the organism may not be great, in which case a more powerful method is inefficient (...) Ideally a method should be selected which matches the information content in the source but is not such a burden as to generate noise."

Measurement procedures with a high channel capacity can be too complicated for the respondents. For example, the current time trade-off protocol for the EQ-5D instrument uses a specialized computer program (EQ-VT) that was developed to conduct time trade-off exercises (graphical bars, instructions, health-state descriptions). However, empirical studies showed that the time trade-off valuation technique based on the EQ-VT alone was producing unexpected results. Therefore, the current time trade-off protocol used for the EQ-5D instrument states that apart from the EQ-VT, trained interviewer assistance is mandatory (Oppe et al., 2014). So respondents are now confronted with complex and strenuous instructions offered by the interviewer on top of the already complex EQ-VT representation. Recent studies showed that the respondents' values vary substantially between interviewers (Shah et al., 2011; Yang, 2015). Interviewer effects have also been traced in an earlier EQ-5D study for values derived with the visual analogue scale (Krabbe, 2002). A legitimate question is whether the evidence for significant interviewer effects can be considered as the final nail in the coffin of the time trade-off valuation technique (Lugnér and Krabbe, 2017).

SYNTHESIZING MEASUREMENT MODELS

Measurement strategies from health economics are not coherent, nor are they based on genuine measurement principles. The conventional approach of classical test theory makes several assumptions that cannot be tested. Clinimetrics has an eye for the practical relevance of the content but has little to say about how to measure it. Item-response theory combines conceptual beauty with analytical power but seems difficult at the development stage. The strengths of each of these approaches should be integrated to arrive at better health measurement instruments.

Classical test theory can apply factor analysis to explore and strengthen the construct validity of concepts such as health status. Preference-based measurement methods do not have access to anything similar. It would be interesting to develop new models that combine such features. What we are looking for are new measurement tools that may be quite technical in the background but should be simple for the patients to assess. One option for developing a more powerful tool would be to combine the strengths of the descriptive measurement approach (as worked out under classical test theory for a small set of items that are focused on one specific health aspect or domain) with the strengths of preference-based measurement approaches. Such a framework should be able to provide a single value to express the overall quality of a patient's health status. In addition, it should provide precise measures for each of the health domains. The next step would be to replace the classical test theory part of this framework by an item-response theory approach.

According to Streiner et al. (2015, p. 299), the reason why item-response theory is not widely used lies largely in its "hard" assumption of unidimensionality. One implication is that item-response theory cannot be used to construct indices based on items that are causal rather than effect indicators (Chapter 5: "Reflective versus informative"). Thus, it would be wrong to use item-response theory to construct indices where the items themselves define the construct rather than being manifestations of an underlying latent trait. A second implication is that item response theory models cannot be used when the underlying construct is itself multifaceted and complex, as are many constructs in the health field. This may be true for the standard item-response theory models applied so far. However, extended item response theory models (Chapter 14) may resolve these issues, thereby offering a simple and novel way of measuring the quality of patients' health.

MEASUREMENT VERSUS DATA PROCESSING

Neither standard Rasch analysis nor other item-response theory models may be suitable for the type of data that can be obtained with profile health instruments. Several researchers have processed their data by applying Rasch analysis to Likert items, even though these items do not usually have the correct response structure to justify the use of the Rasch model. We have noticed this kind of incongruence in many studies on transforming profile health instruments into preference-based health instruments (Young et al., 2009, 2010, 2011; King et al., 2016). Rasch analysis requires a "cumulative" data structure (i.e., if a respondent agrees with a statement of a certain level, this person also agrees with the statements that precede this level). In standard profile health-status instruments, we are dealing with an "ideal-point" or "single-peaked" data structure (Chapter 5). When persons whose attitudes are to be measured agree or disagree with a statement, the implied response function is single-peaked. In other words, it is expected that a person will agree with the statements that are close to the person's own attitude and disagree with those statements (e.g., categories of the item) that are far from the person's location on the scale in either direction. The impact of using a cumulative model when an ideal-point item-response theory model would be more correct is unclear, but the merit of applying the right response model to the right type of response data seems worth reiterating.

IMPORTANCE

Sometimes researchers apply statistical techniques, notably factor analysis, to determine which items are important. Factor analysis is based on subtracting relationship (information) from correlations between the items. If there are items that are not very important but are scored in line with other items, then these items will show up in a factor. If there are items that are very important but are only reported by a small number of patients, the correlation with the other items will be low. Consequently, such important or relevant items will not show up in the factor solution, which is one of the reasons why clinimetrics has emerged. A nice example of a comparison between the psychometric analytical approach based on factor analysis and the expert opinion approach of clinimetrics is the study of Juniper et al. (1997).

Several studies have used the Rasch model to select items from a larger set of Likert items stemming from profile health instruments (Young et al., 2009, 2010, 2011; King et al., 2016). The purpose of this Rasch analysis is to identify problems with items that might justify exclusion. According to King et al., at least two items are needed to estimate each latent variable in Rasch analysis. Simulation studies have revealed the necessary condition for an accurate estimate of the parameters (items, respondents) for the one-, two-, and three-parameter models. However, there are no hard and fast rules for the minimum number of respondents or of items needed to make useful estimates. Only rarely would an analysis produce useful estimates with fewer than, say, 15—20 items and under 200 respondents for the Rasch model (Yen, 1992). These are indications in support of standard scale development. The application of Rasch analysis to select items from a small set of existing items is different from the standard Rasch analysis with quite larger set of items. But it is likely that an analysis with just two items could easily produce spurious results.

Even more pertinent is whether this type of Rasch analysis is appropriate to the goal of the study. When trying to transform lengthy profile health instruments into a condensed classification system by selecting some of the original items from the profile instrument, it is difficult to base decisions on statistical data processing. It would be better to select the items that are most important than to select the ones that best fit a statistical model, with a potential loss of specificity—certainly when performed with very few items.

RESPONSIVENESS REVISITED

A commendable contribution of clinimetrics is its endeavor to give meaning to the measures collected with health-status instruments. A "recently" introduced measurement criterion from this research tradition, responsiveness (to change), has drawn considerable attention from the users and developers of health instruments. Even though many within the field of health measurement and patient-reported outcomes research argue that responsiveness is important, there is little consensus on how to express it (Norman et al., 2007; Revicki et al., 2008).

Conspicuously absent is a grounded theory about how responsiveness is related to the two classic psychometric concepts of reliability and validity. Responsiveness seems to have a bearing on validity because an instrument has to measure what it was designed to measure to measure accurately. Responsiveness also seems to have a bearing on reliability; if an instrument is unreliable, it will not be responsive to changes. To be of value, an instrument should be stable when no change occurs, though it should reveal differences in case of improvement or deterioration of a person's health status. However, a responsive health-status instrument is not necessarily accompanied by high reliability (test—retest). Imprecise measures may be collected with a particular instrument that nevertheless yields the same means on a retest (Hays and Hadorn, 1992). The soundness of responsiveness as a theoretical concept and its analytical strategies is still a matter of debate (Lindeboom et al., 2005). Responsiveness is often regarded as an aspect of longitudinal construct validity

(Hays and Hadorn, 1992; Reeve et al., 2013). The question then arises: What is longitudinal construct validity? There seems to be considerable circularity in the reasoning and explanation of responsiveness. The concept itself seems to be rather esoteric: formulas and statistical procedures for responsiveness are absent or incomplete.

Although responsiveness is hard to define, it may be described in terms of some of its characteristics. One is that each measurement instrument has a range within which it operates (scale). Therefore, its responsiveness in a given situation depends on whether the instrument operates within the range of variation exhibited in the population being studied. For example, an instrument that is appropriate for patients in an ambulatory care setting might lack sensitivity in the range of variation typical of patients in an intensive care unit. Two structural characteristics increase sensitivity (responsiveness) for the detection of changes. First, the more items an instrument represents at one scale, the more responsive to change it will be (because of the measure's increased reliability). Second, offering respondents a wider range of response categories on each item will increase the precision of the responses (at least in theory), thereby making the scale more reliable and more responsive to changes. Of course, an instrument should be valid to a certain level; otherwise the scale will not reveal the differences. These characteristics are related to the granularity of an instrument. In this regard, Hays and Woolley (2000) drew an interesting conclusion: that the quest to identify a clinically meaningful difference (responsiveness) in health measurement research is part of a more general goal, namely to provide familiar anchors for unfamiliar units to aid interpretation. Straightforward interpretation of outcomes is rather difficult on the health scales developed under classical test theory or based on a clinimetric approach. Health economists have tried to solve this problem by fixing the value of health on a predetermined scale from dead (0.0) to full health (1.0).

So, difficulties arise in the assessment of responsiveness. There is even some discussion about the interpretation and meaning of the concept (Aiken, 1977; Norman, 1989; Krabbe et al., 2004; Lindeboom et al., 2005). Responsiveness as a postulated third measurement property has a high intuitive appeal, but so far it seems to lack a firm scientific grounding (FDA, 2009; Revicki et al., 2008).

DEATH

Using the concept of "death" in health measurement methods is controversial. All of the preference-based measurement systems that have been developed so far present very bad hypothetical health states to overtly healthy people who are asked to make a numerical judgment of how bad these health states are. This raises some questions. First of all, can healthy people make a valid judgment in this case? Second, given the analytical framework, it is necessary to assess many so-called "worse-than-dead" health states. But neither in practice nor in the (clinical) studies that use such classification systems do we ever come across these devastating health states. That is easily explained: there are not many patients who have to live in a state worse than death, and those who do are generally not going to live much longer. Therefore, such patients never participate in studies, nor are they asked to fill in a patient-reported instrument.

Including the option of immediate death in measurement tasks is a contentious issue (Flynn, 2010; Norman et al., 2015). Although it may offend our moral sensibilities, that option is embedded in all of the techniques to arrive at index values suitable for cost-effectiveness analysis. The concept of death is so confrontational and hard to grasp that some find it astonishing that health economists make death a central element in their valuation methods (Kamm, 1993). But valuation techniques are based on all kinds of normative assumptions about how people should interpret the presented health situations and how they should make rational trade-offs and choices. Now, the question is not only whether people are rational (or should be) but also whether these valuation techniques can be considered a form of measurement. It is generally believed that true measurement should exclude normative issues. However, to some extent it is understandable that some valuation techniques include being dead, since death is part of the life course (the end of it). It is intuitive to use "dead" as the lowest point on the scale to express the quality of health (and even life). There is also evidence to suggest that "dead" is not the worst state for many of the respondents (Sintonen, 1981; Torrance, 1984). In addition, it seems that death cannot be conceived as a distinct health status but as a totally different state of being. Therefore, using dead as the lower anchor point seems to be conceptually different from using anchors expressed in terms of lasting health states. Nonetheless, the existential entity of "death" is commonly used as the anchoring label in valuation techniques.

Furthermore, it is a matter of preference or convention to anchor the value of "dead" at 0.0 (Weinstein et al., 1980). Inevitably, this convention implies assigning negative values to the worst health states, regardless of the classification system. For quality-adjusted life years, negative values may complicate the computations or even be incongruous (Pliskin et al., 1980; Torrance, 1984; Krabbe and Bonsel, 1998). In sum, it seems rather difficult to deal adequately with these two sets of health states (better and worse than death) in a single measurement procedure.

PATIENT CENTERED

Traditionally, the effectiveness of (curative) interventions is measured with clinical outcome measures such as blood pressure, recurrence of disease activity, or survival. With the shift to patient-centered care and the trend away from cure toward care, measures covering health aspects that are relevant to patients are increasingly sought after in the effort to evaluate treatment effectiveness.

What distinguishes informative health outcomes from all other measures of health is the need to solicit and incorporate patients' values and preferences into the final assessment. Doing it right requires measurement methods that use the patient's own perception as the point of reference. Patients should not only be involved in selecting which health attributes to assess, but they also should provide a value for their own health condition.

There is a concern that existing health instruments, which base their content predominantly on consensus and expert opinion, are insufficiently sensitive to the perspective of the individual patient, in particular that of particular patient groups. When researchers propose items/attributes of health outcomes as candidates, they risk omitting important aspects that may have greater relevance to patients and risk including aspects that have little or no relevance. It is now more or less accepted that the better strategy is to select items/attributes based on patient input (Cella et al., 1993; Carr and Higginson, 2001; Hamming and de Vries, 2007; Ridgeway et al., 2013; Krueger and Stone, 2014).

One of the most challenging tasks in the development of a health instrument is to determine which attributes should be incorporated to capture the full range of health. Moreover, self-reported severity of symptoms and functional limitations do not necessarily reveal the extent to which these concerns are important to patients. The strategy proposed here is different. Systematic empirical research could provide grounds to determine the hierarchical structure of a comprehensive set of health aspects that are relevant for capturing health. We may give credit to Feinstein for stressing the importance of the patient: "Finally, although patients are the only persons who can suitably observe, evaluate, and rate their own quality of life and the important features of their own health status, individual patients have seldom been asked or allowed to indicate their own values and beliefs. The decisions about what is important and how to 'weigh' the components usually emerge either from mathematical calculations or from an assembled committee of pundits" (1994).

When health states are assessed by the patients themselves, appropriate measurement methods are needed to rule out bias due to adaptation, coping, and other mechanisms. This calls for a measurement framework capable of incorporating both the input of patients about their experienced health state and an objective appraisal of this health state.

FINIS

Health-status measures serve several purposes. They may be taken as outcome measures for clinical research or program evaluation; they could be used on a one-time basis to provide a functional and psychosocial profile of individual patients; or they might be applied serially to monitor the natural history of disease or responses to standard interventions. It is interesting to recall what Feinstein said in 1994: "Despite persistent problems in conceptualization and measurement, quality of life is an idea whose time has come; its appraisal is now demanded in many clinical trials from which it was formerly excluded."

Increasingly, governmental agencies use health values to inform policy decisions. However, many flaws and limitations are associated with the dominant research paradigm that underpins the instruments currently available to quantify a subjective phenomenon such as health. For example, when the renowned EQ-5D instrument was introduced almost 20 years ago, it was welcomed as a laudable initiative to collect values for health states. Now, two decades later, the restricted content of this instrument and its rather academic and cumbersome measurement procedure call for something better.

There is another mechanism in play that explains the slow progress in developing better health measurement methods. The narrow scope of journals is at least partially responsible for the publication of articles with a limited and highly specific range of topics and methods. Moreover, most journals have their own pool of reviewers and a self-selection mechanism for appointing their editors. All of these conditions put a specific area of interest in the spotlight but leave underexposed any scientific contributions that do not nicely fit into "the field." Submissions on possibly interesting endeavors (definitely a tiny minority) outside the mainstream research traditions are sidelined as irrelevant, inconvenient, or simply "not good." Innovation is the Holy Grail in science, but more often than not its guiding light is eclipsed by orthodoxy and imitation. Interdisciplinary research that is worked out with an open mind is the exception. The status quo is conducive to repetition and replication. For example, it leads to fruitless discussions about whether responsiveness should be considered part of reliability, validity, or be considered as a separate measurement property. Support for the position taken here can be found in the recent book of Louviere et al. (2015, p. 4) about best−worst scaling. There, Louviere disclosed that during the 1990s all his papers on this topic were rejected by academic reviewers. It took some time, but best−worst scaling is now regarded by many as a promising approach to reveal individual preferences.

The development of instruments or scales can be expensive and time consuming, as it usually involves a number of steps: item generation, data collection from a sample in the target population of interest, item reduction, scale validation, translation, and possibly cultural adaptation. The whole procedure can easily take more than a year. Developing a measurement in-strument is not a task to be taken lightly. If done properly, it could take years because the process is iterative. Therefore, the use of a previously (validated) health-status instrument is typically preferable to the development of a new instrument for the same purpose or a slightly different one. Yet, if there is a clear need for a specific health instrument that does not exist and we want to make a difference, then in most cases it would be better to take heed of all the lessons learned from previous exercises and develop something completely new, from scratch.

REFERENCES

Aiken, L.R., 1977. Note on sensitivity: a neglected psychometric concept. Perception and Motor Skills 45 (Suppl. 3), 1330.

Brown, T.C., Peterson, G.L., 2009. An Enquiry into the Method of Paired Comparison: Reliability, Scaling, and Thurstone's Law of Comparative Judgment. General Technical Report RMRS-gtr-216WWW. U.S. Department of Agriculture, Forest Service, Rocky Mountain Research Station, Fort Collins, CO, p. 98.

Carr, A.J., Higginson, I.J., 2001. Are quality of life measures patient centred? BMJ 322 (7298), 1357–1360.

Cella, D., Tulsky, D., Gray, G., Sarafian, B., Linn, E., Bonomi, A., Silberman, M., Yellen, S.B., Winicour, P., Brannon, J., et al., 1993. The functional assessment of cancer therapy scale: development and validation of the general measure. Journal of Clinical Oncology 11 (3), 570–579.

Coombs, C.H., 1964. A Theory of Data. John Wiley & Sons, New York.

Einstein, A., 1905. Ist die trägheit eines körpers von seinem energieinhalt abhängig? Annalen der Physik 18, 639–641.

FDA, 2009. Guidance for Industry – Patient-reported Outcomes Measures: Use in Medical Product Development to Support Labeling Claims. FDA, Silver Spring, MD.

Feinstein, A.R., 1994. Clinical judgment revisited: the distraction of quantitative models. Annals of Internal Medicine 120 (9), 799–805.

Flynn, T., 2010. Using conjoint analysis and choice experiments to estimate QALY values: issues to consider. PharmacoEconomics 28 (9), 711–722.

Glynn, I., 2010. Elegance in Science: The Beauty of Simplicity. Oxford University Press, New York.

Hamming, J.F., de Vries, J., 2007. Measuring quality of life. British Journal of Surgery 94 (8), 923–924.

Hays, R.D., Hadorn, D., 1992. Responsiveness to change: an aspect of validity, not a separate dimension. Quality of Life Research 1 (1), 73–75.

Hays, R.D., Woolley, J.M., 2000. The concept of clinically meaningful difference in health-related quality of life research. PharmacoEconomics 18 (5), 419–423.

Hunt, S.M., McEwen, J., 1980. The development of a subjective health indicator. Sociology of Health & Illness 2 (3), 231–246.

Juniper, E.F., Guyatt, G.H., Streiner, D.L., King, D.R., 1997. Clinical impact versus factor analysis for quality of life questionnaire construction. Journal of Clinical Epidemiology 50 (3), 233–238.

Kamm, F.M., 1993. Morality, Mortality. Volume I: Death and Whom to Save from It. Oxford University Press, New York.

King, M.T., Costa, D.S.J., Aaronson, N.K., Brazier, J.E., Cella, D.F., Fayers, P.M., Grimison, P., Janda, M., Kemmler, G., Norman, R., Pickard, A.S., Rowen, D., Velikova, G., Young, T.A., Viney, R., 2016. QLU-C10D: a health state classification system for a multi-attribute utility measure based on the EORTC QLQ-C30. Quality of Life Research 25 (3), 625–636.

Krabbe, P.F.M., Bonsel, G.J., 1998. Sequence effects, health profiles and the QALY model: in search of realistic modeling. Medical Decision Making 18 (2), 178—186.

Krabbe, P.F.M., Peerenboom, L., Langenhoff, B.S., Reurs, T.J.M., 2004. Responsiveness of the generic EQ-5D summary measure compared to the disease-specific EORTC QLQ C-30. Quality of Life Research 13 (7), 1247—1253.

Krabbe, P.F.M., 2002. In: Good Day Sunshine: About Biases, Irregularities and Inconsistencies in the Valuation of Health States. EuroQol Proceedings, pp. 139—145.

Krueger, A., Stone, A., 2014. Progress in measuring subjective well-being. Science 346 (6205), 2—3.

Lindeboom, R., Sprangers, M.A., Zwinderman, A.H., 2005. Responsiveness: a reinvention of the wheel? Health and Quality of Life Outcomes 3 (1), 8.

Louviere, J.J., Flynn, T.N., Marley, A.A.J., 2015. Best—Worst Scaling: Theory, Methods and Applications. Cambridge University Press, Cambridge.

Lugnér, A.K., Krabbe, P.F.M., 2017. In: Time Trade-Off Revisited. EuroQol Proceedings, 2016.

Norman, G.R., 1989. Issues in the use of change scores in randomized trials. Journal of Clinical Epidemiology 42 (11), 1097—1105.

Norman, G.R., Wyrwich, K.W., Patrick, D.L., 2007. The mathematical relationship among different forms of responsiveness coefficients. Quality of Life Research 16 (5), 815—822.

Norman, R., Viney, R., Aaronson, N.K., Brazier, J.E., Cella, D., Costa, D.S.J., Fayers, P.M., Kemmler, G., Peacock, S., Pickard, A.S., Rowen, D., Street, D.J., Velikova, G., Young, T.A., King, M.T., 2015. Using a discrete choice experiment to value the QLU-C10D: feasibility and sensitivity to presentation format. Quality of Life Research 25 (3), 637—649.

Nunnally, J.C., Bernstein, I.H., 1994. Psychometric Theory, third ed. McGraw-Hill, New York.

Oppe, M., Devlin, N.J., van Hout, B., Krabbe, P.F.M., de Charro, F., 2014. A program of methodological research to arrive at the new international EQ-5D-5L valuation protocol. Value in Health 17 (4), 445—453.

Pliskin, J.S., Shepard, D.S., Weinstein, M.C., 1980. Utility functions for life years and health status. Operations Research 28 (1), 206—224.

Reeve, B.B., Wyrwich, K.W., Wu, A.W., Velikova, G., Terwee, C.B., Snyder, C.F., Schwartz, C., Revicki, D.A., Moinpour, C.M., McLeod, L.D., Lyons, J.C., Lenderking, W.R., Hinds, P.S., Hays, R.D., Greenhalgh, J., Gershon, R., Feeny, D., Fayers, P.M., Cella, D., Brundage, M., Ahmed, S., Aaronson, N.K., Butt, Z., 2013. ISOQOL recommends minimum standards for patient-reported outcome measures used in patient-centered outcomes and comparative effectiveness research. Quality of Life Research 22, 1889—1905.

Reise, S.P., Waller, N.G., 2009. Item response theory and clinical measurement. Annual Review of Clinical Psychology 5, 27—48.

Revicki, D., Hays, R.D., Cella, D., Sloan, J., 2008. Recommended methods for determining responsiveness and minimally important differences for patient-reported outcomes. Journal of Clinical Epidemiology 61 (2), 102—109.

Ridgeway, J., Beebe, T., Chute, C., Eton, D., Hart, L., Frost, M., Jensen, D., Montori, V.M., Smith, J.G., Smith, S.A., Tan, A.D., Yost, K.J., Ziegenfuss, J.Y., Sloan, J.A., 2013. A brief patient-reported outcomes quality of life (PROQOL) instrument to improve patient care. PLoS Medicine 10, 11.

Shah, K., Lloyd, A., Devlin, N., 2011. Participants' responses to valuation tasks and implications for valuing EQ-5D-5L. EuroQol Proceedings 25—52.

Sintonen, H., 1981. An approach to measuring and valuing health states. Social Science & Medicine, Part C: Medical Economics 15 (2), 55—65.

Streiner, D.L., Norman, G.R., Cairney, J., 2015. Health Measurement Scales: A Practical Guide to Their Development and Use, fifth ed. Oxford University Press, Oxford.

Titchener, E.B., 1901. In: Experimental Psychology: A Manual of Laboratory Practice. Volume 1: Qualitative Experiments. Macmillan, New York.

Torrance, G.W., 1984. Health states worse than death. In: van Eimeren, W., Engelbrecht, R., Flagle, C.H.D. (Eds.), Third International Conference on System Science in Health Care. Springer Verlag, Berlin, pp. 1085–1089.

Weinstein, M.C., Fineberg, H.V., Elstein, A.S., Frazier, H.S., Neuhauser, D., Neutra, R.R., McNeil, B.J., 1980. Clinical Decision Analysis. Saunders, Philadelphia.

Wright, B.D., 1977. Solving measurement problems with the Rasch model. Journal of Educational Measurement 14 (2), 97–116.

Yang, Z., 2015. Inconsistency in the valuations of EuroQol EQ-5D-5L health states in China was more related to interviewer and to interview process than to respondents' characteristics. Value in Health 18 (7), A737–A738.

Yen, W.M., 1992. Item response theory. In: Alkin, M. (Ed.), Encyclopedia of Educational Research. Macmillan, New York, pp. 657–667.

Young, T., Yang, Y., Brazier, J.E., Tsuchiya, A., Coyne, K., 2009. The first stage of developing preference-based measures: constructing a health-state classification using Rasch analysis. Quality of Life Research 18 (2), 253–265.

Young, T.A., Rowen, D., Norquist, J., Brazier, J.E., 2010. Developing preference-based health measures: using Rasch analysis to generate health state values. Quality of Life Research 19 (6), 907–917.

Young, T.A., Yang, Y., Brazier, J.E., Tsuchiya, A., 2011. The use of Rasch analysis in reducing a large condition-specific instrument for preference valuation: the case of moving from AQLQ to AQL-5D. Medical Decision Making 31 (1), 195–210.

Further Reading

Barofsky, I., 2012. Quality: It's Definition and Measurement as Applied to the Medically Ill. Springer, New York.

Bond, T.G., Fox, C.M., 2001. Applying the Rasch Model: Fundamental Measurement in the Human Sciences. Lawrence Erlbaum Associates, Mahwah.

Cappelleri, J.C., Zou, K.H., Bushmakin, A.G., Alvir, J.M., Alemayehu, D., Symonds, T., 2014. Patient-Reported Outcomes: Measurement, Implementation and Interpretation. CRC Press, Boca Raton.

Coombs, C.H., 1964. A Theory of Data. John Wiley & Sons, New York.

Crocker, L., Algina, J., 1986. Introduction to Classical and Modern Test Theory. Holt, Rinehart and Winston, New York.

Engelhard Jr., G., 2013. Invariant Measurement: Using Rasch Models in Social Behavioral, and Health Sciences. Routledge, New York.

Fayers, P.M., Machin, D., 2000. Quality of Life. Assessment, Analysis and Interpretation. Wiley, Chichester.

Gold, M.R., Siegel, J.E., Russell, L.B., Weinstein, M.C., 1996. Cost-Effectiveness in Health and Medicine. Oxford University Press, New York.

Hand, D.J., 2004. Measurement Theory and Practice: The World Through Quantification. Arnold, London.

McDowell, I., 2006. Measuring Health: A Guide to Rating Scales and Questionnaires, third ed. Oxford University Press, Oxford.

Nunnally, J.C., Bernstein, I.H., 1994. Psychometric Theory, third ed. McGraw-Hill, New York.

Streiner, D.L., Norman, G.R., Cairney, J., 2015. Health Measurement Scales: A Practical Guide to Their Development and Use, fifth ed. Oxford University Press, Oxford.

de Vet, H.C.W., Terwee, C.B., Mokkink, L.B., Knol, D.L., 2011. Measurement in Medicine: A Practical Guide. Cambridge University Press, Cambridge.

Glossary

Attribute   A characteristic or property that an instrument seeks to describe (e.g., vitality, depression, mobility).

Concept   An abstraction that we use to express the ideas, people, organizations, events, or things we are interested in.

Construct   A set of operational measures allowing for the study of a theoretical concept.

Dimension   Often used as a synonym for "domain," but with a statistical association.

Domain   A general health aspect, usually having several attributes.

Index   An absolute number that can be expressed in proportions (from 0% to 100%) and can aggregate aspects (e.g., indicators, variables, attributes).

Indicator   (1) A quantifiable characteristic of individuals, providing supporting evidence when describing the health of a population. (2) In statistics and research design, an observed value of a variable; in a restricted sense, a variable in latent statistical models (e.g., factor analysis, SEM).

Item   A response task (question), part of a scale, usually in the form of a linguistic statement consisting of a stem (e.g., in the last 7 days I was …) plus a number of ordered response levels.

Latent trait   A variable that is not directly observed but inferred.

Measure   A number with which arithmetic can be done (at least at interval level).

Preference   The (relative) "desirability" of something or a condition.

Scale   One or multiple items that measure a single domain.

Score   A summary contained in responses to the items that are related to the constructs being measured. There are two types of scores: raw and scaled. The former has not been adjusted, the latter is the result of some transformation of the raw score for the purpose of reporting evidence for all respondents on a consistent scale.

Stimulus   An event or object eliciting a response that is to be measured.

Value   Relative measure that expresses the quality (level of severity) of subjective perceptions and experiences of certain (health) condition.

Index